Lecture Notes in Computer Science 16048

Founding Editors

Gerhard Goos
Juris Hartmanis

AF167653

Editorial Board Members

Elisa Bertino, *Purdue University, West Lafayette, IN, USA*
Wen Gao, *Peking University, Beijing, China*
Bernhard Steffen, *TU Dortmund University, Dortmund, Germany*
Moti Yung, *Columbia University, New York, NY, USA*

The series Lecture Notes in Computer Science (LNCS), including its subseries Lecture Notes in Artificial Intelligence (LNAI) and Lecture Notes in Bioinformatics (LNBI), has established itself as a medium for the publication of new developments in computer science and information technology research, teaching, and education.

LNCS enjoys close cooperation with the computer science R & D community, the series counts many renowned academics among its volume editors and paper authors, and collaborates with prestigious societies. Its mission is to serve this international community by providing an invaluable service, mainly focused on the publication of conference and workshop proceedings and postproceedings. LNCS commenced publication in 1973.

Carson K. Leung · Anton Dignös ·
Gabriele Kotsis · A. Min Tjoa · Ismail Khalil
Editors

Big Data Analytics and Knowledge Discovery

27th International Conference, DaWaK 2025
Bangkok, Thailand, August 25–27, 2025
Proceedings

 Springer

Editors
Carson K. Leung (ID)
University of Manitoba
Winnipeg, MB, Canada

Anton Dignös (ID)
Free University of Bozen-Bolzano
Bolzano, Italy

Gabriele Kotsis (ID)
Johannes Kepler University Linz
Linz, Austria

A. Min Tjoa (ID)
Vienna University of Technology
Vienna, Austria

Ismail Khalil (ID)
Johannes Kepler University Linz
Linz, Austria

ISSN 0302-9743 ISSN 1611-3349 (electronic)
Lecture Notes in Computer Science
ISBN 978-3-032-02214-1 ISBN 978-3-032-02215-8 (eBook)
https://doi.org/10.1007/978-3-032-02215-8

© The Editor(s) (if applicable) and The Author(s), under exclusive license
to Springer Nature Switzerland AG 2026

This work is subject to copyright. All rights are solely and exclusively licensed by the Publisher, whether the whole or part of the material is concerned, specifically the rights of translation, reprinting, reuse of illustrations, recitation, broadcasting, reproduction on microfilms or in any other physical way, and transmission or information storage and retrieval, electronic adaptation, computer software, or by similar or dissimilar methodology now known or hereafter developed.
The use of general descriptive names, registered names, trademarks, service marks, etc. in this publication does not imply, even in the absence of a specific statement, that such names are exempt from the relevant protective laws and regulations and therefore free for general use.
The publisher, the authors and the editors are safe to assume that the advice and information in this book are believed to be true and accurate at the date of publication. Neither the publisher nor the authors or the editors give a warranty, expressed or implied, with respect to the material contained herein or for any errors or omissions that may have been made. The publisher remains neutral with regard to jurisdictional claims in published maps and institutional affiliations.

This Springer imprint is published by the registered company Springer Nature Switzerland AG
The registered company address is: Gewerbestrasse 11, 6330 Cham, Switzerland

If disposing of this product, please recycle the paper.

Preface

DAWAK was established in 1999 as the International Conference on Data Warehousing and Knowledge Discovery. It was run continuously under this name until its 16th edition in 2014. Then, it was renamed in 2015 to the International Conference on Big Data Analytics and Knowledge Discovery, while retaining its DAWAK acronym. This name change was made to better reflect new research directions in the broad and rapidly growing areas of big data analytics and knowledge discovery from massive data.

Since its inception, the DAWAK conference has provided a high-quality platform for researchers, practitioners, and developers worldwide in the areas of database systems, data integration, data warehousing, cloud computing, programming languages, both traditional and big data analytics, and more recently, artificial intelligence (AI) and data science. The main objectives of the conference are to explore, disseminate, and exchange knowledge in these fields through scientific and industrial talks. With big data analytics, knowledge discovery, and AI bsecoming important research frontiers in both academia and the IT industry, DAWAK continues to evolve alongside innovations in algorithms, data processing architectures, and high-performance computing technologies.

Important research topics associated with these major areas include: models for big data; parallel processing and DBMS technology; distributed system architectures; scalability and parallelization; query languages, processing and optimization; semantics for big data intelligence; data warehouse and data lake architectures; data mining and text mining; AI, machine learning and deep learning; pre-processing and data cleaning; polystore and multistore architectures; NoSQL storage systems; cloud infra-structures and metadata for big data; big data storage, indexing, quality, provenance, search, discovery and management; big data analytics on various data (e.g., text, graph, unstructured data, spatio-temporal data, sensor data, real-time events); privacy and security in analytics; data visualisation; big data applications; data science; data engineering, management and analytics.

This book comprises the Proceedings of the 27th International Conference on Big Data Analytics and Knowledge Discovery (DAWAK 2025), held August 25–27, 2025, in Bangkok, Thailand. This marked the second time the DAWAK conference was host-ed outside of Europe—the first being DAWAK 2023 in Penang, Malaysia.

DAWAK 2025 received 62 paper submissions. Of these, the Program Committee selected 12 as regular papers, resulting in an acceptance rate of 19%. Additionally, 14 papers were accepted as short papers, aimed at showcasing pioneering research and innovative projects across various big data analytics and knowledge discovery disciplines. These short papers highlight early-stage research, emerging concepts and pre-liminary findings, fostering meaningful discussions and potential collaborations.

The accepted papers cover a variety of research topics on both theoretical and practical aspects. The program included among others the following topics: (1) data management and indices, (2) data mining and knowledge discovery, (3) graph data processing and analytics, (4) large language models (LLMs), (5) neural networks, as well as (6) sequential data analytics and recommendation systems.

Owing to the reputation of DAWAK, selected best papers from DAWAK 2025 will be invited for a special issue of the Data & Knowledge Engineering (DKE, Elsevier) journal. We extend our sincere thanks to Carson Woo, Editor-in-Chief of DKE, for approving the special issue.

We are also grateful to the members of the Program Committee and external reviewers for their thorough and timely evaluations of the submitted papers. Finally, we express our heartfelt appreciation to the DEXA conference organizers for their continuous support and guidance, with special thanks to Ismail Khalil for his invaluable help in all tasks leading to having the proceedings published.

August 2025

Carson K. Leung
Anton Dignös
Gabriele Kotsis
A. Min Tjoa
Ismail Khalil

Organization

Program Committee Chairs

Carson K. Leung	University of Manitoba, Canada
Anton Dignös	Free University of Bozen-Bolzano, Italy

Publicity Chair

Kanda Saikaew	Khon Kaen University, Thailand

Steering Committee

Gabriele Kotsis Johannes	Kepler University Linz, Austria
A Min Tjoa	Vienna University of Technology, Austria
Lukas Fischer	Software Competence Center Hagenberg, Austria
Bernhard Moser	Software Competence Center Hagenberg, Austria
Christine Strauss	University of Vienna, Austria
Ismail Khalil	Johannes Kepler University Linz, Austria

Program Committee Members

Abhishek Santra	University of Texas at Arlington, USA
Alberto Abello	Universitat Politècnica de Catalunya, Spain
Alejandro Maté	University of Alicante, Spain
Andrea Kő	Corvinus University of Budapest, Hungary
Besim Bilalli	Universitat Politècnica de Catalunya, Spain
Boris Novikov	Finland
Carlos Ordonez	University of Houston, USA
Chaiporn Jaikaeo	Kasetsart University, Thailand
Chitsutha Soomlek	Khon Kaen University, Thailand
Chotirat Ratanamahatana	Chulalongkorn University, Thailand
Christos Doulkeridis	University of Piraeus, Greece
Chutiporn Anutariya	Asian Institute of Technology, Thailand
Cristina Dutra de Aguiar	Universidade de São Paulo, Brazil
Darja Solodovnikova	University of Latvia, Latvia

Dimitri Theodoratos	New Jersey Institute of Technology, USA
Ela Pustulka	University of Applied Sciences and Arts Northwestern Switzerland, Switzerland
Elisa Quintarelli	Università di Verona, Italy
Emanuele Storti	Università Politecnica delle Marche, Italy
Enrico Gallinucci	University of Bologna, Italy
Fadila Bentayeb	Université Lyon 2, France
Franck Ravat, IRIT	Université de Toulouse, France
Frank Höppner	Ostfalia University of Applied Sciences, Germany
Genoveva Vargas-Solar	CNRS-LIRIS, France
Hwan-Seung Yong	Ewha Womans University, South Korea
Isabelle Wattiau	ESSEC & CNAM, France
Jaroslav Pokorný	Charles University in Prague, Czech Republic
Jens Lechtenbörger	University of Münster, Germany
Jérôme Darmont	Université Lyon 2, France
Jorge Bernardino	Polytechnic Institute of Coimbra, Portugal
Jun Miyazaki	Tokyo Institute of Technology, Japan
Kamel Boukhalfa	USTHB University, Algeria
Kanda Saikaew	Khon Kaen University, Thailand
Kazuo Goda	University of Tokyo, Japan
Keun Ho Ryu	Chungbuk National University, South Korea
Kjetil Nørvåg	Norwegian University of Science and Technology, Norway
Ladjel Bellatreche	ENSMA, France
Laurent D'Orazio	University of Rennes, CNRS, IRISA, France
Leonidas Fegaras	University of Texas at Arlington, USA
Luca Gagliardelli	University of Modena and Reggio Emilia, Italy
Maik Thiele	Technische Universität Dresden, Germany
Marcin Gorawski	Silesian University of Technology, Poland
Markus Endres	University of Passau, Germany
Marut Buranarach	National Electronics and Computer Technology Center - NECTEC, Thailand
Matteo Francia	University of Bologna, Italy
Michele Linardi	CY Cergy Paris University, France
Mirjana Ivanovic	University of Novi Sad, Serbia
Mourad Khayati	University of Fribourg, Switzerland
Olivier Teste	IRIT, France
Omar Boussaid	Université Lyon 2, France
Oscar Romero	Universitat Politècnica de Catalunya, Spain
Patrick Marcel	Université d'Orléans, France
Rim Moussa	ENI Carthage, Tunisia
Sana Sellami	Aix-Marseille University, France

Sandro Bimonte	INRAE, France
Selma Khouri	École Nationale Supérieure d'Informatique, Algeria
Silvia Chiusano	Politecnico di Torino, Italy
Sofian Maabout	LaBRI, University of Bordeaux, France
Soror Sahri	Université Paris Cité, France
Soumia Benkrid	Ecole Nationale Supérieure d'Informatique, Algeria
Soumyava Das	Teradata Labs, USA
Stephane Jean	University of Poitiers, ISAE-ENSMA Poitiers, France
Sven Groppe	University of Luebeck, Germany
Trasapong Thaiupathump	Chiang Mai University, Thailand
Vatcharaporn Esichaikul	Asian Institute of Technology, Thailand
Witold Andrzejewski	Poznań University of Technology, Poland
Wookey Lee	Inha University, South Korea
Yinuo Zhang	Teradata, USA
Young-Koo Lee	Kyung Hee University, South Korea

External Reviewers

Imane Hocine	University of Luxembourg, Luxemburg
Javier Sanchis	University of Alicante, Spain
Manel Ben Sassi	Université de la Manouba, ENSI, RIADI, Tunisia

Organizers

Contents

Graph Data Processing and Analytics

Data management and Indices

Keynote Talk

Sparse Matrix Algorithms for Evolving Neural Networks

Carlos Ordonez[✉]

Department of Computer Science, University of Houston, Houston, USA
carlos@central.uh.edu

Abstract. There is significant research into sparse and dense matrix computations, with high-performance computing techniques, reducing time complexity and improving parallel speedup mostly with ample main memory, considering I/O on secondary storage as a less important aspect. On the other hand, there has been important work in the database, data mining and big data communities accelerating the computation of machine learning models on large data sets. However, massive neural networks and constantly changing data sets are pushing matrix computation demands further. We first present a survey on three key problems identifying research issues: maintaining a large data set updated under frequent matrix entry insertions and deletions, sparse matrix addition/multiplication and recomputing a deep neural network when a sparse data set changes frequently. We then propose a research agenda focusing on those three major problems solved with parallel I/O efficient algorithms storing and processing matrices with coordinate tuples (like a database relational table): matrix entry insertion/deletion, matrix addition/multiplication and assembling these algorithms into state of the art neural networks. We argue coordinate tuples complement and can potentially replace established main memory storage mechanisms, like dense arrays and compressed row/column formats. In summary, we believe database-inspired, parallel I/O efficient, algorithms tailored for sparse matrices can help updating, explaining and monitoring evolving neural networks on large dynamic data sets.

1 Introduction

1.1 Motivation

Computing and monitoring dynamic machine learning models plays a crucial role in addressing science and society problems involving evolving matrix data sets and evolving graphs, where matrix entries are inserted and deleted frequently. This challenge will persist as a significant concern in the future due to the versatility of sparse matrices, which can represent interactions or connections among numerous variables, objects, people, devices, and more [12,14]. The dynamic nature of these data structures, characterized by frequent updates and changes, necessitates the development of efficient and adaptive machine learning models that can accurately capture and respond to these shifts. As the complexity and

© The Author(s), under exclusive license to Springer Nature Switzerland AG 2026
C. K. Leung et al. (Eds.): DaWaK 2025, LNCS 16048, pp. 3–17, 2026.
https://doi.org/10.1007/978-3-032-02215-8_1

data scale of these problems continue to grow, the ability to compute and monitor dynamic models will remain essential for extracting meaningful insights and driving scientific advancements.

The Coordinate (COO) format [10] is generally not the preferred choice in most existing High-Performance Computing (HPC) and Artificial Intelligence (AI) libraries (e.g. Python NumPy, SciPy, PyTorch). This is largely because many libraries opt for dense formats to leverage CPU vectorized instructions for vector and matrix operations [24], while other libraries utilize compressed formats for sparse matrices, like Compressed Sparse Row (CSR [22]) and Compressed Sparse Column (CSC [19]) due to their efficiency in data transfer and access through the upper memory hierarchy, including RAM, CPU L1/L2 cache, and registers. The primary reasons for the limited adoption of COO format are its relatively slower performance and redundant coordinate information. However, as the landscape of HPC and AI continues to evolve, this mindset and inferiority may shift, potentially leading to a reevaluation of the COO format's role in AI, databases and HPC.

The gap between dense, compressed formats and coordinate (COO) format for sparse matrices may shrink as advancements in hardware technology continue to evolve. As CPUs increasingly feature more cores, allowing for greater parallelization, and main memory capacities expand, enabling larger datasets to be processed in-memory, the advantages of compressed formats may diminish. On the other hand, there is growing interest in optimizing GPUs for sparse matrices. Furthermore, the emergence of non-volatile memory technologies that increasingly approach the speeds of traditional memory will also help to reduce the performance differences between these formats. As a result, the benefits of using COO format, such as simpler algorithm implementation, easier code verification, straightforward transfer in and out and explainable intermediate results, may become more appealing, potentially making it a viable choice for dynamic applications.

1.2   Adapting Theory, Parallel Computing and Database Algorithms for Evolving Neural Networks

Insertion and deletion of a sparse matrix requires innovation that bridges the fields of algorithms [3], parallel computing [14], and database systems [16]. This is because traditional algorithms for dense matrices are not directly applicable to sparse matrices, and existing algorithms for sparse matrices are often limited to specific storage formats or operations. Furthermore, parallel algorithms for sparse matrix addition and multiplication, which are fundamental operations in many applications, require significant changes if the matrix is stored as a table containing coordinate tuples. While this storage format is slower in main memory compared to compressed formats, it offers broader algorithmic possibilities, easier verification of code correctness, and clear understanding of the algorithm. The time performance gap may shrink or become less important as CPUs and GPUs get faster and RAM grows larger.

As hardware continues to evolve with advancements in CPUs and GPUs [8] featuring increased core counts, alongside novel storage technologies like Non-Volatile Memory (NVM) approaching main memory speeds, the processing capabilities for large-scale data sets will substantially improve. However, the exponential growth of data sets and the increasing prevalence of neural networks will create new challenges, particularly in observing and tuning neural network models as data changes. In this context, innovative parallel algorithms designed to leverage new hardware architectures and efficiently handle dynamic data will play a crucial role in addressing these challenges, enabling faster insights and more accurate predictions.

1.3   Potential Impact on Science and Society

Algorithms optimized for the Coordinate (COO) tuple format have the potential to significantly impact various scientific and societal problems in the future. Sparse matrices are used in several science applications involving large graphs or high-dimensional data for maintaining dynamic neural networks. Biology is one such field, where protein-protein interaction networks can be represented as large graphs with millions of vertices and edges, and dynamic neural networks can be used to predict new interactions or identify patterns in the data (e.g., the BioGRID database). Atmospheric sciences is another field, where climate models can generate large amounts of high-dimensional data that require efficient storage and analysis, and dynamic neural networks can be used to predict weather patterns or identify trends in the data (e.g., the Climate Data Online dataset). Additionally, geology and physics can also benefit from COO format sparse matrices, where seismic data and particle collision data can be represented as large graphs and high-dimensional data that require efficient analysis and updates, and dynamic neural networks can be used to predict earthquakes or identify patterns in particle collisions (e.g., the Incorporated Research Institutions for Seismology dataset and the CERN Open Data Portal).

The efficient representation and analysis of complex data structures is crucial for informed decision-making in several key areas of society, as illustrated in the following examples. In traffic management, for instance, sparse matrices can be used to represent large graphs of traffic flow and congestion, enabling dynamic neural networks to predict traffic patterns and optimize traffic light control (e.g., the Transportation Networks for Research dataset). Similarly, in epidemiology, COO format sparse matrices can be applied to disease transmission networks, allowing dynamic neural networks to predict the spread of diseases and identify key factors in disease transmission (e.g., the Centers for Disease Control and Prevention's (CDC) Influenza dataset). Furthermore, in economics, sparse matrices can be used to represent large graphs of economic transactions and relationships, enabling dynamic neural networks to identify patterns and trends in economic data (e.g., the US Census Bureau's economic datasets).

2   Survey: State of the Art

2.1   Updating a Large Sparse Matrix in Batches

Updating a large sparse matrix in secondary storage requires clever algorithms to minimize the number of I/O operations [18], using optimal space in main memory. Serial algorithms, such as the in-place update, involve reading the matrix (or a matrix block) from secondary storage, updating the elements in main memory, and writing the updated matrix back to secondary storage. This method is simple but may require multiple passes over the data, which can be time-consuming for large matrices. Another serial approach is the buffer-based update algorithm, which uses a buffer in main memory to accumulate updates before writing them to secondary storage. The buffer is flushed when it is full or when a certain change threshold is reached. On the other hand, in parallel computing, algorithms such as parallel in-place update and parallel buffer-based update can be used to speed up updates. These approaches involve dividing the matrix into smaller chunks and updating each chunk in parallel using multiple threads or processes. The log-structured merge (LSM) algorithm, commonly used in databases, can also be parallelized by dividing the log file into smaller chunks and merging each updated chunk with the main matrix in parallel.

From a theory perspective, I/O-efficient algorithms, are designed to minimize the number of I/O operations required to update the matrix. These algorithms can be applied to sparse matrix updates to reduce the number of I/O operations (read access). Cache-oblivious algorithms, which are designed to optimize cache performance, can also be used to improve the efficiency of sparse matrix updates.

In the context of database systems [16], indexed tables can be used to update the matrix, providing atomicity, isolation and consistency guarantees. On the hand, columnar database systems can provide efficient updates and exploration queries with sparse matrices. Finally, array database systems allow manipulating unlimited size arrays on secondary storage.

In conclusion, the three major approaches of serial and parallel I/O efficient algorithms, theoretical foundations, and database solutions must be combined to develop efficient algorithms for updating a large sparse matrix on secondary storage. By following theoretical foundations and combining the strengths of each approach, researchers can develop algorithms that minimize I/O operations, optimize cache performance, and provide atomicity and consistency guarantees.

2.2   Fast Linear Algebra: Matrix Multiplication

Existing algorithms for fundamental matrix operators like matrix addition and matrix multiplication typically utilize dense and sparse compressed formats [17], which are optimized for modern multi-core CPU architectures [20]. Sparse formats, such as Compressed Sparse Row (CSR) or Compressed Sparse Column (CSC), store non-zero elements in a contiguous block of memory, allowing for efficient memory access patterns and minimizing memory bandwidth usage [7,15]. This leads to significant performance improvements on modern CPUs, which are designed to handle large amounts of data in parallel [13]. Moreover, the trend towards using Graphics Processing Units (GPUs) for matrix computations has further emphasized the importance of using dense matrix formats [7]. GPUs, with their massive parallelism and high-bandwidth memory, are particularly well-suited for dense matrix operations. Dense matrix formats can take full advantage of the GPU architecture, allowing for thousands of threads to perform math calculations simultaneously. GPUs have produced significant performance gains in many applications, including machine learning, scientific simulations, and data analytics.

In contrast, the Coordinate Format (COO) [10], which stores matrix elements as a table with tuples containing the row, column, and value, is generally considered slow, redundant, and space-inefficient. Why? Because the coordinate format requires more memory accesses and has a higher overhead due to the need to iterate over a list (sequence) of tuples. Additionally, the coordinate format is more memory-intensive, as it requires storing the row and column indices for each non-zero element, resulting in a slightly larger memory footprint (but still proportional to the number of non-zero entries, known as nnz).

As a result, dense and compressed matrix formats have become the defacto standard for high-performance matrix computations on both CPUs and GPUs. While the COO format may still be useful in certain niche applications, dense and compressed formats are generally the preferred choice for most use cases due to their superior performance and efficiency. Nevertheless, we believe the COO format may become more useful as hardware gets faster, sparse matrices change more frequently and they are read and written on secondary storage.

2.3   Computing Neural Networks with Sparse Input Matrices

The latest research on computing neural networks with sparse input matrices that have periodic changes has focused on developing efficient methods to handle these changes [23]. One approach is iterative linearization, which allows for sparse feature updates and quantifies the frequency of feature learning needed to achieve comparable performance. This method has shown remarkable performance on par with standard training methods, highlighting the importance of feature learning in neural networks.

Most current neural networks use dense matrices during forward and backward propagation. Dense matrices are also widely supported by popular deep learning frameworks and libraries. Additionally, dense matrices take advantage of optimized linear in multi-core CPUs and GPUs. However, some neural networks, such as those used in natural language processing and recommendation systems, often deal with large sparse matrices as input. For example, the Word2Vec model uses a sparse matrix to represent word embeddings, where each row corresponds to a word and each column corresponds to a feature. Similarly, the DeepWalk model uses a sparse matrix to represent graph embeddings, where each row corresponds to a node and each column corresponds to a feature. The GraphSAGE model also uses sparse matrices to represent graph data, where each row corresponds to a node and each column corresponds to a feature. On the other hand, sparse matrices may appear in intermediate results during forward and backward propagation when neurons are dropped to avoid overfit or when gradients vanish. Overall, the development of efficient methods for updating neural networks with sparse input matrices is an active area of research, and new techniques are being explored to take advantage of the benefits of sparse matrices while minimizing their drawbacks.

2.4 Theoretical Models for Parallel Computation

The best theory computation models to study the time and space complexity of parallel matrix computations (to be discussed later) are the Parallel Random Access Machine (PRAM) model and the Bulk Synchronous Parallel (BSP) model, but extensions are needed for newer architectures with the massive parallelism of GPUs. Some alternative parallel model include the CGM (Coarse Grained Multi-computer; distributed, synchronous, each processor has limited memory), LogP model (Latency-overhead-gap-Processors, which may be more practical than PRAM, it assumes short messages among processors) and PP (Pipeline Parallelism; best for GPUs). The PRAM model is a widely used theoretical model for parallel computing that assumes a shared memory and multiple processors that can access and modify the memory simultaneously. It is suitable for studying parallel algorithms on multi-core CPUs. For GPU computing, the BSP model is more suitable as it takes into account the specific characteristics of GPU architectures, such as the bulk-synchronous execution model and the memory hierarchy. However, some researchers argue that a different model, such as the GPU-PRAM model, may be necessary to accurately capture the unique characteristics of GPU architectures, such as the SIMT (Single Instruction, Multiple Thread) execution model and the memory coalescing. It is necessary to use a different theoretical model for GPU compared to CPU because of the following reasons. Different memory hierarchies: GPU has a different memory hierarchy compared to CPU, with a larger register file and a smaller cache. Different execution models: GPU uses a bulk-synchronous execution model, whereas CPU uses a traditional one-task execution model. Different thread scheduling: GPU incorporate a more complex thread scheduling mechanism compared to CPU. Some models for GPU parallel computing worth mentioning include: GPU-PRAM: an extension of the PRAM model for GPU architectures; BSP-GPU: an extension of the BSP model for GPU architectures; SIMT model: a model specifically designed for SIMT architectures. In conclusion, while the PRAM and BSP models can be used to study parallel algorithms on multi-core CPUs and GPUs, a different model may be necessary to accurately capture the unique characteristics of GPU architectures.

3   Future Research

We defend the idea of storing sparse matrices in coordinate format, opposing established compressed formats like CSR and CSC. When a sparse matrix is stored in coordinate format, represented as (i, j, v) tuples, it offers a unique advantage over compressed formats like Compressed Sparse Row (CSR) or Compressed Sparse Column (CSC). Coordinate tuples enable a wide range of database systems and list-based algorithms to compute machine learning models. This flexibility is crucial, as it enables extending existing algorithms, exploiting the structural properties of the matrix and data-oriented parallel processing. Moreover, it provides a deeper understanding of the theoretical implications of sparse matrix computations. By working directly with the (i, j, v) tuples, research can gain insights into the underlying complexity of the algorithms and explore the relationships between the matrix elements. This understanding can lead to the development of more explainable, intuitive algorithms, but also efficient and scalable. The coordinate format uncompressed storage allows for a more direct analysis of computational and space complexity.

Algorithms for sparse matrices in coordinate tuple format are a relatively new area of research compared to the state of the art. This is because classical parallel I/O efficient algorithms have traditionally been designed with arrays in main memory as the preferred data structure, rather than sparse matrices stored in tuple format. As a result, adapting these algorithms to work with COO format sparse matrices has not received significant attention. However, there is a substantial body of existing work on theoretical algorithms for sparse matrices in main memory, as well as I/O efficient algorithms for other data structures, that can be combined and extended to develop new algorithms for sparse matrices in COO format. By building on these foundations, researchers can create novel algorithms that efficiently handle the unique characteristics of sparse matrices in COO format.

We envision future research should study these three major problems below, all of which assume that the input sparse matrix is maintained on secondary storage, intermediate matrices may be maintained on secondary storage and the output matrices (weights) may also need to be stored. Research is needed to address the following key challenges considering data-oriented parallel processing and reducing I/O cost:

1. Algorithms for inserting and deleting batches of matrix entries, which can efficiently handle the dynamic nature of sparse matrices.
2. Algorithms for fundamental matrix operations such as matrix addition and matrix multiplication, which can minimize the overhead of data transfer between secondary storage and main memory, preserving fast transfer in the upper memory hierarchy (registers, L1/L2 cache, RAM).
3. Incorporating the previous algorithms into evolving neural networks that are computed on sparse matrices, which are periodically updated with scattered changes, requiring efficient and scalable solutions to maintain model accuracy and performance.

3.1   Problem I: Insertion and Deletion of Batches of Matrix Entries

Data sets are not static: they continuously change. The Internet is pushing changes further and the rate of change may be unpredictable. Recent research has explored various techniques for maintaining a sparse matrix up to date with batches of insertions and deletions of a few matrix entries, particularly within a time window. Under this window-based approach, parallel aspects such as $O(1)$ data structures, parallel insertion and deletion algorithms, and concurrent access control have been studied. The primary reason for this focus is that most High-Performance Computing (HPC) research assumes the sparse matrix can be stored with arrays maintained in main memory, limiting the applicability of these techniques to problems where the matrix is so large that it cannot fit in main memory. Notable contributions include research by Bender et al. on parallel data structures for dynamic sparse matrices [6], and studies by Liu et al. on efficient sparse matrix multiplication across diverse systems, including database systems [21]. However, a thorough review of research literature reveals a significant gap specifically addressing the challenges of maintaining sparse matrices with unpredictable batches of insertions and deletions, indicating a need for further research in this area.

Optimizing batches of insertions and deletions of sparse matrix entries is a more general problem than a sliding window approach because it can handle a wide range of dynamic scenarios, rather than just a fixed-size window of recent updates (insert new records, delete old records). In many domains, the rate and pattern of insertions and deletions can vary significantly over time, with different time scales, rhythms, and spikes in activity. By providing more flexibility and adaptability, optimizing batches of insertions and deletions can lead to better performance and efficiency in a wide range of applications, from social networks and financial markets to recommendation systems and more. For example, in a social network, the rate of new user sign-ups and friendships may be relatively steady, but there may be sudden spikes in activity around holidays or major events. In contrast, a financial market may experience rapid changes in trading activity during times of economic uncertainty. A sliding window approach may not be able to adapt to these varying patterns with spikes, whereas optimizing batches of insertions and deletions can handle these changes more effectively. Furthermore, optimizing batches of insertions and deletions requires collecting and processing more information than just the recent updates. In addition to the entry coordinates and values of the matrix entries, timestamps must also be collected and considered. This allows the optimization algorithm to take into account the temporal relationships between updates and make more informed decisions to process them in groups to reduce I/O cost.

Modern hardware aspects must be considered. Future research should develop parallel algorithms for multi-core CPU or hybrid CPU/GPU, but excluding GPU-only (since the data set is assumed to be read from secondary storage), for efficient insertion/deletion of a few matrix entries (O(1) or O(log(n))) of a large sparse matrix with O(n) edges stored in Coordinate (COO) format. Two important I/O techniques include matrix tiling to improve data locality and identifying I/O patterns in sparse matrices to exploit buffers and group I/O operations. This research direction aims to leverage the strengths of both multi-core CPUs and hybrid CPU/GPU architectures to achieve significant performance gains in a few updates of a sparse matrix. By focusing on COO format, which is well-suited for parallel processing, future research can explore novel algorithms that minimize the time complexity of insertion and deletion operations, making them suitable for large-scale applications. The development of such algorithms will require careful consideration of data structures, memory management, and synchronization techniques to ensure efficient and scalable performance.

Let X be the input data set, consisting of n column vectors with d dimensions (i.e. a $d \times n$ matrix). We envision two potential parallel algorithms and data structures to insert batches of matrix entries in COO format of a large matrix stored on secondary storage:

1. Batch-Insert-Delete Algorithm:
   We can assume there a single process updating X, to simplify algorithms and avoid concurrency mechanisms. However, concurrent updates should be eventually considered. This algorithm uses a combination of a buffer-based approach and a parallel merge-sort algorithm. The buffer is stored in main memory and it is used to accumulate incoming batches and deleted batches of matrix X entries. When the buffer is full, the algorithm sorts the buffer using a parallel merge-sort algorithm and then merges the sorted buffer with the existing matrix on secondary storage to propagate changes. This algorithm can be improved with subscript ranges to process the matrix in blocks to improve I/O locality.
   Data Structure: A combination of a buffer (in main memory) and a sparse matrix (on secondary storage) stored in coordinate tuple format. Time Complexity derivation: Buffer accumulation: $O(b)$, where b is the size of the buffer. Parallel merge-sort: $O(b \log(b)/p)$, where p is the number of machines (cores). Merge with existing matrix: $O(n_z + b)$, where n_z is the number of non-zero entries in the existing matrix.
   Parallel Speedup: A parallel merge-sort algorithm can potentially achieve a linear speedup of up to p, where p is the number of machines (cores).
2. Log-Structured-Merge (LSM) Algorithm:
   The LSM approach is commonly used in modern NoSQL database systems to handle high insertion rates. Therefore, new matrix update algorithms can use the log-structured merge (LSM) file approach to insert large batches of matrix entries, but with few or no deletions. That is, X is constantly growing, mainly n as data size grows, but also d can grow as more features are added. The algorithm can work with a combination of in-memory buffers and on-disk storage to accumulate and merge batches of matrix entries, similar to the more common algorithm introduced above.
   Data Structure: A combination of in-memory buffers and on-disk storage, with a sparse matrix stored in coordinate format.
   Time Complexity: Buffer accumulation: $O(b)$, where b is the size of the buffer. Parallel merge: $O(b \log(b)/p)$, where p is the number of machines (cores). Merge with existing matrix: $O(n_z + b)$, where n_z is the number of non-zero entries in the existing matrix. Parallel Speedup: a potential speedup of up to p, where p is the number of machines (cores).

There are important differences compared to algorithms used in query processing. Database algorithms typically focus on handling high insertion rates for transactions and demanding queries on normalized tables, whereas our envisioned algorithms focus on inserting batches of matrix entries in parallel, where there is no notion of normalized tables Database algorithms use different data structures, such as B-trees or hash tables on rows, whereas future algorithms will work on matrix sub-blocks, with coarser indexing In summary, our algorithms are designed to efficiently insert batches of matrix entries in parallel, taking into account the characteristics of secondary storage and multi-core CPUs. While database algorithms share similarities, new algorithms will be tailored to the specific requirements of sparse matrix insertion and deletion.

3.2 Problem II: Algorithms for Addition and Multiplication of Sparse Matrices Stored with Coordinate Tuples

The development of parallel, I/O-efficient algorithms for sparse-sparse and sparse-dense matrix operators is crucial to process large matrices. Two fundamental matrix operators must be studied: addition and multiplication. For matrix addition, an algorithm can take advantage of the fact that the coordinate tuples allow efficient insertion and deletion of entries. By utilizing parallel processing on multi-core CPUs and hybrid CPU/GPU architectures, the algorithm can quickly identify and combine corresponding entries from the two input matrices, resulting in a new sparse matrix in COO format. The parallelization of this process can be achieved by dividing the matrices into smaller blocks and processing them concurrently. In contrast, matrix multiplication is a more complex operation that requires careful consideration of I/O patterns to get rows from the left matrix and columns from the right matrix, reading them as coordinate tuples. To avoid recomputing all multiplications, the algorithm can employ techniques such as caching intermediate results and reusing previously computed products. By leveraging the parallel processing capabilities of multi-core CPUs and hybrid CPU/GPU architectures, the algorithm can efficiently perform the necessary multiplications and accumulations to produce the resulting sparse matrix in coordinate format as well. Despite the differences between addition and multiplication, both algorithms share commonalities in their reliance on efficient I/O operations by block and parallel processing. In both cases, the use of coordinate tuples and partitioned storage enables efficient handling of sparse matrices.

HPC research on parallel algorithms for sparse matrix addition and multiplication, combining sparse and dense matrices, has explored various main memory formats, including dense arrays, COO, CSR, and CSC. Several studies have investigated the potential of these formats, such as the work by Bulu et al. on parallel sparse matrix-vector multiplication using compressed sparse blocks (CSB) [4]. Other notable contributions include the study by Kaya on parallel algorithms for computing sparse matrix permanents [3], and the work by Liu et al. on sparse matrix-matrix multiplication using the COO format [21]. The coordinate tuple format opens new possibilities on modern hardware with faster

CPUs and GPUs, offering opportunities for improved performance and efficiency. However, theoretical aspects, such as potential time complexity, I/O optimization, and parallel speedup, have not received sufficient attention, largely due to the efficiency of linear algebra libraries on modern CPUs and GPUs for dense matrices, as noted by Demmel et al. [11] and by Ballard et al. [5]. Parallel aspects of these algorithms have been explored in both multi-core and distributed memory settings. The impact of I/O patterns and matrix structure on performance has also been studied, highlighting the need for careful consideration of these factors in algorithm design.

Future research should focus on developing new parallel algorithms for sparse matrix addition and sparse matrix multiplication, specifically tailored for matrices stored with coordinate (COO) tuples. Let the two input matrices be A, B, compatible for matrix multiplication $A \cdot B$. Matrix addition is an easier and straightforward case since both matrices have equal dimensions. The primary goal should be to achieve ideal time complexity and parallel speedup for these matrix operators in three distinct scenarios: (1) when A, B are resident in main memory, (2) when A, B are read from secondary storage, (3) when A is large and it is read from secondary storage and B is in main memory (or vice-versa), but $\text{size}(B) \ll \text{size}(A)$. Scenario (3) where one matrix is in main memory and the other is on secondary storage can be reduced to Scenario (1) or (2) when both matrices are about the same size. For Scenarios (1) and (2) hashing, sorting can used to merge rows from A with columns from B, similar to a relational join operator. For Scenario (3) B can be converted to a dense representation, accessing B elements directly with subscripts i, j from A. NVM offers significantly faster access times, larger block sizes, and non-volatility, making it a default option for storing and processing large matrices. Therefore, this research direction is particularly relevant, given the increasing adoption of non-volatile memory (NVM) with fast PCI connection, as a replacement for traditional disks and SATA solid-state drives (SSDs). For GPUs A, B entries can be easily sorted by A column, B row and then aligned before transferring to GPU memory for parallel multiplication.

3.3 Problem III: Incorporating Sparse Matrix Algorithms into Evolving Neural Networks

The landscape of deep learning has undergone a significant shift, with Transformers and Graph Convolutional Networks (GCNs) [1] emerging as dominant architectures [2,9], leaving behind plain Deep Learning, Recurrent Neural Networks (RNNs), and Convolutional Neural Networks (CNNs). Current research is now focusing on incorporating sparse matrix algorithms into these two emerging types of neural networks. The first step involves utilizing algorithms that update the input sparse matrix with batches of insertions and deletions, allowing the network to efficiently adapt to changing data. In Transformers, these updated sparse matrices can be leveraged to compute self-attention mechanisms, where algorithms for sparse matrix addition and multiplication play a crucial role in combining and transforming input embeddings. Similarly, in GCNs, these algorithms can be applied to update adjacency matrices and node features, enabling the network to learn from the evolving graph structure. By integrating sparse matrix algorithms, both Transformers and GCNs can benefit from improved efficiency and scalability, particularly when dealing with large and dynamic datasets.

Despite the extensive research on sparse matrices in HPC and parallel computing, there is scarce work on utilizing sparse matrices within Transformers and Graph Convolutional Networks (GCNs) neural networks, where the input data set undergoes periodic batches of insertions and deletions. The vast majority of existing research on sparse matrices has focused on optimizing their use in traditional HPC applications, such as linear algebra operations and scientific simulations. However, the potential benefits of sparse matrices in neural networks, particularly in the context of dynamic and evolving data, remain largely unexplored. As a result, there is a significant gap in the literature regarding the application of sparse matrices in Transformers and GCNs, where the input data is constantly changing due to insertions and deletions. This lack of research presents an opportunity for innovative work that could lead to significant advancements in the efficiency and scalability of these neural networks.

Adapting the Coordinate (COO) format to optimize I/O can yield substantial algorithm efficiency benefits, particularly in scenarios involving frequent matrix updates to recompute neural networks faster. By harnessing the COO format, matrix updates can be executed more rapidly, and neural network models can be recomputed with increased efficiency, thereby avoiding the need for costly recomputations from scratch. Moreover, incremental algorithms can process updates in small batches, facilitating a faster update cycle characterized by a short lag of a few seconds between the insertion or deletion of matrix batches and the subsequent update of the neural network. This database-oriented approach enables more responsive and adaptive modeling, making it well-suited for applications where data is constantly evolving and timely insights are required.

4   Deployment

Since it would be a long-time effort to develop prototypes to solve the three major problems introduced above, here we provide some guidelines for programming and experimental evaluation. To deploy the algorithms and data structures discussed above, Python is a better language than C++ or Java for faster prototype development of sparse matrix algorithms due to its ease of use, flexibility, and extensive community support. Python's syntax and nature make it an ideal language for rapid prototyping and development, allowing developers to focus on the theoretical aspects of the algorithms rather than the implementation details. Additionally, Python's vast array of libraries and frameworks provide a wealth of pre-built functionality, enabling developers to build upon existing work and accelerate their development process. However, low-level languages like C++ will still be needed for specific sparse matrix bottlenecks, where performance is critical and optimization is required. C++'s ability to provide direct memory access and fine-grained control over hardware resources make it an ideal choice for optimizing performance-critical components.

The following popular Python libraries can be used to develop and test the theory of matrix algorithms and neural networks, listed in order of maturity and popularity:

1. NumPy: A mature and widely-used library for efficient numerical computation, including linear algebra.
2. SciPy: A comprehensive library for scientific computing, providing functions for linear algebra, optimization, and more.
3. PyTorch: A popular deep learning framework that provides a dynamic computation graph, tensors and automatic differentiation.
4. scikit-learn: A widely-used library for machine learning, providing tools for data pre-processing, feature selection, and model evaluation.
5. Pandas: A library for data manipulation and analysis, providing data structures and functions for working with structured data.
6. TensorFlow: similar to PyTorch, which was more popular a few years ago.

To experimentally validate time complexity and parallel speedup theory results, the following steps can be followed:

1. Program the algorithms using the chosen Py libraries and frameworks, with a focus on simplicity and readability.
2. Use synthetic data to test the algorithms under various conditions, such as different matrix sizes and numbers of threads.
3. Measure the execution time and parallel speedup of the algorithms using profiling tools and benchmarking techniques.
4. Compare the results to the theoretical predictions and analyze any discrepancies, using statistical methods and data visualization techniques to identify trends and patterns.
5. Iterate on the design and implementation of the algorithms based on the results of the experiments, refining the implementation and testing new hypotheses as needed.

5   Conclusions

We presented a vision paper, with a tentative research agenda. Therefore, neither theory results nor experiments were provided. From a research and code development perspective, the coordinate tuple format holds potential benefits that can streamline the development process and facilitate innovation, bridging database systems and HPC. Specifically, this format can be processed using established I/O-efficient algorithms, which can accelerate computation and reduce overhead. The format's simplicity also makes it easier to analyze space and time complexity, enabling researchers to better understand the computational resources required and optimize their code accordingly. Furthermore, the coordinate format straightforward structure simplifies the process of writing correct code, reducing the likelihood of errors and bugs. Additionally, the explicit representation of coordinates provides query capabilities and storage transparency, making it easier to track and interpret results, and ultimately facilitating the development of more reliable and efficient algorithms.

References

1. Abadal, S., Jain, S., Guirado, R., López-Alonso, J., Alarcón, E.: Computing graph neural networks: a survey from algorithms to accelerators. ACM Comput. Surv. **54**(9), 191:1–191:38 (2022)
2. Aggarwal, C.C.: Neural Networks and Deep Learning - A Textbook. Springer, Cham (2023). https://doi.org/10.1007/978-3-031-29642-0
3. Aho, A., Hopcroft, J.E., Ullman, J.D.: Data Structures and Algorithms, 2nd edn. Addison/Wesley, Redwood City, California (1983)
4. Azad, A., Buluç, A.: A work-efficient parallel sparse matrix-sparse vector multiplication algorithm. In: 2017 IEEE International Parallel and Distributed Processing Symposium, IPDPS 2017, Orlando, FL, USA, May 29 - June 2, 2017, pp. 688–697. IEEE Computer Society (2017)
5. Ballard, G., Demmel, J., Gearhart, A.: Brief announcement: communication bounds for heterogeneous architectures. In: Rajaraman, R., auf der Heide, F.M., eds, SPAA: Proceedings of ACM Symposium on Parallelism in Algorithms and Architectures, pp. 257–258. ACM (2011)
6. Bender, M.A., Stølting Brodal, G., Fagerberg, R., Jacob, R., Vicari, E.: Optimal sparse matrix dense vector multiplication in the i/o-model. In: Gibbons, P.B., Scheideler, C., eds, SPAA 2007: Proceedings of the 19th Annual ACM Symposium on Parallelism in Algorithms and Architectures, San Diego, California, USA, June 9-11, 2007, pp. 61–70. ACM (2007)
7. Brock, B., Buluç, A., Yelick, K.A.: RDMA-based algorithms for sparse matrix multiplication on GPUs. In: Proceedings of the 38th ACM International Conference on Supercomputing, ICS 2024, Kyoto, Japan, June 4-7, 2024, pp. 225–235. ACM (2024)
8. Capra, M., Bussolino, B., Marchisio, A., Masera, G., Martina, M., Shafique, M.: Hardware and software optimizations for accelerating deep neural networks: survey of current trends, challenges, and the road ahead. IEEE Access **8**, 225134–225180 (2020)
9. Chen, C., et al.: A survey on graph neural networks and graph transformers in computer vision: a task-oriented perspective. IEEE Trans. Pattern Anal. Mach. Intell. **46**(12), 10297–10318 (2024)
10. Dang, H.-V., Schmidt, B.: The sliced COO format for sparse matrix-vector multiplication on CUDA-enabled GPUs. In: Proceedings of the International Conference on Computational Science, ICCS, volume 9 of Procedia Computer Science, pp. 57–66. Elsevier (2012)
11. Demmel, J., Grigori, L., Hoemmen, M., Langou, J.: Communication-optimal parallel and sequential QR and LU factorizations. SIAM J. Sci. Comput. **34**(1) (2012)
12. Demmel, J.W.: Applied Numerical Linear Algebra. SIAM, 1st edition (1997)
13. Deveci, M., Trott, C., Rajamanickam, S.: Multithreaded sparse matrix-matrix multiplication for many-core and GPU architectures. Parallel Comput. **78**, 33–46 (2018)

14. Dongarra, J., Duff, I.S., Sorensen, D.C., van der Vost, H.A.: Numerical Linear Algebra for High-Performance Computers. SIAM (1998)
15. Ezouaoui, S., Hamdi-Larbi, O., Mahjoub, Z.: Towards efficient algorithms for compressed sparse-sparse matrix product. In: 2017 International Conference on High Performance Computing & Simulation, HPCS 2017, Genoa, Italy, July 17-21, 2017, pp. 651–658. IEEE (2017)
16. Garcia-Molina, H., Ullman, J.D., Widom, J.: Database Systems: The Complete Book. Prentice Hall, 2nd edition (2008)
17. Huang, H., Chow, E.: Exploring the design space of distributed parallel sparse matrix-multiple vector multiplication. IEEE Trans. Parallel Distrib. Syst. 35(11), 1977–1988 (2024)
18. Jang, M.-H., Ko, Y.-Y., Gwon, H.-M., Jo, I., Park, Y., Kim, S.-W.: SAGE: a storage-based approach for scalable and efficient sparse generalized matrix-matrix multiplication. In: Proceedings of the 32nd ACM International Conference on Information and Knowledge Management, CIKM, pp. 923–933. ACM (2023)
19. Jayakody, S., Wang, J.: EMBARK: memory bounded architectural improvement in CSR-CSC sparse matrix multiplication. In: 9th IEEE International Conference on Collaboration and Internet Computing, CIC 2023, Atlanta, GA, USA, November 1-4, 2023, pp. 8–17. IEEE (2023)
20. Kim, B., et al.: MViD: sparse matrix-vector multiplication in mobile DRAM for accelerating recurrent neural networks. IEEE Trans. Comput. 69(7):955–967 (2020)
21. Liu, W., Vinter, B.: CSR5: an efficient storage format for cross-platform sparse matrix-vector multiplication. In: Bhuyan, L.N., Chong, F., Sarkar, V., eds, Proceedings of ACM on International Conference on Supercomputing (ICS), pp. 339–350. ACM (2015)
22. Nagahara, Y., Yan, J., Kawamura, K., Motomura, M., Van Chu, T.: Efficient COO to CSR conversion for accelerating sparse matrix processing on FPGA. In: IEEE International Conference on Consumer Electronics, ICCE 2024, Las Vegas, NV, USA, January 6-8, 2024, pp. 1–2. IEEE (2024)
23. Soltaniyeh, M., Martin, R.P., Nagarakatte, S.: An accelerator for sparse convolutional neural networks leveraging systolic general matrix-matrix multiplication. ACM Trans. Archit. Code Optim. 19(3), 42:1–42:26 (2022)
24. Zhang, H., Mills, R.T., Rupp, K., Smith, B.F.: Vectorized parallel sparse matrix-vector multiplication in PETSC using AVX-512. In: Proceedings of the 47th International Conference on Parallel Processing, ICPP 2018, Eugene, OR, USA, August 13-16, 2018, pp. 55:1–55:10. ACM (2018)

Invited Talk

Data Integration in the AI Era: Research Trends and Still Open Issues

Robert Wrembel[(✉)] [iD]

Poznan University of Technology, Poznań, Poland
robert.wrembel@cs.put.poznan.pl

Abstract. Data integration (DI) has been an area for intensive research for decades, which resulted in a few acknowledged reference architectures. The architectures can be categorized as supporting: (1) virtual integration (federated and mediated), (2) physical integration (data warehouse), and (3) hybrid (data lake, data lakehouse, data mesh). Regardless of their specific type, all these architectures rely on a complex integration layer. The layer is implemented by a sophisticated software, for designing, orchestrating, and running the so-called DI processes. On the one hand, in all business domains, large volumes of highly heterogeneous data are produced, e.g., medical systems, smart cities, smart agriculture, which require further advancements in the data integration technologies. On the other hand, the widespread adoption of artificial intelligence (AI) solutions is now extending towards DI, offering alternative solutions, opening new research paths, and generating new open problems.

In this talk, I will share my perspective on the application and potential of AI solutions for selected DI problems. I will also highlight still unresolved issues within the field of DI. The talk will be structured into three main parts: (1) an overview of data integration architectures, (2) selected AI techniques for DI (like data wrangling, data quality, schema matching, optimization of systems, and code generation), and (3) still open problems in DI. The findings presented in the talk are based on my experience in running research and development DI projects for various business entities. It offers a concise overview of common DI challenges and potential solutions, serving as a quick-start guide for further exploration.

Keywords: data integration architectures · data integration process · data wrangling · data quality · code generation · data lineage · machine learning · artificial intelligence

1 Preliminaries

A typical data landscape is described by heterogeneity of data sources (DSs)/ data storage systems, their distribution, and their silo architectures. The heterogeneity is manifested by DSs offering different functionalities, data models, and implementation technologies. The distribution means that DSs are physically

© The Author(s), under exclusive license to Springer Nature Switzerland AG 2026
C. K. Leung et al. (Eds.): DaWaK 2025, LNCS 16048, pp. 21–36, 2026.
https://doi.org/10.1007/978-3-032-02215-8_2

deployed in geographically separated hardware (a stand-alone server, a cloud). A silo architecture means that data are stored in large repositories (typically in legacy systems). These repositories are typically difficult to integrate due to technological or policy reasons.

Moreover, for years, complex, data-driven systems have been developed, e.g., for health care, precision/sustainable agriculture, and smart cities. These systems produce huge volumes of highly heterogeneous data (a.k.a. big data) that need to be integrated to feed various applications offering descriptive or predictive analytics.

The need of integrating data in such a landscape is indisputable, but building an integrated system is not trivial. The most challenging issues in data integration include: (1) providing means for efficient access to data, (2) building a homogeneous (a data model and structures) integrated data set, (3) cleaning, homogenizing, and discovering duplicates, (4) providing means for tracing data processing from their source to destination (a.k.a. data lineage, data provenance).

To address this problem, a few (de facto) standard data integration (DI) architectures are inevitable in modern information systems and they are constantly facing new challenges caused by complex, fast arriving, and ample data as well as emerging data engineering technologies. Thus, DI architectures and processes are among very frequently researched topics [91,102,112,123].

Naturally, in recent years, these architectures have been augmented with machine learning (ML) or artificial intelligence (AI) solutions, with the intention to make the integration faster, easier, and cheaper. Typical support from ML/AI concerns: (1) data wrangling (homogenization, error detection, and cleaning), (2) data deduplication, (3) schema matching, (4) systems optimization and self-tuning capabilities, and (5) automatic code generation.

In this talk, I will share my perspective on the application and potential of the aforementioned ML/AI solutions for selected DI problems. The list of the addressed problems will not be excessive, but include the core ones. I will also highlight still unresolved issues within the field of DI, specifically: performance optimization of DI processes, data wrangling, and discovering data lineage.

The talk will be structured into three main parts: (1) an overview of data integration architectures, (2) selected AI techniques for DI, and (3) still open problems in DI. The findings presented in the talk are based on my experience in running research and development DI projects for various business entities, including IT, financial, and farming industries.

2   Data Integration Architectures: Overview

DI aims at making heterogeneous and distributed data available for an end user for further processing (like analytics). Research and development projects on the topic resulted in a few standard DI architectures, namely: (1) federated [14] and mediated [15], (2) data warehouse (DW) [34], (3) lambda [64], (4) data lake (DL) [44], (5) data lakehouse (DLH) [45], (6) polystore [100], and recently (7) data mesh/ data fabric [29].

In all of the aforementioned architectures, the content of DSs is brought together in an integrated system, either as virtual or physical data. To this end, a special purpose layer is placed between DSs and end user applications. This layer is implemented by a sophisticated software, which runs the so-called DI processes [36,94].

DI processes are core elements of all DI architectures. DI processes are complex workflows composed of tasks that are responsible for extracting data from DSs, data wrangling (transforming data into a common model and data structures, cleaning data, removing missing, inconsistent, and redundant data items), integrating data, and loading them into a central repository (i.e., DW, DL, or DLH) or making them available in virtual integration architectures (i.e., federated, mediated, polystore, or data mesh). DI processes are managed by a dedicated software - a DI engine (an ETL engine in a DW architecture). Most of the DI engines support a set of predefined (out of the box) tasks [49].

Virtual DI architectures include *federated databases* [14,33,86] and *mediator*-based systems [15,109]. In these architectures, data are integrated and wrangled on demand by a DI layer. The federated architecture is used to integrate databases built on the same data model (relational) and it uses one access interface (query language). The mediated architecture is applied to integrate not only databases but also other types of DSs.

The first representative of a physical integration is a *data warehouse* (DW) architecture [103], where the integration is implemented by means of DI processes (called ETL in this architecture). Integrated data are persistently stored in a central repository - a data warehouse. This architecture is efficient in application domains like insurance, finances, trading, sales, which process large volumes of simple data, e.g., strings, numbers, and dates.

The standard DW architecture extended with capabilities of collecting data that arrive as streams is called *lambda* [40,64]. It includes two data processing lanes - the standard batch one and the real-time one. The architecture was developed in order to be able to analyze batch-arriving data with stream-arriving data in the same system. Both lanes are integrated using the serving layer, which is typically implemented by means of virtual integration and/or physical integration.

In order to integrate heterogeneous (multimodal data) a data lake architecture was proposed. A *data lake* is a repository that stores heterogeneous data ingested from DSs in their original formats [44,67]. Such data have to be further homogenized by DI processes, to produce data available for applications, e.g., [53]. In a pure data lake architecture, data are unified on-the-fly, like in a mediated system.

In a *data lakehouse* [34,45,92,117] data coming from a data lake are first unified by DI processes and then physically stored in one or more data warehouses, which are part of the whole architecture. Each data warehouse provides data prepared for specific analytical applications.

Recently, a technological concept called data mesh has gained popularity in DI. A *data mesh* defines a data architecture and data governance approach

[12,29], where each component in a mesh is a DS having a dedicated owner. A data mesh architecture is implemented by a set of technologies, which are called *data fabric* [97]. A data fabric includes among others: data storage and data management systems (e.g., databases, distributed file systems), DI architectures, queuing systems and message brokers, data security and governance frameworks, as well as data analytics and visualization.

Further readings on various DI architectures are available in [7,44,94,98].

Methods for developing DI processes have been researched and developed for decades (see [4,94]) and were included in commercial (and some open license) DI design environments and DI engines [39], but still designing efficient DI processes, optimizing their executions, and managing them is challenging and expensive.

The fast advancements in ML/AI techniques makes their application to the design, optimization, and management of DI processes promising. However, research works on DI focus mainly on mappings between values [11] or schemas [30], data cleaning [50], data deduplication [8,118], see Sect. 3. Moreover, even though multiple providers of DI technologies and consulting companies opt for applying ML techniques in data integration, a clear step-by-step and end-to-end approach has not been proposed yet.

3   Selected ML/AI Techniques for Data Integration

In recent years, the application of ML/AI techniques, including LLMs, has rapidly emerged as a cutting-edge trend in data engineering. Traditional ML/AI techniques for data engineering are mostly based on supervised learning, which requires substantial amount of training data to produce models of acceptable quality (typically measured by precision, recall, F1). Moreover, multiple problems require specific domain knowledge from human experts.

Whereas it is claimed that LLMs, being trained on massive data volumes, already possess some parts of this knowledge. On top of it, their understanding of natural language and generative capabilities, offer alternative solutions to data integration. LLMs also alow user interactions with natural language prompts to provide solutions to simple data integration tasks. Finally, LLMs come out of the box with encapsulated knowledge from prior training. For this reason, tuning their capabilities to a given problem at hand requires (a) few training examples - this is the so-called few-shot learning or even zero-shot learning.

The most common applications of ML/AI techniques include: data wrangling, data deduplication, schema matching, systems optimization, and automatic code generation. Further reading on ML/AI/LLMs applied to various data integration tasks is available among others in [9,25,50,51,66].

3.1   Data Wrangling

Data wrangling is a costly process of preparing data for analytics [81], which includes tasks like data transformation, homogenization, error detection and removal, imputation. ML solutions proved to be efficient in some of these tasks.

For example, multiple works for error and anomaly detection apply: (1) clustering (e.g., [35, 90, 96, 108]), where correct values form clusters whereas incorrect values lay outside of these clusters or (2) neural networks, mostly for time series (e.g., [24, 38, 114]).

Data imputation also profits from ML (e.g., [42, 57, 61, 113, 116]). For example, in [116] the authors proposed an attention mechanism to input data, based on discovered data distributions. The neural network learns by observing what value was observed and what was imputed. In *HoloClean* the originally developed methods for data imputation based on statistical modeling [85] were extended with attention mechanism [113]. The attention mechanism learns patterns of data distribution in attribute tables. Based on these patterns, the framework is able to propose values for imputation. Their experimental evaluations show that the attention mechanism is able to outperform state-of-the-art methods based on ensemble tree models and deep learning.

[42] proposes multiple imputation models based on deep denoising autoencoders.

In *ActiveClean* [57] an ML model is trained on clean data to infer repair actions. [61] presents a framework for discovering data repair actions. The system is based on a set of cleaning rules. The set of adequate rules to be fired for a given cleaning problem is suggested by an ML algorithm. To this end, the proposed solution first collects training data, created by a user who solves a few cleaning examples. After a training step, the system automatically infers whether a rule should be fired for a given dirty row.

Recent advancements in LLMs resulted in the application of this technology to data wrangling. [51] and [66] present the results of assessing LLMs for basic data wrangling tasks, like (1) data transformation, (2) imputation, and (3) detecting errors and anomalies. Based on conducted experiments on various versions of *GPT3* the authors showed that *GPT3* was able to correctly run most of the tested tasks on simple data, based on a few learning examples provided in prompts (few-shot learning).

3.2   Deduplication

Applying ML for deduplication started with classification methods for dividing record pairs into classes of: duplicates, probably duplicates, and non-duplicates (e.g., [32, 87, 93]). These methods were subsequently extended with ML methods for learning schemes for record comparisons (e.g., [10, 37, 55]) and learning record matching rules (e.g., [26]), supported by active/self learning approaches (e.g., [22, 52, 54, 105]). In these solutions, deduplication tasks are encapsulated in an ML model, learned from data.

More advanced ML methods apply complex neural networks (e.g., [20, 31, 63, 101, 118, 120]) as the main component of pre-trained language models (PtLMs) or large language models (LLMs). For example, *DeepER* [31] contributes a method for encoding records into vectors for subsequent similarity comparison. The encoding is based on recurrent neural networks extended with long short term memory capabilities. The contribution also draws upon the existence of pre-trained word embeddings, if available. [63] reports the comparison of two classes of the state-of-the-art solutions: (C1) based on statistical modeling extended with ML with (C2) solutions based on deep learning. The comparison was concluded that C2 outperforms C1 for textual and dirty data.

The application of PtLMs to data deduplication tasks gained popularity (e.g., [60,77,79,106]) as it alleviated the need for explicitly implementing multiple deduplication tasks. However, a significant limitation of PtLMs is: (1) their need for large volumes of training data and (2) the observed decline in performance (typically measured by F1) when comparing records that were not included in the training set, as reported in [79]. These limitations have driven significant recent interest in LLMs (e.g., [48,70,78,80,107]). The evaluations of different types of LLMs is available in [16,52,78].

3.3   Schema Matching

Schema matching is one of the core tasks in DI. Its goal is to identify semantically corresponding database objects (e.g., tables and attributes) in different DSs. In the past, traditionally, schema matching has relied on a combination of user-defined mappings and rules as well as on similarity analysis of object structures, types, names, and data. For numerous DSs and large schemas these methods are time-consuming and error-prone. This is where ML/AI techniques were employed as well, first - for learning patterns and relationships from existing data and mappings, and second - for discovering these patterns and matching rules in unseen schemas [83]. The recent solutions apply neural networks [71], pre-trained language models [73,121,122], and LLMs for schema matching [75,88].

3.4   Systems Optimization

Various ML techniques have already been successfully applied to optimizing system performance. They build performance models that are based on performance characteristics (typically CPU, I/O, and memory usage), collected during a normal runtime of a system or during excessive testing phases.

For example, in [47,82,99] regression methods are proposed: in [82] - to discover the most influential DBMS parameters and their values, with the goal of minimizing latency and maximizing throughput of query workloads; in [47] - to recommend optimal parallelization parameters for a given code; in [99] - to optimize resource allocation for virtual machines.

The works reported in [56,76] focus on applying ML techniques to provide auto-tuning capabilities in the so-called self-driving database management systems. The solution proposed in [56] learns features of data engines from offline testing and online monitoring. Based on these insights, the system is able to route queries to adequate engines for efficient executions. Other learned models are used for automatic configuration tuning, scaling, and data migration. In [76] the authors apply ML to predict the behaviour of a DBMS (by learning from incoming workloads) and to adjust systems' parameters to handle a predicted workload.

[119] proposes a solution to tune multiple components of a system, by means of a encoder - decoder architecture implemented as neural networks. [104] uses reinforcement learning for finding good system configurations. To this end, the solution iteratively explores the space of possible settings: first, selecting a given setting (including additional database objects, values of various system parameters) and evaluating their performance on a workload sample and second, evaluating the performance on a test workload. [84] presents a system for supporting the process of designing data integration pipelines. To this end, it uses: (1) a repository of historical pipelines augmented with multiple metadata and (2) a recommender engine that for a given data integration at hand proposes the most suitable pipeline. Further readings on ML solutions for systems tuning can be found in [3,41,89,110].

3.5   Automatic Code Generation

Methods to convert sentences in a natural language into SQL have proliferated within recent years. They are supported by LLMs. Since data integration tasks are often expressed by means of SQL commands, these approaches are tempting to be used in designing DI pipelines.

However, there are two main challenges in applying LLMs to SQL code generation. First, efficient implementations of SQL commands for a given source DBMS (using advanced features of SQL like the *with* clause, optimization hints). Second, understanding the semantics of table and columns in source systems in the context of the semantics of prompts. To this end, some solutions apply the attention mechanism to figure out adequate columns for inclusion in SQL queries [58], various techniques of sentence embeddings [43], or reinforcement learning [69]. [72] provides a study on the efficiency of LLMs for SQL code generation. Its final remark states that LLMs poses *"remarkable capacity of both proprietary and advanced open-source models to accurately translate natural language questions into executable SQL code"*.

Applying LLMs to generating code snippets in languages other than SQL has became a hot research topic as well, see for example [2,17,21,59,68,95].

4   Still Open Problems in Data Integration

This section outlines my subjective view on still open research challenges in data integration, with the focus on: (1) optimizing performance of DI processes, (2) data wrangling techniques, and (3) discovering lineage links. At this stage of development of ML/AI and LLMs it is tempting to apply these techniques to the aforementioned open problems.

4.1 Performance Optimization of DI Processes

In order to reduce the execution time of a DI process, a few classes of solutions have been proposed (see [111] for a brief overview), like scaling of a DI server, parallel processing of DI tasks, re-ordering of DI tasks. The last technique has been well researched and resulted in a few approaches [4]. A special case of re-ordering is called *push-down* optimization. Its principle is to move some DI tasks into a data source, to be executed there. It is typically applicable to tasks at the beginning of a DI process. Push-down is available in a few DI engines, but only for relational DSs and for a few non-relational DSs.

In this context, the issues that needs further research include: (1) building models to decide which DI tasks can be pushed down to contribute to the improvement of performance of a DI process and (2) how to efficiently implement a given pushed down task in a DS, leveraging the functionality and internals of the DS.

As discussed earlier, the application of ML/AI techniques for systems optimization and the application of LLMs to code generation proved to provide promising results. For this reason, a natural extension is to apply these techniques to the push-down optimization. ML/AI techniques should decide whether a given DI task should be pushed down into a DS to increase the performance of a DI process, whereas LLMs should generate an efficient implementation of the task in the DS.

DI processes apply not only predefined tasks available in design tools, but also require the deployment of **user defined functions** (UDFs), in order to implement specific tasks [19,27]. UDFs can be implemented in various programming languages and typically they are treated by a DI engine as **black-boxes** (BBUDFs). For this reason, optimizing the execution of DI processes with BBUDFs is more than challenging.

To be able to apply optimization techniques for DI process execution, one must know performance characteristics of BBUDFs and (if possible) their semantics. Here ML/AI come with potential solutions. Assuming that classification models can be learned by ML/AI from historical performance data of known UDFs, these models can be used to classify BBUDFs, based on their performance characteristics. This way, some insights into these BBUDFs can be discovered (as in the initial approach reported in [13]).

4.2 Data Wrangling

Most of the solutions to data wrangling support standard data types, mostly text strings, numbers, and dates. The application of IT to all fields of industry (e.g., farming, healthcare, manufacturing) requires cleaning heterogeneous (a.k.a. multi-modal data), like spatial data, spatio-temporal time series, 2D and 3D images, audio (see for example projects in the intelligent/eco farming industry [23,62]).

Typically, in this application area, data are produced by robotic devices, which constitute the so-called Internet of Robotic Things (IoRT). IoRT produce multi-modal data that arrive to an integration system as streams. For this reason, it is necessary to wrangle such data in real-time. This, however, needs further development of edge-fog-cloud architectures with adequate efficient data wrangling solutions.

4.3 Discovering Data Lineage

To supervise data integration, data lineage (DL) is used to track the full data life cycle - from their origin (i.e., a DS) to their destination (often a data warehouse) [1,28,115]. This tracking information is vital for governance, compliance, and quality assessment and allows organizations to validate data and understand object dependencies. A special case of DL is object lineage, where links between database objects are discovered. DL links are often unavailable and have to be discovered (see. [5]).

To discover missing object lineage, in our preliminary work we applied ML techniques on available metadata. Feature vectors were built and a classification model was employed to determine whether one database object is a source for another. Initial experiments on large database schemas showed that the discovery of broken lineage links was possible at an acceptably high probability. However, this research direction has just started, opens new possibilities, and needs further solid investigation.

5 Final Comments

The discussion presented in this paper on the application of ML/AI/LLM technologies to solving data integration challenges should be extended towards a broader scope, in particular:

- whether these technologies can revolutionize the development and deployment methods of efficient DI pipelines,
- how to build end-to-end DI pipelines with the support of these technologies,
- how to assure and verify the quality of data produced by such pipelines,
- how to leverage the technologies for building complex DI architectures with appropriately co-designed software and hardware,
- how to mitigate bias in ML/AI/LLM used to build DI pipelines,
- whether the technologies could help solving the still unsolved challenge of managing the evolution of ETL processes (e.g., [6,18,74,94]), which is a totally unexplored research topic.

Furthermore, recent years have seen a growing application of LLMs to various data engineering problems. Research, including studies referenced in this paper, has demonstrated the capacity of LLMs to deliver promising results for specific data integration tasks. For example, [46,65,66] illustrated that carefully designed prompt engineering can substantially enhance performance. Complementing this, [66] provided experimental proof that LLMs not only outperform state-of-the-art pre-trained language models in data deduplication, imputation, and error detection but also show promising results with few-shot learning.

Even though such promising results were obtained, a fundamental key towards a successful application of ML/AI/LLM for data integration is the ability to tightly couple these technologies with database management systems, other storage systems, and DI tools (engines), so that they have less restricted access to data, schemas, performance characteristics, and other metadata. Moreover, LLMs (or rather large technological models - LTMs) should become specialized modules for particular technologies like data integration, systems optimization, or data wrangling. Specialized LTMs should be made pluggable into systems, where they could further be trained and tuned to support specific tasks.

While LLMs offer significant promise, their application, including to data integration, still faces considerable challenges. First, they generate non-deterministic outputs (as they learn constantly). Second, the verification of correctness of the generated results, especially ML models and code snippets is costly or impossible (the lack of explainability of AI models). Third, as mentioned before, constructing adequate prompts seems to have a great impact on the results generated by LLMs. For this reason, another field of research is emerging - it is automatic prompt engineering (tuning, optimization).

References

1. What is data lineage? IBM documentation. https://www.ibm.com/topics/data-lineage
2. Akella, A., Narayanam, K.: Data wrangling task automation using code-generating language models. In: AAAI-25, Association for the Advancement of Artificial Intelligence (AAAI) (2025)
3. Aken, D.V., Pavlo, A., Gordon, G.J., Zhang, B.: Automatic database management system tuning through large-scale machine learning. In: International Conference on Management of Data (SIGMOD), pp. 1009–1024 (2017)
4. Ali, S.M.F., Wrembel, R.: From conceptual design to performance optimization of ETL workflows: current state of research and open problems. VLDB J. **26**(6), 777–801 (2017). https://doi.org/10.1007/s00778-017-0477-2
5. Andrzejewski, W., Boiński, P., Wrembel, R.: On fixing broken lineage. In: Provenance Week @ ACM SIGMOD/PODS Conference (2025)
6. Awiti, J.: Algorithms and architecture for managing evolving ETL workflows. In: European Conference on Advances in Databases and Information Systems. CCIS, vol. 1064 (2019)
7. Azzini, A., et al.: Advances in data management in the big data era. In: Goedicke, M., Neuhold, E., Rannenberg, K. (eds.) Advancing Research in Information and Communication Technology. IAICT, vol. 600, pp. 99–126. Springer, Cham (2021). https://doi.org/10.1007/978-3-030-81701-5_4
8. Barlaug, N., Gulla, J. A.: Neural networks for entity matching: a survey. ACM Trans. Knowl. Disc. Data, **15**(3) (2021)
9. Berti-Équille, L., Bonifati, A., Milo, T.: Machine learning to data management: a round trip. In: IEEE International Conference on Data Engineering ICDE. IEEE Computer Society (2018)
10. Bianco, G.D., Gonçalves, M.A., Duarte, D.: BLOSS: effective meta-blocking with almost no effort. Inf. Syst. **75** (2018)

11. Birgersson, M., Hansson, G., Franke, U.: Data integration using machine learning. In: IEEE International Enterprise Distributed Object Computing Workshop (EDOC), pp. 1–10 (2016)
12. Bode, J., Kühl, N., Kreuzberger, D., Hirschl, S., Holtmann, C.: Data mesh: best practices to avoid the data mess. *CoRR*, abs/2302.01713 (2023)
13. Bodziony, M., Ciesielski, B., Lehnhardt, A., Wrembel, R.: On reasoning about black-box UDFs by classifying their performance characteristics. In *Harnessing Opportunities: reshaping ISD in the post-COVID-19 and Generative AI Era (ISD)* (2024)
14. Bouguettaya, A., Benatallah, B., Elmargamid, A.: Interconnecting Heterogeneous Information Systems. Kluwer Academic Publishers (1998). ISBN 0792382161
15. Brezany, P., Tjoa, A.M., Wanek, H., Wöhrer, A.: Mediators in the architecture of grid information systems. In: Wyrzykowski, R., Dongarra, J., Paprzycki, M., Waśniewski, J. (eds.) PPAM 2003. LNCS, vol. 3019, pp. 788–795. Springer, Heidelberg (2004). https://doi.org/10.1007/978-3-540-24669-5_103
16. Brunner, U., Stockinger, K.: Entity matching with transformer architectures - a step forward in data integration. In: International Conference on Extending Database Technology (EDBT), pp. 463–473. OpenProceedings.org (2020)
17. Busch, D., Bainczyk, A., Smyth, S., Steffen, B.: LLM-based code generation and system migration in language-driven engineering. Int. J. Softw. Tools Technol. Transf. **27**(1) (2025)
18. Butkevicius, D., Freiberger, P.D., Halberg, F.M.: MAIME: A maintenance manager for ETL processes. In: Workshops of the EDBT/ICDT Joint Conference. CEUR Workshop Proceedings, vol. 1810. CEUR-WS.org (2017)
19. Chen, Q., Wu, R., Hsu, M., Zhang, B.: Extend core UDF framework for GPU-enabled analytical query evaluation. In: International Database Engineering and Applications Symposium (IDEAS), pp. 143–151 (2011)
20. Chen, R., Shen, Y., Zhang, D.: GNEM: a generic one-to-set neural entity matching framework. In: The Web Conference (WWW). ACM (2021)
21. Chen, W., Tong, W., Case, A., Zhang, T.: Dango: a mixed-initiative data wrangling system using large language model. In: Conference on Human Factors in Computing Systems (CHI). ACM (2025)
22. Chen, X., Xu, Y., Broneske, D., Durand, G.C., Zoun, R., Saake, G.: Heterogeneous committee-based active learning for entity resolution (HeALER). In: Welzer, T., Eder, J., Podgorelec, V., Kamišalić Latifić, A. (eds.) ADBIS 2019. LNCS, vol. 11695, pp. 69–85. Springer, Cham (2019). https://doi.org/10.1007/978-3-030-28730-6_5
23. Chist-Era. Development of a plug-and-play middleware for integrating robot sensor data with gis tools in a cloud environment (GIS4IoRT). Chist-Era Project Call 2023. https://www.chistera.eu/projects-call-2023, https://ncn.gov.pl/sites/default/files/pliki/chistera-wrembel-en.pdf
24. Cho, Y., Lee, J., Ham, G., Jang, D., Kim, D.: Generality-aware self-supervised transformer for multivariate time series anomaly detection. Appl. Intell. **55**(7) (2025)
25. Chu, X., Ilyas, I.F., Krishnan, S., Wang, J.: Data cleaning: overview and emerging challenges. In: International Conference on Management of Data (SIGMOD) (2016)
26. Cohen, W.W., Richman, J.: Learning to match and cluster large high-dimensional data sets for data integration. In: ACM SIGKDD International Conference on Knowledge Discovery and Data Mining (KDD). ACM (2002)

27. Crotty, A., et al.: An architecture for compiling UDF-centric workflows. VLDB Endowment **8**(12), 1466–1477 (2015)
28. Cui, Y., Widom, J.: Lineage tracing for general data warehouse transformations. VLDB J. **12**(1) (2003)
29. Dehghani, Z.: Data Mesh: Delivering Data-Driven Value at Scale. O'Reilly (2022). ISBN 1492092398
30. Dong, L., Rekatsinas, T.: Data integration and machine learning: a natural synergy. VLDB Endowment, **11**(12) (2018)
31. Ebraheem, M., Thirumuruganathan, S., Joty, S., Ouzzani, M., Tang, N.: Distributed representations of tuples for entity resolution. VLDB Endowment, **11**(11) (2018)
32. Elfeky, M.G., Verykios, V.S., Elmagarmid, A.K.: Tailor: a record linkage tool box. In: International Conference on Data Engineering (ICDE). IEEE Computer Society (2002)
33. Elmagarmid, A., Rusinkiewicz, M., Sheth, A.: Management of Heterogeneous and Autonomous Database Systems. Morgan Kaufmann Publishers (1999). ISBN 1-55860-216-X
34. Errami, S.A., Hajji, H., Kadi, K.A.E., Badir, H.: Spatial big data architecture: from data warehouses and data lakes to the lakehouse. J. Parallel Distrib. Comput. **176**, 70–79 (2023)
35. Fumanal-Idocin, J., Rodríguez-Martínez, I., Indurain, A., Minárová, M., Bustince, H.: Almost aggregations in the gravitational clustering to perform anomaly detection. Inf. Sci. **612** (2022)
36. Furche, T., Gottlob, G., Libkin, L., Orsi, G., Paton, N.W.: Data wrangling for big data: challenges and opportunities. In: International Conference on Extending Database Technology (EDBT), pp. 473–478 (2016)
37. Gagliardelli, L., Papadakis, G., Simonini, G., Bergamaschi, S., Palpanas, T.: Generalized supervised meta-blocking. VLDB Endowment, **15**(9) (2022)
38. Gao, C., Ma, H., Pei, Q., Chen, Y.: Dynamic graph-based graph attention network for anomaly detection in industrial multivariate time series data. Appl. Intell. **55**(6) (2025)
39. Gartner. Magic quadrant for data integration tools (2022)
40. Gillet, A., Leclercq, É., Cullot, N.: Lambda+, the renewal of the lambda architecture: category theory to the rescue. In: La Rosa, M., Sadiq, S., Teniente, E. (eds.) CAiSE 2021. LNCS, vol. 12751, pp. 381–396. Springer, Cham (2021). https://doi.org/10.1007/978-3-030-79382-1_23
41. Golfarelli, M., Graziani, S., Rizzi, S.: An active learning approach to build adaptive cost models for web services. Data Knowl. Eng. **119**, 89–104 (2019)
42. Gondara, L., Wang, K.: MIDA: multiple imputation using denoising autoencoders. In: Phung, D., Tseng, V.S., Webb, G.I., Ho, B., Ganji, M., Rashidi, L. (eds.) PAKDD 2018. LNCS (LNAI), vol. 10939, pp. 260–272. Springer, Cham (2018). https://doi.org/10.1007/978-3-319-93040-4_21
43. Guo, C., Tian, Z., Tang, J., Li, S., Wang, T.: Multi-pattern retrieval-augmented framework for text-to-sql with poincaré-skeleton retrieval and meta-instruction reasoning. Inf. Process. Manage. **62**(3) (2025)
44. Hai, R., Koutras, C., Quix, C., Jarke, M.: Data lakes: a survey of functions and systems (2023)
45. Harby, A.A., Zulkernine, F.H.: From data warehouse to lakehouse: a comparative review. In: IEEE Big Data, pp. 389–395 (2022)

46. He, Z., Naphade, S., Huang, T.K.: Prompting in the dark: assessing human performance in prompt engineering for data labeling when gold labels are absent. In: Conference on Human Factors in Computing Systems (CHI). ACM (2025)
47. Hernández, Á.B., Pérez, M.S., Gupta, S., Muntés-Mulero, V.: Using machine learning to optimize parallelism in big data applications. Futur. Gener. Comput. Syst. **86**, 1076–1092 (2018)
48. Huang, Z.: Disambiguate entity matching using large language models through relation discovery. In: Conference on Governance, Understanding and Integration of Data for Effective and Responsible AI (GUIDE-AI). ACM (2024)
49. IBM. Product documentation: Infosphere information server 11.3 (2023). https://www.ibm.com/docs/en/iis/11.3?topic=jobs-processing-data
50. Ilyas, I.F., Rekatsinas, T.: Machine learning and data cleaning: which serves the other? ACM J. Data Inf. Q. **14**(3), 13:1–13:11 (2022)
51. Jaimovitch-López, G., Ferri, C., Hernández-Orallo, J., Martínez-Plumed, F., Ramírez-Quintana, M.J.: Can language models automate data wrangling? Mach. Learn. **112**(6) (2023)
52. Jain, A., Sarawagi, S., Sen, P.: Deep indexed active learning for matching heterogeneous entity representations. VLDB Endowment, **15**(1) (2021)
53. Jemmali, R., Abdelhédi, F., Zurfluh, G.: DLToDW: Transferring relational and NoSQL databases from a data lake. SN Comput. Sci. **3**(5), 381 (2022)
54. Jurek, A., Hong, J., Chi, Y., Liu, W.: A novel ensemble learning approach to unsupervised record linkage. Inf. Syst. **71** (2017)
55. Kejriwal, M., Miranker, D.P.: A two-step blocking scheme learner for scalable link discovery. In: International Workshop on Ontology Matching @ISWC, vol. 1317, pp. 49–60. CEUR-WS.org (2014)
56. Kraska, T., et al.: Check out the big brain on BRAD: simplifying cloud data processing with learned automated data meshes. VLDB Endowment **16**(11), 3293–3301 (2023)
57. Krishnan, S., Wang, J., Wu, E., Franklin, M.J., Goldberg, K.: Activeclean: interactive data cleaning for statistical modeling. VLDB Endowment, **9**(12) (2016)
58. Li, R., Chen, Y., Zhang, H., Yang, J., Xiao, Q., Jiang, S.: A question-aware few-shot text-to-SQL neural model for industrial databases. Int. J. Intell. Syst. **1**, 2025 (2025)
59. Li, X., Döhmen, T.: Towards efficient data wrangling with LLMs using code generation. In: Workshop on Data Management for End-to-End Machine Learning (DEEM) @SIGMOD. ACM (2024)
60. Li, Y., Li, J., Suhara, Y., Doan, A., Tan, W.: Deep entity matching with pre-trained language models. VLDB Endowment, **14**(1) (2020)
61. Mecca, G., Papotti, P., Santoro, D., Veltri, E.: BUNNI: learning repair actions in rule-driven data cleaning. ACM J. Data Inf. Qual. **16**(2) (2024)
62. MSCA. IoRT data management and analysis for sustainable agriculture (Green-FieldData). HORIZON-MSCA-2024-DN-01 (Marie Skłodowska-Curie Actions Doctoral Networks (2024). https://maiage.inrae.fr/node/3242
63. Mudgal, S., et al.: Deep learning for entity matching: a design space exploration. In: SIGMOD International Conference on Management of Data. ACM (2018)
64. Munshi, A.A., Mohamed, Y.A.I.: Data lake lambda architecture for smart grids big data analytics. IEEE Access **6**, 40463–40471 (2018)
65. Nananukul, N., Sisaengsuwanchai, K., Kejriwal, M.: Cost-efficient prompt engineering for unsupervised entity resolution in the product matching domain. Discov. Artif. Intell. **4**(1) (2024)

66. Narayan, A., Chami, I., Orr, L.J., Ré, C.: Can foundation models wrangle your data? VLDB Endowment, **16**(4) (2022)
67. Nargesian, F., Zhu, E., Miller, R.J., Pu, K.Q., Arocena, P.C.: Data lake management: challenges and opportunities. VLDB Endowment **12**(12), 1986–1989 (2019)
68. Nejjar, M., Zacharias, L., Stiehle, F., Weber, I.: LLMs for science: usage for code generation and data analysis. J. Softw. Evol. Process, **37**(1) (2025)
69. Nguyen, X., Phan, X., Piccardi, M.: Fine-tuning text-to-SQL models with reinforcement-learning training objectives. Nat. Lang. Process. J. **10** (2025)
70. Nuntachit, N., Sugannasil, P.: Can chatgpt outperform other language models? an experiment on using chatgpt for entity matching versus other language models. In: Barolli, L. (ed.) 3PGCIC 2023. LNDECT, vol. 189, pp. 14–26. Springer, Cham (2023). https://doi.org/10.1007/978-3-031-46970-1_2
71. Oh, H., Kulvatunyou, B.S., Jones, A.T., Finin, T.: Employing word-embedding for schema matching in standard lifecycle management. J. Ind. Inf. Integr. **38** (2024)
72. Ojuri, S., Han, T.A., Chiong, R., Stefano, A.D.: Optimizing text-to-SQL conversion techniques through the integration of intelligent agents and large language models. Inf. Process. Manage. **62**(5) (2025)
73. Pan, Z., Yang, M., Monti, A.: Schema matching based on energy domain pre-trained language model. Energy Inform. **6**(1) (2023)
74. Papastefanatos, G., Vassiliadis, P., Simitsis, A., Vassiliou, Y.: Policy-regulated management of ETL evolution. J. Data Semant. **13**, 147–177 (2009)
75. Parciak, M., Vandevoort, B., Neven, F., Peeters, L.M., Vansummeren, S.: Schema matching with large language models: an experimental study. In: Workshops @VLDB (2024)
76. Pavlo, A., et al.: Make your database system dream of electric sheep: towards self-driving operation. VLDB Endowment **14**(12), 3211–3221 (2021)
77. Peeters, R., Bizer, C.: Dual-objective fine-tuning of BERT for entity matching. VLDB Endowment, **14**(10) (2021)
78. Peeters, R., Bizer, C.: Using chatGPT for entity matching. In: New Trends in Database and Information Systems - ADBIS Short Papers, Doctoral Consortium and Workshops. CCIS, vol. 1850. Springer, Cham (2023)
79. Peeters, R., Der, R.C., Bizer, C..: WDC products: a multi-dimensional entity matching benchmark. In: International Conference on Extending Database Technology (EDBT). OpenProceedings.org (2024)
80. Peeters, R., Steiner, A., Bizer, C.: Entity matching using large language models. In: International Conference on Extending Database Technology (EDBT). OpenProceedings.org (2025)
81. Petricek, T., van den Burg, G.J.J., Nazábal, A., Ceritli, T., Jiménez-Ruiz, E., Williams, C.K.I.: AI assistants: a framework for semi-automated data wrangling. IEEE Trans. Knowl. Data Eng. **35**(9) (2023)
82. Pumma, S., Feng, W., Phunchongharn, P., Chapeland, S., Achalakul, T.: A runtime estimation framework for ALICE. Futur. Gener. Comput. Syst. **72**, 65–77 (2017)
83. Rahm, E., Bernstein, P.A.: A survey of approaches to automatic schema matching. VLDB J. **10**(4) (2001)
84. Redyuk, S., Kaoudi, Z., Schelter, S., Markl, V.: Assisted design of data science pipelines. VLDB J. **33**(4) (2024)
85. Rekatsinas, T., Chu, X., Ilyas, I.F., Ré, C.: Holoclean: holistic data repairs with probabilistic inference. VLDB Endowment, **10**(11) (2017)

86. Rusinkiewicz, M., Czejdo, B.D., Embley, D.W.: An implementation model for muldidatabase queries. In: International Conference on Database and Expert Systems Applications (DEXA). Springer, Cham (1991)

87. Sarawagi, S., Bhamidipaty, A.: Interactive deduplication using active learning. In: ACM SIGKDD International Conference on Knowledge Discovery and Data Mining (KDD). ACM (2002)

88. Seedat, N., van der Schaar, M.: Matchmaker: self-improving large language model programs for schema matching. *CoRR*, abs/2410.24105 (2024)

89. Sellami, R., Defude, B.: Complex queries optimization and evaluation over relational and NoSQL data stores in cloud environments. IEEE Trans. Big Data **4**(2), 217–230 (2018)

90. Shi, P., Zhao, Z., Zhong, H., Shen, H., Ding, L.: An improved agglomerative hierarchical clustering anomaly detection method for scientific data. Concurrency Comput. Pract. Experience, **33**(6) (2021)

91. Siddiqi, S., Kern, R., Boehm, M.: SAGA: a scalable framework for optimizing data cleaning pipelines for machine learning applications. SIGMOD, **1**(3) (2023)

92. Sienkiewicz, M., Wrembel, R.: Managing data in a big financial institution: Conclusions from a r&d project. In: Workshops of the EDBT/ICDT Joint Conference. CEUR Workshop Proceedings, vol. 2841 (2021)

93. Silva, J.A., Pereira, D.A.: A multiclass classification approach for incremental entity resolution on short textual data. Int. J. Bus. Intell. Data Min. **18**(2) (2021)

94. Simitsis, A., Skiadopoulos, S., Vassiliadis, P.: The history, present, and future of ETL technology (invited). In: International Workshop on Design, Optimization, Languages and Analytical Processing of Big Data (DOLAP) @EDBT/ICDT. CEUR Workshop Proceedings, vol. 3369, pp. 3–12 (2023)

95. Sobo, A., Mubarak, A., Baimagambetov, A., Polatidis, N.: Evaluating LLMs for code generation in HRI: a comparative study of chatgpt, gemini, and claude. Appl. Artif. Intell. **39**(1) (2025)

96. Song, S., Li, C., Zhang, X.: Turn waste into wealth: on simultaneous clustering and cleaning over dirty data. In: ACM SIGKDD International Conference on Knowledge Discovery and Data Mining. ACM (2015)

97. Strengholt, P.: Data management at scale: modern data architecture with data mesh and data fabric. O'Reilly (2023). ISBN 1098138864

98. Friedman, N.H.T.: Data hubs, data lakes and data warehouses: How they are different and why they are better together (2020). Gartner

99. Taheri, J., Zomaya, A.Y., Kassler, A.: vmbbprofiler: a black-box profiling approach to quantify sensitivity of virtual machines to shared cloud resources. Computing **99**(12), 1149–1177 (2017)

100. Tan, R., Chirkova, R., Gadepally, V., Mattson, T.G.: Enabling query processing across heterogeneous data models: a survey. In: IEEE Big Data, pp. 3211–3220 (2017)

101. Thirumuruganathan, S., et al.: Deep learning for blocking in entity matching: a design space exploration. VLDB Endowment, **14**(11) (2021)

102. Timakum, T., Lee, S., Hu, H., Song, I., Song, M.: DOLAP: a 25 year journey through research trends and performance (invited talk). In: International Workshop on Design, Optimization, Languages and Analytical Processing of Big Data (DOLAP), vol. 3653. CEUR-WS.org (2024)

103. Vaisman, A.A., Zimányi, E.: Data Warehouse Systems - Design and Implementation, 2nd edn. Data-Centric Systems and Applications. Springer, Cham (2022)

104. Wang, J., Trummer, I., Basu, D.: UDO: universal database optimization using reinforcement learning. VLDB Endowment, **14**(13) (2021)

105. Wang, Q., Vatsalan, D., Christen, P.: Efficient interactive training selection for large-scale entity resolution. In: Cao, T., Lim, E.-P., Zhou, Z.-H., Ho, T.-B., Cheung, D., Motoda, H. (eds.) PAKDD 2015. LNCS (LNAI), vol. 9078, pp. 562–573. Springer, Cham (2015). https://doi.org/10.1007/978-3-319-18032-8_44

106. Wang, R., Zhang, Y.: Pre-trained language models for entity blocking: a reproducibility study. In: Conference the North American Chapter of the Association for Computational Linguistics. ACL (2024)

107. Wang, T., et al.: Match, compare, or select? an investigation of large language models for entity matching. CoRR, abs/2405.16884 (2024)

108. Wenz, V., Kesper, A., Taentzer, G.: Clustering heterogeneous data values for data quality analysis. ACM J. Data Inf. Qual. 15(3) (2023)

109. Wiederhold, G.: Mediators in the architecture of future information systems. Computer, 25(3) (1992)

110. Witt, C., Bux, M., Gusew, W., Leser, U.: Predictive performance modeling for distributed batch processing using black box monitoring and machine learning. Inf. Syst. 82, 33–52 (2019)

111. Wrembel, R.: On three missing pieces in the data integration puzzle. In: Workshops of the EDBT/ICDT Joint Conference, vol. 3946. CEUR-WS.org (2025)

112. Wrembel, R., Abelló, A., Song, I.: DOLAP data warehouse research over two decades: trends and challenges. Inf. Syst. 85, 44–47 (2019)

113. Wu, R., Zhang, A., Ilyas, I.F., Rekatsinas, T.: Attention-based learning for missing data imputation in HoloClean. In: Conference on Machine Learning and Systems, (MLSys) (2020)

114. Xu, D., Xia, T., Hou, J., Xiang, Y., Xuan, Q.: MSR-GAN: multi-scales decomposition representations for unsupervised anomaly detection. Appl. Intell. 55(8) (2025)

115. Yamada, M., Kitagawa, H., Amagasa, T., Matono, A.: Augmented lineage: traceability of data analysis including complex UDF processing. VLDB J. 32(5) (2023)

116. Yoon, J., Jordon, J., van der Schaar, M.: GAIN: missing data imputation using generative adversarial nets. In: International Conference on Machine Learning (ICML). Proceedings of Machine Learning Research, vol. 80 (2018)

117. Zaharia, M., Ghodsi, A., Xin, R., Armbrust, M.: Lakehouse: a new generation of open platforms that unify data warehousing and advanced analytics. In: CIDR (2021)

118. Zeakis, A., Papadakis, G., Skoutas, D., Koubarakis, M.: Pre-trained embeddings for entity resolution: an experimental analysis. VLDB Endowment, 16(9) (2023)

119. Zhang, W., Lim, W.S., Butrovich, M., Pavlo, A.: The holon approach for simultaneously tuning multiple components in a self-driving database management system with machine learning via synthesized proto-actions. VLDB Endowment, 17(11) (2024)

120. Zhang, W., Wei, H., Sisman, B., Dong, X.L., Faloutsos, C., Page, D.: Autoblock: a hands-off blocking framework for entity matching. In: International Conference on Web Search and Data Mining (WSDM). ACM (2020)

121. Zhang, Y., Di, M., Luo, H., Xu, C., Tsai, R.T.: SMUTF: schema matching using generative tags and hybrid features. Inf. Syst. 133 (2025)

122. Zhang, Y., et al.: Schema matching using pre-trained language models. In: International Conferenc on Data Engineering (ICDE) (2023)

123. Zhu, J., Mao, Y., Chen, L., Ge, C., Wei, Z., Gao, Y.: Fusionquery: on-demand fusion queries over multi-source heterogeneous data. VLDB Endowment, 17(6) (2024)

Tutorial

Leveraging Machine Learning Techniques for Customer Data Deduplication - Hard-Won Lessons from a Real-World Project in the Financial Industry

Robert Wrembel$^{(\boxtimes)}$ ⓘ, Witold Andrzejewski ⓘ, Pawel Boiński ⓘ, and Bartosz Bębel ⓘ

Poznan University of Technology, Poznań, Poland
{robert.wrembel,witold.andrzejewski,pawel.boinski,
bartosz.bebel}@cs.put.poznan.pl

Abstract. This paper is associated with a tutorial presented at DEXA 2025 Conferences and Workshops. The tutorial shares the practical experience gained from a 3-year R&D project for a big financial institution in Poland. The project aimed at developing deduplication pipelines for customer records. It involved the development of two distinct end-to-end deduplication pipelines that are based on (1) statistical/probabilistic modeling and on (2) machine learning. This tutorial focuses on lessons learned from developing the *machine learning pipeline*, within the context of a real-world industrial setting. Moreover, this tutorial provides an overview of approaches to data deduplication, including the traditional state-of-the-art baseline deduplication pipeline, solutions based on machine learning and neural networks that apply pre-trained and large language models.

Keywords: data quality · data deduplication · entity resolution · entity matching · statistical modeling · machine learning · classification · neural networks · pre-trained language model · large language model

1 Introduction to Data Deduplication

Data stored in repositories of companies are often erroneous. Typical data errors include: missing, wrong, outdated, and misspelled values. On top of it, companies typically face the problem of multiple database records describing the same physical entity - these multiple records will further be called duplicates. Duplicates result among others from: (1) storing data in multiple but not synchronized data repositories, (2) using applications that do not check for duplicates while inserting or updating data, and (3) from erroneous data inserted into a system, i.e., when monitoring of data quality is missing. Being able to discover duplicates is of particular importance while dealing with personal data, e.g., in

© The Author(s), under exclusive license to Springer Nature Switzerland AG 2026
C. K. Leung et al. (Eds.): DaWaK 2025, LNCS 16048, pp. 39–52, 2026.
https://doi.org/10.1007/978-3-032-02215-8_3

healthcare, insurance, and financial industries. Such duplicates cause economic loses of an institution (e.g., the same content mailed multiple times to the same person), distort the results of analyses, increase customer dissatisfaction, and as a consequence, deteriorate the reputation of the institution. For this reason, multiple approaches to data deduplication (a.k.a. entity resolution, entity matching, entity reconciliation, record linkage) have been proposed in the research literature.

Techniques for discovering duplicates have been intensively researched over decades, resulting in advanced solutions that leverage: (1) rules, (2) statistical/probabilistic methods, (3) machine learning (ML), and (4) neural networks (NN), i.e., pre-trained language models, and recently, large language models (see for example [6,17,49] for the state-of-the-art solutions).

Despite these solutions have been developed, discovering duplicate data in large repositories is still very challenging. The challenge results form: (1) a large number of data to be searched and compared for duplicates, (2) the quality of data to be analyzed and deduplicated that is far from being perfect, and (3) the need of a human domain expert knowledge in the deduplication process. Moreover, for large data repositories, the exact number of duplicates is unknown, which further challenges the deduplication process, e.g., [40] and makes its verification difficult.

2   Data Deduplication Pipeline and Tasks

In general, in a data deduplication process, pairs of records are compared and their similarities are computed. In the naive approach, record comparison has a quadratic computational complexity, which is unacceptable for real applications. To handle a deduplication process more efficiently, a *base-line data deduplication pipeline* (BLDDP) was developed [17,19,23,36,49,50], see Fig. 1.

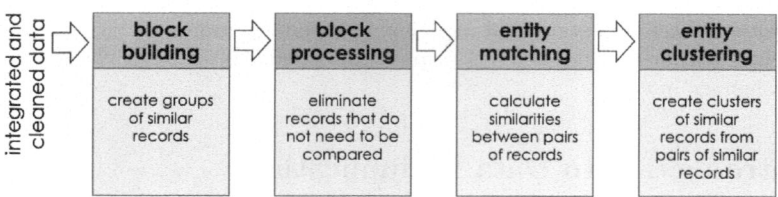

Fig. 1. The base-line data deduplication pipeline

The pipeline is equipped with multiple complex algorithms, supporting the main tasks in the pipeline. The tasks include: *block building* (a.k.a. blocking) - it organizes records into groups, such that each group includes records that seem to be similar; *block processing* - its goal is to eliminate records from groups created in the previous task that do not need to be compared, because they are likely not to be duplicates; *entity matching* - it computes similarity values between records compared in pairs; *entity clustering* - it aims at creating groups of similar records (that represent the same real-world entity) from the results of the previous task.

While building a deduplication pipeline, its designer has to provide solutions to the following detailed tasks [2]:

- T1: selecting the most adequate set of attributes A_i^G (i=1, ..., n) for dividing records into smaller groups G_i,
- T2: selecting a method for dividing records into groups G_i,
- T3: selecting an algorithm for comparing pairs of records in G_i,
- T4: tuning parameters of the selected algorithm for comparing records (e.g., window width in the sorted neighborhood algorithm),
- T5: selecting the most adequate attributes A_m^C (m = 1, ..., k) whose values will be compared in pairs of records in G_i,
- T6: selecting the most adequate similarity measures to compare values of A_m^C,
- T7: finding adequate weights of attributes A_m^C, as not every attribute is equally important in the comparison, i.e., not equally contributes to the overall similarity of pairs of records in G_i,
- T8: finding an overall similarity formula for a pair of records G_i being compared,
- T9: finding adequate similarity thresholds to distinguish between similarity classes: duplicates, probably duplicates, not duplicates,
- T10: building pairs of similar records,
- T11: building an overall graph of similarities between the compared pairs of records,
- T12: dividing the graph of similarities into sub-graphs, each of which represents groups of similar records, as there are typically more than two records referring to the same real-world object.

3   Solutions for Building Data Deduplication Pipelines

Multiple open-source and commercial systems supporting the development of data deduplication pipelines are available, see [34] for their short overviews. Nonetheless, their functionality, towards a fully automated development of tasks T1-T12 is limited and still needs to be researched. Much more advanced solutions come from the research; these solutions can be classified as based on: (1) rules, (2) statistical/probabilistic modeling, (3) machine learning, and (4) neural networks.

A **rule** system for deduplication (e.g., citeFJLM09,SM+17) is the most intuitive solution and the most easy to understand by a human. Typically, designing deduplication rules involves human experts and therefore is a time-consuming task. Furthermore, to deduplicate industry size data sets (with millions of rows and dozens of attributes), the set of rules becomes huge. Such a set is impossible to be managed without a dedicated software. Interesting challenges in this context include: detecting rule subsumptions and designing the sequence of rules to be fired when checking data. For these reasons, other approaches to deduplication have been proposed, as discussed further in this section.

The **statistical/probabilistic modeling** is the foundation of the other approaches. It requires that tasks T1-T12 are explicitly designed, implemented, and run. Even though the BLDDP was intensively researched (e.g., [5,15,17,19, 49]), easy and ready-to-use methods for tasks T1, T4, T5, T6, T7, T8, T9 still need to be investigated.

Basic **ML deduplication models** are based on standard classifiers (e.g., decision trees, logistic regression, SVM) to divide record pairs into classes of: duplicates, probably duplicates, and non-duplicates (e.g., [14,22,56,57]). ML algorithms are also used for learning blocking schemes - T2 (e.g., [33]), block processing - T3, T4 (e.g., [8,25]), and entity matching - T5-T10 (e.g., [18]). In active/self learning approaches (e.g., [14,31,64]), the tasks of *block building* and *entity matching* are often executed implicitly by a given ML model, learned from training data. The training data are labeled either by human experts or by a labeling ML model, like the *Snorkel* model available in Python.

More advanced ML methods apply complex **neural networks** (NNs), mainly to tasks T1, T2, T5-T9 (e.g., [12,13,30,39,60,68,69]) as the main component of **pre-trained language models** (PtLMs) or **large language models** (LLMs). PtLMs and LLMs use a data processing pipeline different than the BLDDP. First, data are transformed into a format suitable for a neural network. Second, the data are transformed into embeddings. Third, the embeddings are compared (i.e., *entity matching*) to discover duplicates. The matched pairs of entities must further be processed by non-neural-networks algorithms within the *entity clustering* step (from the BLDDP).

From methods based on complex NNs, *DeepER* [21] is the first solution which uses deep learning (DL) techniques. It applies recurrent neural networks (RNN) and word embeddings as entity representations, to discover similarities between entities. *DeepMatcher* [39] contributes the architectural template with three main modules for: (1) attribute embedding, (2) attribute similarity representation, and (3) classification. The authors compared the efficiency of various implementations of the modules by means of various DL techniques (e.g., word-level embeddings vs. character-level embeddings, RNN vs. Smooth Inverse Frequency vs. Attention vs. Hybrid) to implement modules for computing attribute similarities.

The application of PtLMs to data deduplication tasks gained popularity (e.g., [37,51,53,65]) as it alleviated the need for explicitly implementing the aforementioned tasks T5, T6, T7, T8, T9, and T10. A special type of a NN, called a transformer [61], became a core component natural language processing systems. In [12], the authors compared four transformer architectures, namely *BERT*, *XLNet*, *DistilBERT*, and *RoBERTa* on the task of deduplication. Experimental results showed the advantages of these solutions over earlier solutions, namely *Magellan* [34] and *DeepMatcher* [67]. [60] compared the use of different DL techniques like *Autoencoder*, *Seq2seq*, *CTT*, and *Sentence-BERT* for block building. The experiments showed that DL-based solutions outperformed non-DL solutions (*Blocking Scheme Learner* [38] and *Token Blocking* [44]).

A significant limitation of PtLMs is: (1) their need for extensive training data and (2) the observed decline in performance (typically measured by F1) when comparing records that were not included in the training set, as reported in [53]. These limitations have driven significant recent interest in LLMs (e.g., [29,42,52,54,66]).

LLMs can be categorized as: (1) static models, which associate every token with a fixed embeddings vector; (2) models based on *BERT*, which vectorize every token based on its context, and (3) *Sentence-BERT* models, which associate every sequence of tokens with a context-aware embeddings vector. [68] extended [12] by using this categorization to compare the efficiency of 12 LLMs. The experiments showed that *Sentence-BERT* models outperformed the two other models. [30] presented the first proposal for integrating an entity blocker and entity matcher into a single module using set of transformer-based PtLMs (e.g., BERT), to form a committee of encoders.

The main focus of research on LLMs for data deduplication has been on experimenting with various types of prompts. These prompts commonly employ strategies such as: (1) simple to complex text task descriptions (e.g., [52,54,66]), (2) the inclusion of learning examples (e.g., [52]), (3) the provision of explicit rules for comparing records (e.g., [52]), and (4) instructions to identify relationships between entities to support duplicate identification [29]. As reported in [52], the experimental comparisons of the performance (often by means of F1) of PtLMs and LLMs showed that LLMs alleviate the drawbacks of PtLMs and that LLMs offer better performance (typically measured by F1).

The approaches to deduplication based on ML and complex neural networks - PtLMs and LLMs face limitations in real-world scenarios. First, the availability of large, high-quality training datasets remains a challenge. Second, the opaqueness of neural network models limits their interpretability, making it difficult to understand and debug a deduplication process. Moreover, LLMs introduce another pitfall of time performance for large data sets being deduplicated, and as a consequence, additional monetary costs.

4   Our Approach to Data Deduplication

Drawing upon our project experience, this section contrasts research solutions for deduplication with business expectations. Furthermore, we present two deduplication pipelines developed within our work: the first leverages statistical/probabilistic modeling, and the second utilizes machine learning.

4.1   Data Deduplication: Research Vs. Business Projects

The execution of a 3-year data cleaning and deduplication project in the financial sector demonstrate that research solutions often face significant challenges in practical application, especially within the financial domain, as discussed below.

- The quality of data being deduplicated - in most of the approaches it is assumed that data were cleaned before delivering them into a deduplication system, regardless of the type of the system. However, in practice, data that arrive to the deduplication system are partially dirty. It is because for large data volumes, e.g., millions or dozens of millions of rows, it is infeasible to perfectly clean such massive data sets within the given project time frame.
- The sizes of deduplicated data - the research solutions were tested on much smaller data sets (e.g., [35,53,55] than in real business data deduplication projects that process millions or dozens of millions of rows.
- The availability of labeled training data for ML algorithms - in business projects such labeled data have to be created by human experts. For large data sets, it is impossible to label a training data set large enough for a data volume at hand, within the given project time frame. A promising labeling technique - active learning still requires substantial effort from experts. Auto-labeling solutions, like *Snorkel*, need human verification of labeled data as well.
- Available development environments - financial institutions are permitted to employ only software that has undergone strict internal certification processes, driven by security regulations. This, in turn, restricts development teams to a limited portfolio of off-the-shelf solutions.
- The performance of a deduplication pipeline - deduplication processes are time-consuming, and their runtime performance can become critical for massive datasets. Consequently, complex deduplication pipelines with advanced algorithms may not be suitable for multiple business scenarios.
- Understandability of deduplication processes - the developed solutions must be easy to understand, thus to maintain and extend, by the IT staff of a company. For this reason, intuitive and simple solutions are preferred.

4.2   Statistical/Probabilistic Pipeline

The statistical/probabilistic pipeline that we proposed is visualized in Fig. 2. It explicitly implements tasks T1-T12 (outlined in Sect. 2). The solutions implemented in the tasks are based on expert knowledge, rules, statistical/probabilistic modeling, and graph processing algorithms (see [4,5,11] for details).

4.3   ML Pipeline

As an alternative to the solution outlined in Sect. 4.2, we developed also the machine learning pipeline. Is is composed of seven tasks [3], as shown in Fig. 3.

In *task 1*, 1000 pairs of records were manually labeled by three domain experts. A label indicated one of the three classes, namely: true duplicates - T, probably duplicates - P, and non-duplicates - N. To automatically label a larger training set of pairs, in *task 2* the set of rules was built manually with the support of domain experts. The rules were next applied in *task 3* to automatically label a set of 2.5 million of pairs of customer records. In *task 4*, from the set

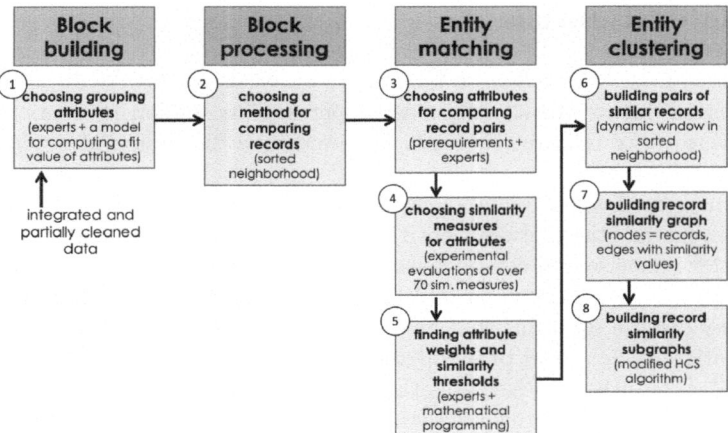

Fig. 2. The overall statistical/probabilistic data deduplication pipeline used in the business project reported in this tutorial

of pairs built in *task 3*, a training and a testing data sets were created by stratified sampling. Having created these two subsets, standard feature engineering techniques were applied to these subsets in *task 5*.

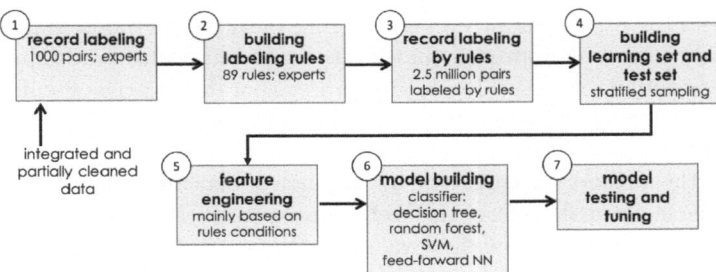

Fig. 3. The ML pipeline used in the business project reported in this tutorial

In *Task 6*, four distinct classification models were developed using the training set: a decision tree, a random forest, a support vector machine (SVM), and a feed-forward neural network. These models were built using Python package *sklearn*. Hyperparameters of the models were adjusted manually or (if needed) with the help of the *optuna* package, based on the dataset labeled by domain experts. Finally, in *task 7*, the quality of all the produced models was tested based on: (1) the testing set obtained from the automatically labeled data and (2) the set of 1000 pairs manually labeled by domain experts. The model quality was measured by means of precision, recall, accuracy, and F1.

5   Related Tutorials

To the best of our knowledge, so far over 10 tutorials on data deduplication was proposed and delivered by world class experts. This section presents a chronological overview of relevant tutorials, followed by a discussion of our proposal's unique contributions.

[27] surveys classical techniques for entity resolution (ER), algorithms for computing similarities between entities as well as rule-based and probabilistic methods for pairwise and cluster-based ER. Subsequently, the authors recognize the importance of solutions to multi-entity ER, which are important for Web data. They discuss solutions that are based on clustering algorithms, probabilistic generative models, and probabilistic logical languages.

In [26], the author investigates the sources of uncertainty inherent in entity resolution. The work categorizes these uncertainties, identifies their impact across various entity resolution tasks, and presents a set of algorithms designed to address them. These algorithms leverage the probability theory in various tasks of an entity resolution pipeline.

[46] focuses on novel blocking algorithms designed for deduplicating data from the Web. The authors discuss also performance issues of these algorithms in the context of processing large volumes of data. They present two techniques for improving the performance, i.e., meta-blocking and parallel processing. The tutorial was enhanced with the inclusion of new blocking algorithms and presented by the same research team two years later [47].

In [32], the authors begin by presenting and analyzing the advantages and disadvantages of five popular privacy-preserving blocking methods. Subsequently, they introduce a novel approach that leverages locality-sensitive hashing (LSH) to identify similar records. This approach incorporates a privacy enhancement component, where data are anonymized (encoded) by means of the Bloom filter, prior to applying the LSH.

In [20], the authors recognize the importance of benchmarks for entity matching, with the focus on data from the Web. Subsequently, they discuss benchmarks for linked data and cover real-data vs. synthetic-data benchmarks.

The challenges posed by data from the Web are recognized also in [58]. The tutorial focuses on ER for data in the RDF format, by means of: (1) Web-scale blocking and meta-blocking techniques, (2) iterative techniques, and (3) progressive techniques, which attempt to discover as many matches as possible, within a given limited computing budget.

[48] focuses also on ER for data from the Web. It discusses challenges introduced by Web data (such as heterogeneity, volume, and quality) in the design of a data deduplication pipeline. For such data, schema-agnostic algorithms are needed - the tutorial introduces a suite of such algorithms. Furthermore, the tutorial covers tasks of the BLDDP and examines parallelization strategies to improve their efficiency.

[7] surveys ML techniques applied to various data management tasks, with a particular focus on data integration challenges. This includes data integration at both the data and schema levels, as well as error detection, cleaning, and

deduplication. Its focus on data deduplication covers: exploring the use of (semi-) supervised learning models for learning similarity and blocking functions as well as the role of active learning in ER.

[28] focuses on the following aspects of ER: injecting domain knowledge in the process of discovering matching entities, automation, and explainability. The first aspect is covered by the survey of crowd-sourcing techniques, the second - by active learning and deep learning, the third one - by discussing techniques for explainable ER.

[16] is the most comprehensive tutorial on ER from the ones reviewed here. It covers: (1) an introduction to data linkage, (2) the data linkage process with the focus on cleaning, blocking, similarity, linkage complexity, and accuracy, (3) advanced data linkage techniques like tuning blocking parameters, training data, processing graphs of similar entities, and (4) privacy.

[45] includes a systematic overview of the evolution of entity resolution pipelines. The pipelines are divided into four generations, each of which focuses on a different challenge posed by data. The first generation (G1) addresses data veracity and adequate deduplication techniques. G2 adds volume - in order to handle it, the focus is put on parallelization. G3 adds variety - techniques for ER on data from the Web are discussed. G4 adds velocity - time constraints are defined, and solutions like progressive and real-time ER are presented. Finally, G5 covers novel techniques for ER, like crowd-sourcing and deep learning. This tutorial was enhanced with the inclusion of new algorithms and the integration of LLMs and it was presented in [41].

The aforementioned tutorials cover theoretical and practical solutions to ER. Most of the solutions use publicly available data sets. [7,28,41,45] address ER techniques supported by ML. However, none of the tutorials listed in this section reports findings from a real R&D project done for a big company from the financial industry.

The main **differences** between our tutorial and the aforementioned ones include:

- Our tutorial presents **practical solutions** to tasks T1-T12, which have been verified on a large real data set of customer records and deployed in the production infrastructure of a large financial institution. Notice that: (1) this customer data set is larger than reported in the related research literature and (2) the quality of deduplicated data was worse than assumed in the related research literature (which introduced yet another challenge to the already difficult problem).
- The tutorial shares experience that we gained in the project, where we implemented two alternative deduplication pipelines, i.e., *statistical/probabilistic* and *ML*. Both pipelines are comparatively assessed in our tutorial, identifying their advantages and limitations. We also discuss challenges in annotating data for ML algorithms.
- Lastly, the tutorial provides a *unique comparative analysis* of the underlying assumptions in research projects and the specific assumptions and requirements inherent in the real-world R&D project referred to.

6   Tutorial Scope and Focus

This tutorial shares our experience in developing and deploying two alternative end-to-end deduplication pipelines for customer records, capable of processing over 20 million records per batch. The experience is based on a R&D project conducted in years 2020–2023 for a big financial institution in Poland. This tutorial is an extended version of [2], where we covered the *statistical/probabilistic pipeline*. This tutorial focuses mainly on the *ML pipeline*.

The tutorial is divided into three parts. The *first* one introduces the fundamental concepts and techniques for deduplication followed by a brief description of the BLDDP. The *statistical/probabilistic pipeline* developed within the project, extends the BLDDP in order to address specific project goals by means of tasks T1–T12, which we briefly discuss. We also point out to the problem of partially dirty data in the deduplication pipeline.

In the *second* part, we outline our *ML pipeline* applied to deduplicating large customer data sets. We show how we coped with the limited availability of labeled data for training and testing ML models, and share our findings on difficulty of properly detecting different similarity classes. We compare results of applying the most popular classification techniques to the same deduplication problem and compare them with the results obtained from the *statistical/probabilistic pipeline*. Finally, we present a is a short survey of the most important AI/ML techniques for data deduplication, like:

- active learning (e.g., [14,56,59]),
- learning similarity measures (e.g., [10,18]),
- entity blocking (e.g., [9,24,38]),
- generalized entity matching (e.g., [1,62,63]),
- AutoML in entity matching (e.g., [43]),
- deep learning (e.g., [13,68,69]),
- PtLMs and LLMs for ER (e.g., [12,52,60,66]).

The *third* part of this tutorial details key lessons learned from designing the two deduplication pipelines outlined in Sect. 4, with a particular focus on the *ML pipeline*. It further discusses critical project findings, considering both technical challenges and the non-functional requirements imposed by the financial institution. Finally, we present a comparison between the actual requirements of a real-world data deduplication project for a large financial institution and those typically assumed in research (e.g., data quality, data volume, performance, deployment architecture, and software), demonstrating their frequent incompatibility in practice.

References

1. Ahmadi, N., Sand, H., Papotti, P.: Unsupervised matching of data and text . In: International Conference on Data Engineering (ICDE). IEEE Computer Society (2022)

2. Andrzejewski, W., Bębel, B., Boiński, P., Wrembel, R.: On customer data deduplication - research vs. industrial perspective:. In: Tekli, J., et al. (eds.) ADBIS 2024. Communications in Computer and Information Science, vol. 2186, pp. 392-400. Springer, Cham (2024). https://doi.org/10.1007/978-3-031-70421-5_37
3. Andrzejewski, W., Bębel, B., Boiński, P., Kowalewska, J., Marszalek, A., Wrembel, R.: Statistical modeling vs. machine learning for deduplication of customer records. In: Workshops of the EDBT/ICDT Joint Conference. CEUR Workshop Proceedings, vol. 3651. CEUR-WS.org (2024)
4. Andrzejewski, W., Bębel, B., Boiński, P., Sienkiewicz, M., Wrembel, R.: Text similarity measures in a data deduplication pipeline for customers records. In: International Workshop on Design, Optimization, Languages and Analytical Processing of Big Data (DOLAP) @(EDBT/ICDT. CEUR Workshop Proceedings, vol. 3369, pp. 33–42. CEUR-WS.org (2023)
5. Andrzejewski, W., Bębel, B., Boiński, P., Wrembel, R.: On tuning parameters guiding similarity computations in a data deduplication pipeline for customers records: experience from a R&D project. Inf. Syst. **121** (2024)
6. Barlaug, N., Gulla, J.A.: Neural networks for entity matching: a survey. ACM Trans. Knowl. Discov. Data **15**(3) (2021)
7. Berti-Équille, L., Bonifati, A., Milo, T.: Machine learning to data management: a round trip. In: IEEE International Conference on Data Engineering (ICDE) (2018)
8. Bianco, G.D., Gonçalves, M.A., Duarte, D.: BLOSS: effective meta-blocking with almost no effort. Inf. Syst. **75** (2018)
9. Bilenko, M., Kamath, B., Mooney, R.J.: Adaptive blocking: learning to scale up record linkage. In: International Conference on Data Mining (ICDM). IEEE Computer Society (2006)
10. Bilenko, M., Mooney, R.J.: Adaptive duplicate detection using learnable string similarity measures. In: ACM SIGKDD International Conference on Knowledge Discovery and Data Mining (KDD). ACM (2003)
11. Boiński, P., Andrzejewski, W., Bębel, B., Wrembel, R.: On tuning the sorted neighborhood method for record comparisons in a data deduplicaton pipeline: industrial experience report. In: Strauss, C., Amagasa, T., Kotsis, G., Tjoa, A.M., Khalil, I. (eds.) DEXA 2023. LNCS, vol. 14146, pp. 164–178. Springer, Cham (2023). https://doi.org/10.1007/978-3-031-39847-6_11
12. Brunner, U., Stockinger, K.: Entity matching with transformer architectures - a step forward in data integration. In: International Conference on Extending Database Technology (EDBT), pp. 463–473. OpenProceedings.org (2020)
13. Chen, R., Shen, Y., Zhang, D.: GNEM: a generic one-to-set neural entity matching framework. In: The Web Conference (WWW). ACM (2021)
14. Chen, X., Xu, Y., Broneske, D., Durand, G.C., Zoun, R., Saake, G.: Heterogeneous committee-based active learning for entity resolution (HeALER). In: Welzer, T., Eder, J., Podgorelec, V., Kamišalić Latifić, A. (eds.) ADBIS 2019. LNCS, vol. 11695, pp. 69–85. Springer, Cham (2019). https://doi.org/10.1007/978-3-030-28730-6_5
15. Christen, P.: A survey of indexing techniques for scalable record linkage and deduplication. IEEE Trans. Knowl. Data Eng. **24**(9) (2012)
16. Christen, P.: Record linkage: introduction, recent advances, and privacy issues (part 1, 2, 3, 4) (2019). youtube.com. Accessed December 2024
17. Christophides, V., Efthymiou, V., Palpanas, T., Papadakis, G., Stefanidis, K.: An overview of end-to-end entity resolution for big data. ACM Comput. Surv. **53**(6) (2021)

18. Cohen, W.W., Richman, J.: Learning to match and cluster large high-dimensional data sets for data integration. In: ACM SIGKDD International Conference on Knowledge Discovery and Data Mining (KDD). ACM (2002)
19. Colyer, A.: The morning paper on An overview of end-to-end entity resolution for big data (2020). https://blog.acolyer.org/2020/12/14/entity-resolution/
20. Daskalaki, E., Saveta, T., Fundulaki, I., Herschel, M.: A tutorial on instance matching benchmarks. In: European Semantic Web Conference (ESWC) (2016)
21. Ebraheem, M., Thirumuruganathan, S., Joty, S., Ouzzani, M., Tang, N.: Distributed representations of tuples for entity resolution. VLDB Endow. **11**(11) (2018)
22. Elfeky, M.G., Verykios, V.S., Elmagarmid, A.K.: Tailor: a record linkage tool box. In: International Conference on Data Engineering (ICDE). IEEE Computer Society (2002)
23. Elmagarmid, A.K., Ipeirotis, P.G., Verykios, V.S.: Duplicate record detection: a survey. IEEE Trans. Knowl. Data Eng. **19**(1) (2007)
24. Evangelista, L.O., Cortez, E., da Silva, A.S., Jr, W.M.: Adaptive and flexible blocking for record linkage tasks. J. Inf. Data Manage. **1**(2) (2010)
25. Gagliardelli, L., Papadakis, G., Simonini, G., Bergamaschi, S., Palpanas, T.: Generalized supervised meta-blocking. VLDB Endow. **15**(9) (2022)
26. Gal, A.: Tutorial: uncertain entity resolution. Re-evaluating entity resolution in the big data era. VLDB Endow. **7**(13) (2014)
27. Getoor, L., Machanavajjhala, A.: Entity resolution: theory, practice & open challenges. VLDB Endow. **5**(12) (2012)
28. Gurajada, S., Popa, L., Qian, K., Sen, P.: Learning-based methods with human-in-the-loop for entity resolution. In: International Conference on Information and Knowledge Management (CIKM). ACM (2019)
29. Huang, Z.: Disambiguate entity matching using large language models through relation discovery. In: Conference on Governance, Understanding and Integration of Data for Effective and Responsible AI (GUIDE-AI). ACM (2024)
30. Jain, A., Sarawagi, S., Sen, P.: Deep indexed active learning for matching heterogeneous entity representations. VLDB Endow. **15**(1) (2021)
31. Jurek, A., Hong, J., Chi, Y., Liu, W.: A novel ensemble learning approach to unsupervised record linkage. Inf. Syst. **71** (2017)
32. Karapiperis, D., Verykios, V.S., Katsiri, E., Delis, A.: A tutorial on blocking methods for privacy-preserving record linkage. In: Karydis, I., Sioutas, S., Triantafillou, P., Tsoumakos, D. (eds.) ALGOCLOUD 2015. LNCS, vol. 9511, pp. 3–15. Springer, Cham (2016). https://doi.org/10.1007/978-3-319-29919-8_1
33. Kejriwal, M., Miranker, D.P.: A two-step blocking scheme learner for scalable link discovery. In: International Workshop on Ontology Matching @ISWC, vol. 1317, pp. 49–60. CEUR-WS.org (2014)
34. Konda, P., et al.: Magellan: toward building entity matching management systems. Technical report, University of Wisconsin-Madison (2016)
35. Konda, P., et al.: Magellan: toward building entity matching management systems. VLDB Endow. **9**(12) (2016)
36. Köpcke, H., Rahm, E.: Frameworks for entity matching: a comparison. Data Knowl. Eng. **69**(2) (2010)
37. Li, Y., Li, J., Suhara, Y., Doan, A., Tan, W.: Deep entity matching with pre-trained language models. VLDB Endow. **14**(1) (2020)
38. Michelson, M., Knoblock, C.A.: Learning blocking schemes for record linkage. In: National Conference on Artificial Intelligence and Innovative Applications of Artificial Intelligence. AAAI Press (2006)

39. Mudgal, S., et al.: Deep learning for entity matching: a design space exploration. In: SIGMOD International Conference on Management of Data. ACM (2018)
40. Nanayakkara, C., Christen, P., Christen, V.: Unsupervised evaluation of entity resolution. J. Data Inf. Qual. **17**(1) (2025)
41. Nikoletos, K., Ioannou, E., Papadakis, G.: The five generations of entity resolution on web data. In: Stefanidis, K., Systä, K., Matera, M., Heil, S., Kondylakis, H., Quintarelli, E. (eds.) ICWE 2024. LNCS, vol. 14629, pp. 469–473. Springer, Cham (2024)
42. Nuntachit, N., Sugannasil, P.: Can ChatGPT outperform other language models? An experiment on using ChatGPT for entity matching versus other language models. In: Barolli, L. (eds.) 3PGCIC 2023. LNDECT, vol. 189, pp. 14–26. Springer, Cham (2023). https://doi.org/10.1007/978-3-031-46970-1_2
43. Paganelli, M., Buono, F.D., Pevarello, M., Guerra, F., Vincini, M.: Automated machine learning for entity matching tasks. In: International Conference on Extending Database Technology (EDBT). OpenProceedings.org (2021)
44. Papadakis, G., Ioannou, E., Niederée, C., Fankhauser, P.: Efficient entity resolution for large heterogeneous information spaces. In: International Conference on Web Search and Data Mining (WSDM). ACM (2011)
45. Papadakis, G., Ioannou, E., Palpanas, T.: Entity resolution: past, present and yet-to-come. In: International Conference on Extending Database Technology (EDBT) (2020)
46. Papadakis, G., Palpanas, T.: Blocking techniques for web-scale entity resolution. In: International Conference on Web Information System Engineering (WISE) (2014)
47. Papadakis, G., Palpanas, T.: Blocking for large-scale entity resolution: challenges, algorithms, and practical examples. In: IEEE International Conference on Data Engineering (ICDE) (2016)
48. Papadakis, G., Palpanas, T.: Web-scale, schema-agnostic, end-to-end entity resolution. In: The WEB Conference (2018)
49. Papadakis, G., Skoutas, D., Thanos, E., Palpanas, T.: Blocking and filtering techniques for entity resolution: a survey. ACM Comput. Surv. **53**(2) (2020)
50. Papadakis, G., Tsekouras, L., Thanos, E., Giannakopoulos, G., Palpanas, T., Koubarakis, M.: Domain- and structure-agnostic end-to-end entity resolution with Jedai. SIGMOD Rec. **48**(4) (2019)
51. Peeters, R., Bizer, C.: Dual-objective fine-tuning of BERT for entity matching. VLDB Endow. **14**(10) (2021)
52. Peeters, R., Bizer, C.: Using ChatGPT for entity matching. In: Abelló, A., et al. (eds.) ADBIS 2023. CCIS, vol. 1850, pp. 221–230. Springer, Cham (2023). https://doi.org/10.1007/978-3-031-42941-5_20
53. Peeters, R., Der, R.C., Bizer, C.: WDC products: a multi-dimensional entity matching benchmark. In: International Conference on Extending Database Technology (EDBT). OpenProceedings.org (2024)
54. Peeters, R., Steiner, A., Bizer, C.: Entity matching using large language models. In: International Conference on Extending Database Technology (EDBT). OpenProceedings.org (2025)
55. Primpeli, A., Bizer, C.: Profiling entity matching benchmark tasks. In: International Conference on Information and Knowledge Management (CIKM). ACM (2020)
56. Sarawagi, S., Bhamidipaty, A.: Interactive deduplication using active learning. In: ACM SIGKDD International Conference on Knowledge Discovery and Data Mining (KDD). ACM (2002)

57. Silva, J.A., Pereira, D.A.: A multiclass classification approach for incremental entity resolution on short textual data. Int. J. Bus. Intell. Data Min. **18**(2) (2021)
58. Stefanidis, K., Christophides, V., Efthymiou, V.: Web-scale blocking, iterative and progressive entity resolution. In: IEEE International Conference on Data Engineering (ICDE) (2017)
59. Tejada, S., Knoblock, C.A., Minton, S.:. Learning domain-independent string transformation weights for high accuracy object identification. In: ACM SIGKDD International Conference on Knowledge Discovery and Data Mining (KDD). ACM (2002)
60. Thirumuruganathan, S., et al.: Deep learning for blocking in entity matching: a design space exploration. VLDB Endow. **14**(11) (2021)
61. Vaswani, A., et al.: Attention is all you need. In International Conference on Neural Information Processing Systems (NIPS). Curran Associates Inc. (2017)
62. Wang, J., Li, Y., Hirota, W.: Machamp: a generalized entity matching benchmark. In: International Conference on Information & Knowledge Management (CIKM). ACM (2021)
63. Wang, P., Zeng, X., Chen, L., Ye, F., Mao, Y., Zhu, J., Gao, Y.: Promptem: prompt-tuning for low-resource generalized entity matching. VLDB Endow. **16**(2) (2022)
64. Wang, Q., Vatsalan, D., Christen, P.: Efficient interactive training selection for large-scale entity resolution. In: Cao, T., Lim, E.-P., Zhou, Z.-H., Ho, T.-B., Cheung, D., Motoda, H. (eds.) PAKDD 2015. LNCS (LNAI), vol. 9078, pp. 562–573. Springer, Cham (2015). https://doi.org/10.1007/978-3-319-18032-8_44
65. Wang, R., Zhang, Y.: Pre-trained language models for entity blocking: a reproducibility study. In: Conference the North American Chapter of the Association for Computational Linguistics. ACL (2024)
66. Wang, T., et al.: Match, compare, or select? An investigation of large language models for entity matching. CoRR, abs/2405.16884 (2024)
67. Xie, T., Dai, K., Wang, K., Li, R., Zhao, L.: DeepMatcher: a deep transformer-based network for robust and accurate local feature matching. Expert Syst. Appl. **237** (2024)
68. Zeakis, A., Papadakis, G., Skoutas, D., Koubarakis, M.: Pre-trained embeddings for entity resolution: an experimental analysis. VLDB Endow. **16**(9) (2023)
69. Zhang, W., Wei, H., Sisman, B., Dong, X.L., Faloutsos, C., Page, D.: Autoblock: a hands-off blocking framework for entity matching. In: International Conference on Web Search and Data Mining (WSDM). ACM (2020)

Data Mining and Knowledge Discovery

Data Mining and Automated Discovery

FairFES - Fast Exact Sampling for Fair Classification

Manh Khoi Duong$^{(\boxtimes)}$ ⓘ, Nina A. Liebrand ⓘ, and Stefan Conrad ⓘ

Heinrich Heine University, Universitätsstraße 1, 40225 Düsseldorf, Germany
{manh.khoi.duong,nina.liebrand,stefan.conrad}@hhu.de

Abstract. While traditional fairness metrics like statistical disparity evaluate equal treatment across social groups, they fail to account for real-world constraints, such as in hiring, where a predefined acceptance rate across all groups is demanded. To achieve fair datasets for machine learning, we propose two fast and exact sampling methods, FairFESDown and FairFESUp, that are capable of aligning datasets with a specified targeted fairness goal. Unlike existing methods, our approaches have a linear time complexity regarding the datasets' sizes and can handle non-binary protected attributes. We evaluate our methods on several popular classifiers and datasets from the fairness literature, achieving optimal fairness with statistical disparity scores close to zero while maintaining classification performances similar to the original datasets. Our pre-processing methods outperform existing approaches, including FairGAN, FairSMOTE, and FairUS, regarding statistical disparity, classification accuracy, and runtime.

Keywords: Machine Learning · Fairness · Pre-processing · Sampling Methods · AI Act

1 Introduction

With the increasing use of machine learning in sensitive applications such as criminal justice, hiring, or healthcare, fairness has become a critical concern [3,11,24,35,36]. Since algorithmic decision-making systems are shaped by their training data, pre-processing has emerged as a strategy to enhance fairness. Pre-processing techniques reduce biases at an early stage by ensuring that outcomes in the original data are equal across different social groups. These techniques aim to reduce discrimination that fairness metrics such as statistical disparity [6] capture.

While numerous fairness criteria have been proposed over the years for classification [36], regression [1], ranking [2,28], and more, we argue that these criteria are too imprecise for several real-world applications. They overlook the role of under- or overachieving societal constraints, such as legally mandated quotas, which decision-makers must often adhere to. We consider a scenario in which a

---

M. K. Duong and N. A. Liebrand—Equal contribution.

© The Author(s), under exclusive license to Springer Nature Switzerland AG 2026
C. K. Leung et al. (Eds.): DaWaK 2025, LNCS 16048, pp. 55–69, 2026.
https://doi.org/10.1007/978-3-032-02215-8_4

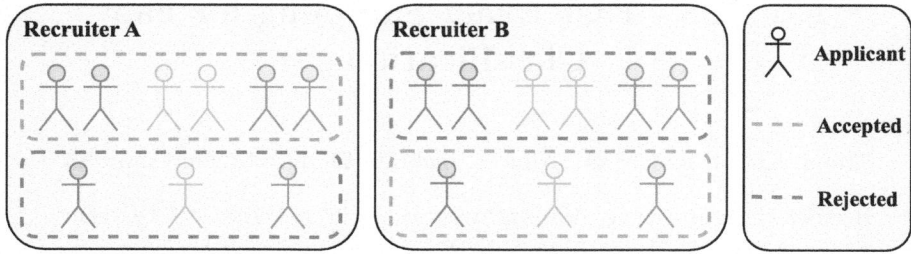

Fig. 1. Both recruiters A and B treat all groups (red, blue, and yellow) equally and do not exhibit statistical disparity. However, both recruiters have different acceptance rates ($A : 66.\overline{3}\%, B : 33.\overline{3}\%$).

company employs two recruiters, A and B, both tasked with selecting candidates from a diverse applicant pool consisting of red, blue, and yellow candidates, as shown in Fig. 1. If we measure the pairwise statistical disparity score between all groups [36], no group is favored by either recruiter. However, both recruiters have different acceptance rates, and this is not captured by known fairness metrics. The company could favor one of the recruiters depending on the number of available positions.

Another critical limitation of fairness metrics in machine learning is their limitation on binary protected attributes [7,8,10,25,33]. Binarized attributes categorize individuals into two broad groups (e.g., white/non-white, young/old), yet real-world demographics are often more complex [14]. Methods limited to binarized attributes ignore discrimination that small and marginalized groups may experience [14]. Thus, such simplifications even risk reinforcing existing inequalities for minority groups. Recent research has begun to address fairness beyond binary group classifications, including the introduction of generalized fairness definitions to account for multiple protected groups [15,18,21,22,31]. To address these limitations, this paper comes with a threefold contribution:

- Motivated by the work of Pleiss et al. [29], we introduce the concept of *targeted fairness* as a flexible fairness parameter that explicitly specifies the proportion of favorable outcomes that all groups must fulfill.
- The criteria are generalized to account for protected attributes with more than two groups. This approach avoids the need to binarize protected attributes, ensuring marginalized groups are not merged with others.
- We present two methods for solving the proposed fairness problem, a downsampling and an upsampling approach, released on GitHub (https://github.com/mkduong-ai/fairfes).

2   Related Work

Bias mitigation techniques in machine learning include pre-processing methods that deal with bias before the training process [3]. This section reviews existing pre-processing methods that are sampling-based and hence are directly comparable with our proposed methods.

One of the earliest works that addressed this topic was proposed by Kamiran and Calders [20]. In their work, they aim to reduce statistical disparity in datasets, that is, the extent to which privileged and unprivileged groups receive different outcomes. To achieve this, they proposed methods based on label modification, resampling, and sample weighting. Recent developments have proposed methods that use synthetic samples to balance outcomes [8,10,13,27].

FairSMOTE [8] utilizes the popular *Synthetic Minority Over-sampling Technique* (SMOTE) [9] for generating synthetic samples to balance outcomes of groups. The dataset is first divided into four partitions based on combinations of the protected attribute (privileged and unprivileged) and binary label (favorable and unfavorable). The number of samples in smaller partitions is increased to match the size of the largest partition, ensuring that all groups have an equal probability (50%) of receiving a favorable outcome.

FairGAN by Patrikar et al. [27] uses a *Conditional Tabular Generative Adversarial Network* (CTGAN) [34] to generate synthetic tabular datasets. The methodology of FairGAN is the same as FairSMOTE [8].

The approach of FairUS [10] differs from the previous methods in that the machine learning model's utility is explicitly considered in the optimization process in addition to fairness. This requires training a machine learning model on an upsampled dataset. Similar to the work of Patrikar et al. [27], CTGANs are also used for the generation of samples. The relative amount of samples to add (λ-values) is found using a *Tree-structured Parzen Estimator* (TPE) [5]. This meta-optimization technique aims to minimize model bias and maximize model utility. The result is a Pareto front of λ-solutions.

All of these methods come with limitations, such as not being able to deal with non-binary protected attributes and not being able to specify a target probability for receiving a favorable outcome.

3 Preliminaries

Definition 1 (Fairness Criterion [15]). *Let E_1, E_2 be events and Z be a random variable representing the protected attribute with values in G. The treatment of group $i \in G$ is defined as $Pr[E_1 \mid E_2, Z = i]$. A (group) fairness criterion requires equal treatment across groups:*

$$Pr[E_1 \mid E_2, Z = i] = Pr[E_1 \mid E_2, Z = j] \quad \forall i, j \in G.$$

E_1 typically denotes a favorable outcome, while E_2 often represents a qualification condition.

Definition 2 (Statistical Disparity (SDP) [36]). *Statistical disparity quantifies the absolute difference between the treatment probabilities of two groups $i, j \in G$:*

$$SDP := |Pr[E_1 \mid Z = i] - Pr[E_1 \mid Z = j]|.$$

Remark 1. Definition 1 only requires probabilities to be equal for all groups. This may not be useful in some practical scenarios. For instance, rejecting all applicants for a job would not be different from accepting all of them, according to this definition. Yet, both options may be undesirable in real-world settings, where policies could dictate a minimum acceptance rate.

Definition 3 (Targeted Fairness). *Motivated by Pleiss et al. [29], we define a targeted fairness criterion that requires treatment probabilities to be equal to a specific target $t_p \in [0, 1]$:*

$$Pr[E_1 \mid E_2, Z = i] = t_p,$$

for all $i \in G$.

Remark 2. It is easy to see that a targeted fairness criterion is a special case of a fairness criterion as it refines its definition. A stakeholder can set, for example, an acceptance rate of $t_p = 0.5$ for each social group when applying for a job.

Definition 4 (Maximum Target Disparity (MTD)). *A possible discrimination measure that aims for the satisfaction of targeted fairness is given by:*

$$MTD := \max_{i \in G} |Pr[E_1 \mid E_2, Z = i] - t_p|.$$

Remark 3. The goal is to minimize the maximum disparity between the observed probabilities and the target t_p. If the maximum target disparity is zero, then the targeted fairness criterion is satisfied. In order for this to happen, there must exist a fraction for all groups, under the given events, that is equal to the target t_p. However, this is rarely the case. Therefore, it is enough to aim for the lowest feasible maximum target disparity.

Remark 4. Minimizing MTD is equivalent to solving following optimization problems:
$$\min |Pr[E_1 \mid E_2, Z = i] - t_p| \quad \forall i \in G.$$

4   Fair Fast Exact Sampling

The objective of this work is to address the problem of achieving fairness in decision-making by systematically resampling a given dataset $\mathcal{D}$. Specifically, we aim to modify $\mathcal{D}$ such that all groups achieve a uniform, predefined target probability t_p for receiving a specified outcome. This process involves either upsampling or downsampling individuals. Let $\hat{p}_i$ describe the observed acceptance rate for group i in dataset $\mathcal{D}$, then the goal is to add or remove samples from the dataset such that $\hat{p}_i = t_p$ for all $i \in G$.

4.1   Empirical Treatment Probability

To calculate MTD, we first need to calculate group treatments. In frequentist statistics, this can be done using empirical counts. For this, let us denote the number of applicants in group i as:

$$n_i = |\{Z = i\} \cap E_2|, \tag{1}$$

and the number of applicants in group i that are accepted as:

$$k_i = |\{Z = i\} \cap E_1 \cap E_2|. \tag{2}$$

The frequentist probabilities of the treatments are then given by:

$$\hat{p}_i := \Pr[E_1 \mid E_2, Z = i] = \frac{k_i}{n_i}. \tag{3}$$

4.2   Sampling Methods

We present two methods for solving the fairness problem: a downsampling and an upsampling approach. The number of observations to remove or add is determined based on the initial acceptance rate and the desired target. For both methods, we consider the following three cases based on $\hat{p}_i$:

1. $\hat{p}_i > t_p$: The group has a higher acceptance rate than desired.
   - Downsampling: Remove favorable outcomes.
   - Upsampling: Add unfavorable outcomes.
2. $\hat{p}_i < t_p$: The group has a lower acceptance rate than desired.
   - Downsampling: Remove unfavorable outcomes.
   - Upsampling: Add favorable outcomes.
3. $\hat{p}_i = t_p$: Nothing needs to be done as the acceptance rate is already met.

Downsampling. The downsampling method adjusts the dataset by selectively removing observations. In the first case $(\hat{p}_i > t_p)$, favorable observations are removed. We denote its count as $m_{i+} \in \{1, \dots, k_i\}$. By removing m_{i+} favorable observations, the total number of samples in group i is reduced as well. Specifically, we aim to find a value m_{i+} such that the acceptance rate achieves the target:

$$\frac{k_i - m_{i+}}{n_i - m_{i+}} \approx t_p. \tag{4}$$

Conversely, when the initial acceptance rate of the group is below the target, we remove unfavorable outcomes in order to increase the acceptance rate. In this scenario, the adjusted acceptance rate is given by:

$$1 - \frac{(n_i - k_i) - m_{i-}}{n_i - m_{i-}} \approx t_p, \tag{5}$$

where $m_{i-} \in \{1, \ldots, (n_i - k_i)\}$ is the number of removed observations for unfavorable outcomes. We note that $\frac{(n_i - k_i)}{n_i}$ describes the rejection rate for group i, i.e., the fraction of those who received an unfavorable outcome. Thus, m_{i-} is subtracted from both the total number of unfavorable outcomes for group i and group size n_i, resulting in a new rejection rate. Subtracting this rejection rate from 1 yields the updated acceptance rate.

The respective values for m_{i+} and m_{i-} can be derived using simple arithmetical operations. They are given by:

$$m_{i+} = \left\lfloor \frac{(n_i \cdot t_p) - k_i}{t_p - 1} \right\rfloor, \tag{6}$$

$$m_{i-} = \left\lfloor \frac{(n_i \cdot t_p) - k_i}{t_p} \right\rfloor. \tag{7}$$

Since m_{i+} and m_{i-} represent counts, the results are rounded down to the nearest integer.

Upsampling. In a similar manner, we can duplicate specific observations to decrease or increase the acceptance rates. In the first case $(\hat{p}_i > t_p)$, we add m_{i-} unfavorable outcomes to reduce the acceptance rate:

$$1 - \frac{(n_i - k_i) + m_{i-}}{n_i + m_{i-}} \approx t_p, \tag{8}$$

and in the second case $(\hat{p}_i < t_p)$, m_{i+} favorable outcomes are added to increase the acceptance rate:

$$\frac{k_i + m_{i+}}{n_i + m_{i+}} \approx t_p. \tag{9}$$

The number of observations to add are respectively given by:

$$m_{i+} = \left\lfloor \frac{(n_i \cdot t_p) - k_i}{1 - t_p} \right\rfloor, \tag{10}$$

$$m_{i-} = \left\lfloor -\frac{(n_i \cdot t_p) - k_i}{t_p} \right\rfloor. \tag{11}$$

Algorithm. Building on these calculations, the *Fair Fast Exact Sampling* algorithm (FairFES) adjusts the distribution of favorable and unfavorable outcomes across protected groups to match the target probability. Algorithm 1 shows both down- and upsampling methods, referred to as FairFESDown and FairFESUp, respectively.

Algorithm 1. Fair Fast Exact Sampling (FairFESDown/FairFESUp)

Require: Dataset $\mathcal{D}$, target probability t_p.
Ensure: Down- or upsampled dataset $\mathcal{D}$.
1: Compute n_i, k_i and $\hat{p}_i$ for each group $i \in G$
2: **for** each $i \in G$ **do**
3:   **if** $\hat{p}_i > t_p$ **then**
4:     FairFESDown: randomly remove m_{i+} favorable samples of i in $\mathcal{D}$
5:     FairFESUp: randomly replicate m_{i-} unfavorable samples of i in $\mathcal{D}$
6:   **end if**
7:   **if** $\hat{p}_i < t_p$ **then**
8:     FairFESDown: randomly remove m_{i-} unfavorable samples of i in $\mathcal{D}$
9:     FairFESUp: randomly replicate m_{i+} favorable samples of i in $\mathcal{D}$
10:   **end if**
11: **end for**
12: **return** $\mathcal{D}$

---

5 Experiments

To evaluate the effectiveness of FairFES, we conduct three experiments. First, we test FairFES on a synthetic dataset to assess its ability to meet our proposed fairness objectives. In the second experiment, we evaluate the runtime of FairFES to determine its efficiency in handling large datasets. Lastly, we compare FairFES against existing pre-processing methods on real-world datasets to evaluate its impact on classifiers' fairness and performance. Overall, we aim to answer the following research questions with our experiments:

- **RQ1:** Can FairFES accurately adjust the distribution of group outcomes to meet any target probability?
- **RQ2:** Does FairFES effectively handle fairness constraints for group sizes greater than two?
- **RQ3:** How does FairFES compare to existing methods in terms of fairness improvement, classification performance, and runtime efficiency?

5.1 Experiment 1: Achieving Targeted Fairness

To assess whether our proposed FairFES methods successfully achieve the specified target probabilities for all groups (**RQ1, RQ2**), we conducted an experiment by generating a dataset. The dataset consists of 10^6 samples and is partitioned into 10 groups. Each group was assigned an initial acceptance rate ranging from 0.25 to 0.70 with evenly spaced 0.05 intervals. The target probability was set to $t_p = 0.4$, meaning that after applying FairFES, each group should ideally have an acceptance rate close to t_p.

Table 1. FairFES down- and upsampling results on a synthetic dataset.

Baseline				FairFESDown			FairFESUp		
Group	k_i	n_i	$\hat{p}_i$	m_{i+}	m_{i-}	$\hat{p}_i$	m_{i+}	m_{i-}	$\hat{p}_i$
0	24 921	99 822	0.25	0	37 520	0.40	25 013	0	0.40
1	29 951	99 851	0.30	0	24 974	0.40	16 649	0	0.40
2	35 245	100 102	0.35	0	11 990	0.40	7 993	0	0.40
3	40 028	99 830	0.40	160	0	0.40	0	240	0.40
4	44 908	99 746	0.45	8 349	0	0.40	0	12 524	0.40
5	50 573	100 740	0.50	17 128	0	0.40	0	25 692	0.40
6	54 873	99 830	0.55	24 902	0	0.40	0	37 352	0.40
7	60 062	100 049	0.60	33 404	0	0.40	0	50 106	0.40
8	65 062	100 018	0.65	41 758	0	0.40	0	62 637	0.40
9	70 163	100 012	0.70	50 264	0	0.40	0	75 395	0.40

Fairness Results. Table 1 shows the down- and upsampling results for each group. The baseline column reports the initial treatment of each group. If we consider the discrimination between groups, we observe a maximum statistical disparity score of 0.45 (group 0 vs. group 9). Group 9 is farthest from the target probability of 0.4, resulting in an MTD of 0.30. The debiased datasets after down- and upsampling successfully adjusted all groups' acceptance rates to the target probability. Although the results are shown rounded to two decimals, rounding to five decimals does not alter the outcome, MTD remains 0 for both methods. These results successfully answer the research questions **RQ1** and **RQ2** that FairFES aligns datasets for more than two groups to a specified target probability with negligible errors.

5.2   Experiment 2: Runtime Analysis

While achieving fairness is important, we cannot ignore the runtime of FairFES (**RQ3**). In this experiment, numerous datasets with varying sizes n and numbers of groups $|G|$ were synthetically generated. To assess the runtimes, either n or $|G|$ was fixed and the other parameter was varied. For these parameters, the following values were used:

- **Dataset Sizes:** Logarithmically evenly spaced from 10^3 to 10^7 samples.
- **Number of Groups:** Logarithmically evenly spaced from 2 to 10^3.

To ensure reliability of the results, all experiments were repeated for 10 trials.

Runtime Results. Before diving into the results of the empirical runtime experiments, we formally prove the runtime complexity of the down- and upsampling methods. The empirical experiments then serve as a validation of our theoretical analysis.

Theorem 1. *Let n be the number of samples in the dataset and $|G|$ be the number of groups in the protected attribute. Both FairFES down- and upsampling algorithms have a worst-case time complexity of $\mathcal{O}(n)$, if we disregard the number of groups $|G|$ as it is often a small number.*

Proof. Both methods loop through the dataset once to count specific occurrences of each group. Counting the occurrences of all groups takes $\mathcal{O}(n)$ operations. With the occurrences counted, the number of modifications (m_{i+} or m_{i-}) to adjust the dataset is calculated for each group. This requires basic arithmetic operations in $\mathcal{O}(1)$, resulting in a time complexity of $\mathcal{O}(|G|)$, as we iterate over all groups. After the number of modifications is calculated, the dataset is adjusted by adding or removing observations using random sampling. Modifying the dataset accordingly can be done in $\mathcal{O}(1)$ operations if a list is used. Altogether, this results in a runtime complexity of $\mathcal{O}(n + |G|)$. Because $|G|$ is often a small number and fixed, we can disregard $|G|$, which results in a runtime complexity of $\mathcal{O}(n)$. □

The results, as shown in Fig. 2a and Fig. 2b, confirm that the runtime of down- and upsampling increases linearly with both the dataset size n and the number of groups $|G|$. The plots show the average runtime and standard deviations over the 10 trials. Both proposed methods exhibit nearly identical runtimes across all tested configurations. Notably, for a dataset size of 10^7, both algorithms took less than one second to finish, demonstrating their efficiency in handling large datasets. Even with a dataset size of 10^6 and 10^3 groups, both methods completed the pre-processing in 5.32 seconds. The standard deviations remain consistently low, with the maximum observed fluctuation being below $0.05\,\text{s}$, ensuring stable performance across different settings.

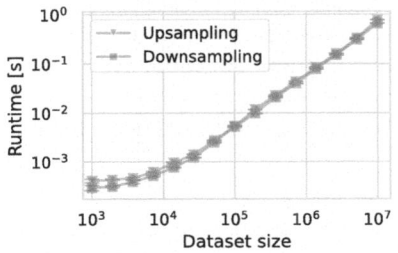 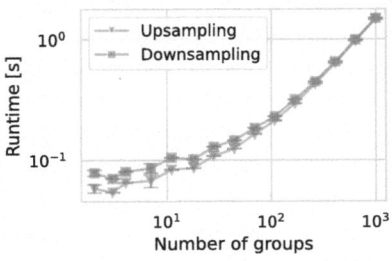

(a) Mean runtimes for different dataset sizes (fixed groups: 2).

(b) Mean runtimes for different number of groups (fixed dataset size: 10^6).

Fig. 2. Log-log plots of the mean runtimes of FairFES methods.

Table 2. Overview of real-world datasets used in experiment 3.

Dataset	Label	Protected Attribute	#Features	#Sample
Adult	Income	Sex: Male, Female Race: American-Indian-Eskimo, Others	22	32 561
Bank	Term Deposit	Job: Entrepreneur, Others	53	41 188
COMPAS	Recidivism	Race: Asian, Others	8	7 214

5.3 Experiment 3: Real-World Datasets

In the following experiment, we compare the performances of the FairFES methods against established fair pre-processing techniques. The experimental pipeline consists of the following steps. Each dataset is initially split randomly into a training set (80%) and a test set (20%). Only the training set is then processed using pre-processing methods. Next, multiple classifiers are trained on both the pre-processed datasets and the original dataset, which serves as a baseline. After training, the models on the unaltered test set are evaluated, where both fairness and classification performance are assessed.

Dataset. To ensure a comprehensive evaluation across different domains, we utilize three popular datasets from the fairness literature: Adult [23], Bank [26], and COMPAS [24] (Table 2). Adult is commonly used to predict income, Bank to predict the success of bank marketing campaigns, and COMPAS for the prediction of recidivism outcomes. It is important to note that all of these datasets are prepared to consist of only binary protected attributes due to the limitations of the established pre-processing methods. This allows for a meaningful comparison of FairFES with existing methods. For the protected attributes, we use the attributes that are known to be discriminatory in these datasets. We intentionally decide to use different groups as the privileged and unprivileged groups than the ones used in previous studies [4,10] to acknowledge understudied minorities.

Methods. The three pre-processing methods FairGAN [33], FairSMOTE [8], and the recently introduced method FairUS [10] are selected for comparison. Because the authors of FairUS compared themselves against FairSMOTE and FairGAN, and made their implementation publicly available, we adopted their code and kept all hyperparameters of the pre-processing methods as provided. For comparative reasons, for FairUS, we consider the Pareto front solution that is closest to the ideal solution using linear scalarization with equal weights to balance fairness and accuracy. This approach ensures that the best trade-off between minimizing statistical disparity and maintaining classification performance is achieved (no bias, maximum theoretical utility). We standardize the experimental setup by using a target probability of $t_p = 0.5$ for FairFES, as other methods naturally aim for this target.

Table 3. Overview of the results of pre-processing methods on real-world datasets trained on the MLP classifier.

Dataset	Method	AUROC	F1	SDP Before	SDP After	SDP Pred	Runtime [s]
Adult (R)	Baseline	0.73 ± 0.03	0.60 ± 0.05	0.17 ± 0.00	0.17 ± 0.00	0.20 ± 0.05	0.00 ± 0.00
	FairFESDown	0.80 ± 0.01	0.65 ± 0.02		$\mathbf{0.00 \pm 0.00}$	0.31 ± 0.05	$\mathbf{0.11 \pm 0.08}$
	FairFESUp	$\mathbf{0.81 \pm 0.01}$	$\mathbf{0.66 \pm 0.01}$		$\mathbf{0.00 \pm 0.00}$	0.27 ± 0.04	0.22 ± 0.14
	FairGAN	0.73 ± 0.03	0.60 ± 0.05		0.43 ± 0.01	$\mathbf{0.16 \pm 0.04}$	4088.40 ± 1363.90
	FairSMOTE	0.80 ± 0.02	0.65 ± 0.02		0.78 ± 0.02	0.24 ± 0.03	1787.86 ± 785.82
	FairUS	0.74 ± 0.02	0.62 ± 0.02		0.17 ± 0.01	$\mathbf{0.16 \pm 0.02}$	7789.36 ± 2792.73
Adult (S)	Baseline	0.72 ± 0.04	0.59 ± 0.07	0.20 ± 0.00	0.20 ± 0.00	0.15 ± 0.04	0.00 ± 0.00
	FairFESDown	0.79 ± 0.02	0.64 ± 0.02		$\mathbf{0.00 \pm 0.00}$	0.19 ± 0.04	0.11 ± 0.04
	FairFESUp	$\mathbf{0.80 \pm 0.01}$	0.65 ± 0.01		$\mathbf{0.00 \pm 0.00}$	0.21 ± 0.03	$\mathbf{0.08 \pm 0.08}$
	FairGAN	0.74 ± 0.04	0.61 ± 0.04		$\mathbf{0.00 \pm 0.00}$	0.18 ± 0.06	1463.04 ± 753.26
	FairSMOTE	0.79 ± 0.02	$\mathbf{0.65 \pm 0.02}$		$\mathbf{0.00 \pm 0.00}$	0.23 ± 0.07	534.57 ± 137.55
	FairUS	0.73 ± 0.04	0.59 ± 0.07		0.07 ± 0.01	$\mathbf{0.15 \pm 0.03}$	1454.86 ± 819.68
Bank	Baseline	0.70 ± 0.10	0.45 ± 0.12	0.03 ± 0.00	0.03 ± 0.00	0.03 ± 0.02	0.00 ± 0.00
	FairFESDown	0.77 ± 0.09	0.45 ± 0.11		$\mathbf{0.00 \pm 0.00}$	$\mathbf{0.02 \pm 0.02}$	$\mathbf{0.24 \pm 0.15}$
	FairFESUp	$\mathbf{0.81 \pm 0.05}$	$\mathbf{0.49 \pm 0.09}$		$\mathbf{0.00 \pm 0.00}$	0.03 ± 0.02	0.35 ± 0.18
	FairGAN	0.73 ± 0.11	0.43 ± 0.15		$\mathbf{0.00 \pm 0.00}$	0.09 ± 0.05	5360.25 ± 2229.55
	FairSMOTE	0.73 ± 0.07	0.43 ± 0.09		$\mathbf{0.00 \pm 0.00}$	0.12 ± 0.11	55655.03 ± 14168.47
	FairUS	0.64 ± 0.08	0.38 ± 0.16		0.07 ± 0.00	0.03 ± 0.02	9898.54 ± 3690.13
COMPAS	Baseline	0.71 ± 0.01	0.67 ± 0.02	0.17 ± 0.00	0.17 ± 0.00	0.18 ± 0.11	0.00 ± 0.00
	FairFESDown	$\mathbf{0.71 \pm 0.02}$	$\mathbf{0.68 \pm 0.01}$		$\mathbf{0.00 \pm 0.00}$	0.17 ± 0.14	0.08 ± 0.08
	FairFESUp	$\mathbf{0.71 \pm 0.01}$	$\mathbf{0.68 \pm 0.02}$		$\mathbf{0.00 \pm 0.00}$	0.16 ± 0.11	$\mathbf{0.03 \pm 0.04}$
	FairGAN	0.70 ± 0.02	0.66 ± 0.03		$\mathbf{0.00 \pm 0.00}$	0.14 ± 0.12	1064.21 ± 123.03
	FairSMOTE	$\mathbf{0.71 \pm 0.01}$	0.67 ± 0.02		$\mathbf{0.00 \pm 0.00}$	0.23 ± 0.10	242.35 ± 128.97
	FairUS	0.70 ± 0.02	0.65 ± 0.03		0.03 ± 0.03	0.15 ± 0.10	2843.16 ± 921.66

Classifier. The debiased datasets, as well as the baseline, are used to train a range of classifiers, including Logistic Regression [12], Decision Tree [30], Random Forest [19], Multi-Layer Perceptron (MLP) [32], and k-Nearest Neighbors [17]. Each classifier is evaluated using Monte Carlo cross-validation over 10 independent trials to account for the variability in the results. This means that the split of the data into training and test sets is done randomly for each trial.

Evaluation Metrics. The fairness of the generated datasets is quantified using statistical disparity. Here, we measure the statistical disparity score before (SDP Before) and after (SDP After) pre-processing as well as the statistical disparity of the classifiers' predictions (SDP Pred). For each classifier, we measure the classification performance using Area Under Curve Receiver Operating Characteristics (AUROC) [16] and F1-score. These metrics are chosen to capture different aspects of model performance. AUROC provides insight into the overall discriminative power of the classifier, while the F1-score balances precision and recall, which is particularly important for imbalanced datasets. Finally, we evaluate the runtime of each pre-processing method to measure computational efficiency.

This is particularly important for larger datasets, where preprocessing time can become a bottleneck in real-world applications.

Results on Real-World Datasets. The primary objective of this comparison is to evaluate the relative efficiency of FairFES in terms of computational runtime, while also assessing its ability to maintain fairness and prediction performance in classification tasks.

Due to brevity reasons, we display the results for the classifier MLP in Table 3. Our proposed FairFESDown and FairFESUp methods demonstrated a significant advantage in terms of computational efficiency. Their runtimes were nearly as fast as not performing any pre-processing method and are in the range of milliseconds. In contrast, other methods, such as FairGAN, FairSMOTE, and FairUS, took minutes to several hours to complete. In terms of predictive performance, FairFESUp led to a clear improvement in AUROC compared to the baseline and all other methods. FairFESDown also performed well and comes with minimal or no sacrifices in terms of both AUROC and F1-score.

From a fairness perspective, our methods achieved optimal SDP and MTD scores of 0.0 while keeping classification performance high. In contrast, FairGAN and FairSMOTE sometimes led to worsened fairness after preprocessing.

6   Conclusion

In this work, we proposed *FairFESDown* and *FairFESUp*, a pair of sampling-based techniques to address the new fairness challenge of *targeted fairness*, which is more specific than traditional fairness criteria. The targeted fairness criterion makes it possible to specifically demand a certain probabilistic outcome for all social groups, e.g., a predefined acceptance rate in a hiring scenario. Further, our definition covers group sizes greater than two, which is more practical in real-world scenarios where protected attributes, such as ethnicity, often comprise multiple categories. The experiments demonstrated that our methods are capable of optimally dealing with the stated fairness issues as measured by our introduced *maximum target disparity* score as well as by statistical disparity. By comparing against recent pre-processing methods with real-world data, we further validated the efficacy of our methods. The prediction performance of classifiers when trained on the *FairFES* pre-processed datasets was maintained or even improved compared to the unprocessed datasets in terms of AUROC and F1-score. Our work comes with a theoretical analysis of the runtime, which was proven to be linear regarding the dataset size. Numerous empirical experiments on synthetic and real-world datasets validated our theoretical analysis. This makes FairFESDown and FairFESUp particularly appealing as efficient and effective pre-processing techniques for fairness-aware machine learning.

References

1. Agarwal, A., Dudík, M., Wu, Z.S.: Fair regression: quantitative definitions and reduction-based algorithms. In: International Conference on Machine Learning, pp. 120–129. PMLR (2019)
2. Asudeh, A., Jagadish, H.V., Stoyanovich, J., Das, G.: Designing fair ranking schemes. In: Proceedings of the 2019 International Conference on Management of Data, pp. 1259–1276 (2019). https://doi.org/10.1145/3299869.3300079
3. Barocas, S., Hardt, M., Narayanan, A.: Fairness and Machine Learning. fairmlbook.org (2019)
4. Bellamy, R.K., et al.: Ai fairness 360: an extensible toolkit for detecting and mitigating algorithmic bias. IBM J. Res. Dev. **63**(4/5), 4:1–4:15 (2019)
5. Bergstra, J., Bardenet, R., Bengio, Y., Kégl, B.: Algorithms for hyper-parameter optimization. In: Advances in Neural Information Processing Systems, vol. 24 (2011)
6. Calders, T., Kamiran, F., Pechenizkiy, M.: Building classifiers with independency constraints. In: 2009 IEEE International Conference on Data Mining Workshops, pp. 13–18 (2009). https://doi.org/10.1109/ICDMW.2009.83
7. Caton, S., Haas, C.: Fairness in machine learning: a survey. ACM Comput. Surv. (2023). https://doi.org/10.1145/3616865
8. Chakraborty, J., Majumder, S., Menzies, T.: Bias in machine learning software: Why? How? What to do? In: Proceedings of the 29th ACM Joint Meeting on European Software Engineering Conference and Symposium on the Foundations of Software Engineering, pp. 429–440 (2021)
9. Chawla, N.V., Bowyer, K.W., Hall, L.O., Kegelmeyer, W.P.: SMOTE: synthetic minority over-sampling technique. J. Artif. Int. Res. **16**(1), 321–357 (2002)
10. Cohen-Inger, N., Rozenblatt, G., Cohen, S., Rokach, L., Shapira, B.: Fairusupsampling optimized method for boosting fairness. In: 27th European Conference on Artificial Intelligence, ECAI 2024, pp. 962–970 (2024)
11. Corbett-Davies, S., Pierson, E., Feller, A., Goel, S., Huq, A.: Algorithmic decision making and the cost of fairness. In: Proceedings of the 23rd ACM SIGKDD International Conference on Knowledge Discovery and Data Mining, pp. 797–806 (2017)
12. Cox, D.R.: The regression analysis of binary sequences. J. Roy. Stat. Soc.: Ser. B (Methodol.) **20**(2), 215–232 (1958)
13. Duong, M.K., Conrad, S.: Dealing with data bias in classification: can generated data ensure representation and fairness? In: Wrembel, R., Gamper, J., Kotsis, G., Tjoa, A.M., Khalil, I. (eds.) Big Data Analytics and Knowledge Discovery. Lecture Notes in Computer Science, vol. 14148, pp. 176–190. Springer, Cham (2023). https://doi.org/10.1007/978-3-031-39831-5_17
14. Duong, M.K., Conrad, S.: Towards fairness and privacy: a novel data pre-processing optimization framework for non-binary protected attributes. In: Benavides-Prado, D., Erfani, S., Fournier-Viger, P., Boo, Y.L., Koh, Y.S. (eds.) Data Science and Machine Learning: 21st Australasian Conference, AusDM 2023, Auckland, New Zealand, December 11–13, 2023, Proceedings, pp. 105–120. Springer Nature Singapore, Singapore (2024). https://doi.org/10.1007/978-981-99-8696-5_8
15. Duong, M.K., Conrad, S.: Measuring and mitigating bias for tabular datasets with multiple protected attributes. In: Calegari, R., Dignum, V., O'Sullivan, B. (eds.) Proceedings of the 2nd Workshop on Fairness and Bias in AI co-located with 27th European Conference on Artificial Intelligence (ECAI 2024), Santiago de Compostela, Spain, October 20th, 2024. CEUR Workshop Proceedings, vol. 3808 (2024)

16. Fawcett, T.: An introduction to roc analysis. Pattern Recogn. Lett. **27**(8), 861–874 (2006). https://doi.org/10.1016/j.patrec.2005.10.010
17. Fix, E., Hodges, J.L.: Discriminatory analysis. nonparametric discrimination: consistency properties (1951)
18. Ghosh, A., Genuit, L., Reagan, M.: Characterizing intersectional group fairness with worst-case comparisons. In: Artificial Intelligence Diversity, Belonging, Equity, and Inclusion, pp. 22–34. PMLR (2021)
19. Ho, T.K.: Random decision forests. In: Proceedings of 3rd International Conference on Document Analysis and Recognition, vol. 1, pp. 278–282. IEEE (1995)
20. Kamiran, F., Calders, T.: Data preprocessing techniques for classification without discrimination. Knowl. Inf. Syst. **33**(1), 1–33 (2012)
21. Kang, J., Xie, T., Wu, X., Maciejewski, R., Tong, H.: InfoFair: information-theoretic intersectional fairness. In: 2022 IEEE International Conference on Big Data (Big Data), pp. 1455–1464 (2022)
22. Kearns, M., Neel, S., Roth, A., Wu, Z.S.: An empirical study of rich subgroup fairness for machine learning. In: Proceedings of the Conference on Fairness, Accountability, and Transparency, pp. 100–109. FAT* '19, Association for Computing Machinery, New York, NY, USA (2019). https://doi.org/10.1145/3287560.3287592
23. Kohavi, R.: Scaling up the Accuracy of Naive-Bayes Classifiers: a Decision-Tree Hybrid, pp. 202–207. KDD'96, AAAI Press (1996)
24. Larson, J., Angwin, J., Mattu, S., Kirchner, L.: Machine bias (2016). www.propublica.org/article/machine-bias-risk-assessments-in-criminal-sentencing
25. Mehrabi, N., Morstatter, F., Saxena, N., Lerman, K., Galstyan, A.: A survey on bias and fairness in machine learning. ACM Comput. Surv. (CSUR) **54**(6), 1–35 (2021)
26. Moro, S., Cortez, P., Rita, P.: A data-driven approach to predict the success of bank telemarketing. Decis. Support Syst. **62**, 22–31 (2014)
27. Patrikar, A.M., Mahenthiran, A., Said, A.: Leveraging synthetic data for AI bias mitigation. In: Synthetic Data for Artificial Intelligence and Machine Learning: Tools, Techniques, and Applications, vol. 12529, pp. 185–190. SPIE (2023)
28. Patro, G.K., Porcaro, L., Mitchell, L., Zhang, Q., Zehlike, M., Garg, N.: Fair ranking: a critical review, challenges, and future directions. In: Proceedings of the 2022 ACM Conference on Fairness, Accountability, and Transparency, pp. 1929–1942. FAccT '22, Association for Computing Machinery (2022)
29. Pleiss, G., Raghavan, M., Wu, F., Kleinberg, J., Weinberger, K.Q.: On fairness and calibration. In: Guyon, I., Luxburg, U.V., Bengio, S., Wallach, H., Fergus, R., Vishwanathan, S., Garnett, R. (eds.) Advances in Neural Information Processing Systems, vol. 30. Curran Associates, Inc. (2017)
30. Quinlan, J.R.: Induction of decision trees. Mach. Learn. **1**(1), 81–106 (1986)
31. Strotherm, J., Ashraf, I., Hammer, B.: Fairness-enhancing classification methods for non-binary sensitive features: how to fairly detect leakages in water distribution systems. PeerJ Comput. Sci. **10**, e2317 (2024)
32. Werbos, P.J.: Backpropagation through time: what it does and how to do it. Proc. IEEE **78**(10), 1550–1560 (1990)
33. Xu, D., Yuan, S., Zhang, L., Wu, X.: FairGAN: fairness-aware generative adversarial networks. In: 2018 IEEE International Conference on Big Data (big data), pp. 570–575. IEEE (2018)
34. Xu, L., Skoularidou, M., Cuesta-Infante, A., Veeramachaneni, K.: Modeling tabular data using conditional GAN. In: Advances in Neural Information Processing Systems, vol. 32 (2019)

35. Zafar, M.B., Valera, I., Gomez Rodriguez, M., Gummadi, K.P.: Fairness beyond disparate treatment & disparate impact: Learning classification without disparate mistreatment. In: Proceedings of the 26th International Conference on World Wide Web (2017). https://doi.org/10.1145/3038912.3052660
36. Žliobaitė, I.: Measuring discrimination in algorithmic decision making. Data Min. Knowl. Disc. **31**(4), 1060–1089 (2017). https://doi.org/10.1007/s10618-017-0506-1

Autism Detection by Analyzing Handwriting Characteristics of Chinese Characters via Deep Learning Models

Yunrui Li[1], Jasin Wong[1], Eva E. Chen[1], Syauki Aulia Thamrin[2],
and Arbee L. P. Chen[1,2]($\boxtimes$)

[1] National Tsing Hua University, Hsinchu, Taiwan
[2] Asia University, Taichung, Taiwan
arbee@asia.edu.tw

Abstract. Autism is a neurodevelopmental disorder that often manifests in childhood, characterized by social difficulties and repetitive behaviors. Motor impairments are also common, with autistic children experiencing challenges in converting sequential actions into integrated movements, affecting fine motor skills and daily activities like handwriting. This study explores the handwriting characteristics of autistic children through an analysis of Chinese characters. Our research has two objectives.First, we identify the handwriting characteristics of autistic children and typically-developing childrenthrough an analysis of Chinese characters. Second, we establish neatness criteria to assess whether a Chinese character is written neatly, aiming to understand handwriting neatness in both groups and to train a classification model using only neatly written characters. The dataset is derived from elementary school workbooks, reflecting real-life situations. To analyze handwriting features, we employ various machine learning/deep learning models equipped with the Class Activation Map (CAM) technique to provide the interpretability of the models by the visualization of the models' focal points. By applying oversampling techniques for data balancing, our model achieves an F1-score of 0.9720, surpassing previous studies on classifying autistic children and typically-developing children. Our findings contribute to a deeper understanding of autistic children's handwriting characteristics, providing insights that can support early detection and intervention.The models can therefore serve as a valuable tool for educators and parents to assess the children's autistic tendencies and support their handwriting development.

Keywords: Handwriting characteristics · Chinese characters · Classification models · Class Activation Map · Model interpretability

1 Introduction

Autism spectrum disorder (ASD) is a neurodevelopmental disorder that is typically identified in childhood. ASD is characterized by signs such as social chal-

© The Author(s), under exclusive license to Springer Nature Switzerland AG 2026
C. K. Leung et al. (Eds.): DaWaK 2025, LNCS 16048, pp. 70–85, 2026.
https://doi.org/10.1007/978-3-032-02215-8_5

lenges, language difficulties, repetitive behaviors, restricted interests, and over-sensitivity or insensitivity to sensory stimuli [1–3]. In addition to the cardinal symptoms of autism, autistic children often experience challenges in motor coordination and planning, resulting in clumsiness [4]. Autistic children may exhibit difficulties in various daily life movements, including maintaining posture [5], playing sports [6], and grasping or manipulating materials with different weights and shapes [7]. Specifically, autistic children often experience challenges in fore-seeing motor outcomes, conceiving goal-directed motor acts, integrating the sensory stimuli followed by movements, conducting sequential movements, and forming coordinated motor responses [8]. Moreover, deficiencies in hand-eye coordination can affect autistic children's learning performance which requires motor precision and accuracy, such as reading and handwriting [9]. Numerous studies have explored the handwriting difficulties experienced by autistic children [10–12]. Previous research has indicated various deviated handwriting performances among autistic children. As illustrated in Fig. 1, (A) shows a template, and (B) demonstrates more disconnected strokes, inconsistent letter sizes, and irregular shapes than (C) (both are from autistic children). It is therefore important to note that there is a large within-group difference among the autism population in their handwriting abilities and performance.

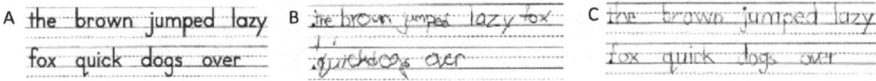

Fig. 1. A is the template, and B and C are both written by individuals with ASD [10].

The differences between Chinese characters and English letters span multiple aspects, with a significant distinction in their glyph structures [13]. For instance, the structure of English letters is relatively simple, typically composed of a small number of curves and straight lines. In contrast, Chinese characters are characterized by rich strokes and intricate structures, incorporating various substructures and organizational patterns that deeply impact the meaning and context of the characters. On the other hand, the common curves and circles found in English letters are relatively less prevalent in Chinese characters, where the predominant elements are straight lines and horizontal strokes. Such distinctions also affect the motor planning and coordination in the process of handwriting, presenting a potential challenge for the learners, particularly for individuals with autism. Therefore, investigating the specific manifestations of Chinese handwriting is a challenging yet meaningful task for research in autistic children.

In this study, we first examined how neatly the Chinese characters were written. Here, the *neatness* refers to the cleanliness of handwriting, encompassing the consistency of character shapes and the organized arrangement of characters. In educational settings, teachers often place significant emphasis on the neatness of handwriting when assessing a child's writing performance. Neatness is not only associated with the aesthetic aspect of writing, it is also crucial for readability. If a child's handwriting is neat, with well-organized text and consistent character forms, the reading experience becomes smoother, and comprehension becomes more accessible. In this research, we have developed a classification model designed specifically to evaluate the neatness of Chinese character handwriting. By introducing the neatness label, the model learns to distinguish between neat and non-neat Chinese characters. If the model's predictions align with the assessment of the neatness from teachers, it can then be used to assist teachers in grading students' writing assignments. This makes the model a practical tool, supporting teachers in more effectively evaluating and guiding students' writing abilities, ultimately enhancing learning outcomes.

Writing is a crucial aspect of language learning for children, and Chinese character writing places higher demands on hand-eye coordination and motor control. To delve further into this matter, we chose to analyze and compare the Chinese handwriting of autistic and TD children. We developed a classification model to decide whether a Chinese handwriting is from autistic or typically-developing (TD) children with the following objectives. Our first objective is to design neatness criteria to decide whether a Chinese character is written neatly. The purpose of this is to track the neatness of handwriting for both autistic and TD children. By using only neatly written Chinese characters, we plan to train the classification model for the writing from autistic and TD children. This constitutes a more challenging task, as distinguishing the handwriting characteristics between autistic and TD children becomes even more intricate when all Chinese characters are written neatly. The purpose of this approach is to further assess the model's ability to differentiate the handwriting styles of autistic children from TD children when only neatly written Chinese characters are considered.

CAM (Class Activation Map) [15] is a visualization technique that assists in understanding the focus of the model. This method allows us to gain insights into which parts of the input contribute significantly to the model's decision, offering a more interpretable perspective on how the model processes and distinguishes writing characteristics from the autistic children. However, direct observation of the results of CAM for thousands of handwriting Chinese characters is both time-consuming and subjective. To address this issue, in our second objective we designed a method of encoding the results of CAM such that the differences between autistic and TD children's handwritings can be objectively and swiftly observed. This approach enhances the efficiency of analyzing Chinese handwriting characteristics associated with the autistic and TD children.

2   Related Work

ASD, being a neurodevelopmental disorder, is associated with developmental alterations in both brain structure and facial tissues [16]. Accordingly, studies have delved into utilizing brain imaging techniques [17,18] and facial image analysis [16,19] to contribute to ASD detection efforts. Moreover, ASD frequently coexists with language impairments, prompting investigations into linguistic aspects for diagnostic insights. Researchers have leveraged speech spectrograms to discern patterns indicative of ASD [20], and natural language processing techniques have been applied to analyze narrative expressions for potential ASD markers [21]. This multifaceted approach recognizes the heterogeneity of ASD symptoms and aims to capture the disorder's varied manifestations through a combination of data sources and analytical methods. Over the years, researchers have extensively explored the potential of these advanced technologies to enhance the understanding of ASD and contribute to early and accurate diagnosis.

Autistic children encounter difficulties initiating actions that result in subsequent movements or ultimate goals, making them notably challenged in tasks involving sequential motor skills [6]. Handwriting is one such task. The subtleties in stroke irregularities, letter shapes, and overall handwriting style may serve as potential indicators of neurodevelopmental variations that are characteristics of ASD. By focusing on this relatively observable aspect of a child's behavior, our study aims to contribute to the growing body of research exploring unconventional yet promising avenues for ASD detection. In the study conducted by Beversdorf et al. [11], an examination of handwriting samples from adults with and without ASD revealed a noteworthy difference in letter size between the ASD group and the TD group. Specifically, individuals with ASD tended to produce significantly larger letters compared to their counterparts. Another relevant study by Johnson et al. [12] focused on autistic children and shed light on spatial aspects of their handwriting. The research disclosed that autistic children exhibited poorer spatial arrangement in their handwriting when compared to TD children. This indicates challenges in organizing and spacing characters appropriately on the writing surface, suggesting that spatial arrangement might be a distinctive feature in the handwriting of individuals with ASD.

Yen et al. [14] collected a handwriting dataset to classify whether the handwriting characters were produced by individuals with ASD or TD, and reveals that autistic children exhibit variations in arcs and spatial distribution for Chinese character writing. Due to the uneven distribution of data (more data from TD individuals than ASD individuals), there was an issue of data imbalance. The study addressed data imbalance through undersampling, achieving an F1-score of 0.954. However, the approach of performing undersampling before splitting the dataset into training and testing sets resulted in inconsistent testing sets across different experiments.

In our study, we aim to investigate whether incorporating phonetic notation data contributes to improving model performance. Moreover, we split the dataset into training and testing sets and applied both undersampling and oversampling exclusively on the training set to address the issue of inconsistent testing sets.

3   Datasets

We collaborated with a local elementary school and an association of autism to recruit participants. We asked students to provide their handwriting workbooks

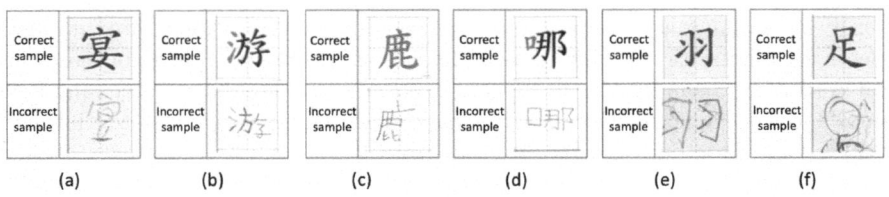

Fig. 2. Example aspects not satisfied by the neatness criteria.

in which they practiced Chinese characters and phonetic notations in class and at home. We were able to collect handwriting that is relatively natural, providing us with potentially more accurate insights into the handwriting differences between ASD and TD children. In total, the dataset comprises handwritings from 23 children: 6 ASD children (average age: 10.5 years; all boys) and 17 TD peers (average age: 8.67 years, eight boys and nine girls). Among them, there are a total of 4 children with mild autism, 1 individual with moderate autism, and 1 individual with severe autism. We used three types of datasets in the analysis: (1) the Chinese character-only dataset, (2) the phonetic notation-only dataset, and (3) the Chinese character + phonetic notation dataset where each Chinese character was written with its corresponding phonetic notation. Due to the page limit, we focus on the use of the Chinese character-only dataset in this paper.

The Chinese character-only dataset comprises 17,950 words, with 14,173 from TD children and 3,777 from autistic children. We considered neatness of writing Chinese characters by defining the neatness criteria based on relevant literature [22,28] and discussions with experienced elementary school teachers. The neatness criteria were divided into two levels: stroke and component, and three factors: position, size, and correctness. The division into stroke and component levels allows for a more nuanced and comprehensive assessment of neatness. The stroke level pertains to the individual strokes that constitute a character, examining their position, size, and correctness. On the other hand, the component level considers the overall arrangement and coherence of the entire character, providing a holistic perspective on neatness. This multi-level approach ensures a detailed and nuanced evaluation, facilitating a more precise understanding of handwriting neatness in the context of our study. Therefore, we access the following six aspects: *stroke position, stroke size, stroke correctness, component position, component size*, and *component size* for each word. The neatness is labeled as "1" if the individual word satisfies five or more aspects, and "0" if it satisfies less than five aspects. Figure 2 exemplifies how we evaluate and label the neatness of each word.

Figure 2(a) demonstrates aspects not satisfied in *stroke position* and *component correctness*. The horizontal stroke in the component "女" is shifted downward, and the "日" component is miswritten. Therefore, it is labeled 0. Figure 2(b) demonstrates only one aspect not satisfied in *component position* where the rightmost component of the character is shifted downward, and is labeled 1. Figure 2(c) demonstrates an aspect not satisfied in *stroke size*, as there is an overly long horizontal stroke at the top, and another aspect not satisfied in *component position* where the "比" component is displaced. It is labeled 0. Figure 2(d) demonstrates only one aspect not satisfied in *component position* where the "口" component is written too large, and is labeled 1. Figure 2(e) demonstrates an aspect not satisfied in *stroke correctness* in the upper-left horizontal stroke and the right vertical stroke. Moreover, another aspect is not satisfied in *stroke position* since some strokes in Fig. 2(e) exceed the grid. Therefore, it is labeled 0. Finally, Fig. 2(f) demonstrates an aspect not satisfied in *component correctness* where the "口" is written as a circle, and another aspect not satisfied in *stroke position* since strokes in Fig. 2(f) exceed the grid. It is labeled 0.

Table 1. Detailed information of the Chinese character-only dataset.

	TD	ASD	Total
Neatness = 1	13,659	1,181	14,840
Neatness = 0	514	2,596	3,110
Total	14,173	3,777	17,950

To assess the reliability of the above criteria, we employ the Fleiss' Kappa value [23], which is used to measure the degree of agreement between raters. In our study, we randomly selected 3600 pictures, which constitute 20% of the Chinese character-only dataset. The labels were made by three raters according to the neatness criteria, and a majority vote was taken. The label was then compared to the author's label. The final calculated Kappa value was 0.7354, falling within the range of 0.61 to 0.80, indicating substantial agreement [24]. In conclusion, the author labeled all the pictures in the Chinese character-only dataset with 14,840 having a neatness label of 1 and 3,110 having a neatness label of 0. Detailed information is presented in Table 1. The value of the Chi-square test is 8824.91, with a p-value < 0.01, which shows TD children were more likely to produce neat Chinese characters than autistic children.

4   Method

The flowchart of building classification models has the following two types: a 5-fold flowchart, as shown in Fig. 3, and the CAM flowchart, depicted in Fig. 4. The 5-fold flowchart signifies the utilization of 5-fold cross-validation during model training. On the other hand, the CAM flowchart employs the CAM results for identifying the handwriting characteristics. In the following, we introduce each step in the flowchart in order.

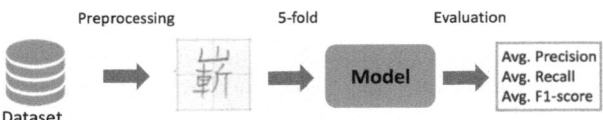

Fig. 3. The 5-fold flowchart.

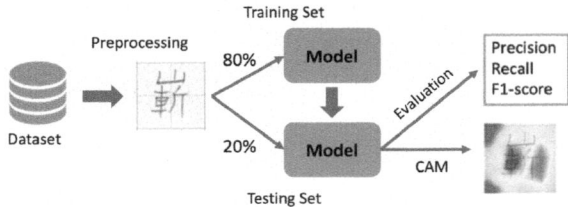

Fig. 4. The CAM flowchart.

4.1   Data Preprocessing and Dataset Balance

Both flowcharts employ the same data preprocessing. First, we convert the image to grayscale. The second step involves resizing the image to 224×224. This is because ResNet-18 [25] requires an input image of size 224×224. As seen in Table I, our data is imbalanced. To address this issue, we apply oversampling and undersampling techniques to the data in the training set. In the case of the 5-fold flowchart, oversampling or undersampling was applied to the training set in each fold. When undertaking undersampling, we randomly select a number of the majority dataset to match the size of the minority dataset. As depicted in Fig. 5, we assumed that the training set contains 14,360 images (which is 80% of the total 17,950 images), comprising 11,338 TD and 3,022 ASD images. Consequently, the final undersampled training set consists of a total of 6,044 images. When conducting oversampling, we employed duplicate oversampling, wherein images are duplicated from the minority dataset to a multiple close to, but not exceeding the majority dataset. As illustrated in Fig. 5, there are 14,360 images in the training set, consisting of 11,338 TD and 3,022 ASD images. We duplicate the ASD images to three times of their original size. Consequently, the final oversampled training set comprised a total of 20,404 images.

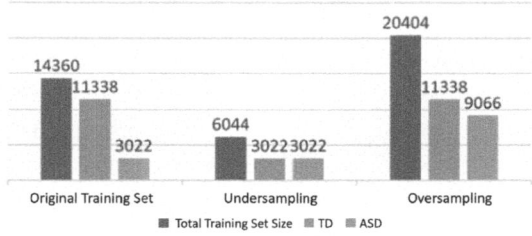

Fig. 5. Data balancing by undersampling and oversampling.

4.2 Model

We employ support vector machine (SVM), decision tree (DT), K-nearest neighbor (KNN), logistic regression (LR), and the ResNet-18 [25] model pretrained on ImageNet [26] for the classification tasks. The ResNet architecture's residual connection design addresses the challenge of training deep convolutional neural networks by mitigating issues associated with excessive layer depth, facilitating the training of deeper networks without suffering from degradation issues.

4.3 Class Activation Map

The CAM technique [27] serves as a crucial visualization tool that facilitates a deeper understanding of a model's focal points. It involves establishing a connection with the Global Average Pooling (GAP) layer after the final convolutional layer. Following this connection, it captures the weights associated with the GAP layer output and linearly combines them with the corresponding feature map to generate the results. The conventional approach outlined above mandates the utilization of the GAP layer, imposing constraints on the overall flexibility of the network architecture. To overcome this limitation, Grad-CAM [15] introduces an innovative solution by incorporating the partial differential of the feature map in the relevant category to supplant the weight output derived from the GAP layer. This modification enables Grad-CAM to be applied across a broader spectrum of CNN architectures, and therefore offers enhanced adaptability. In the context of this study, we adopt the Grad-CAM approach to leverage its proven effectiveness in visualizing and interpreting the focus areas of our model.

5 Experiments

In this section, we analyze the differences in the model focus between the training sets in Chinese character-only dataset and neat Chinese character-only dataset.

5.1 Neatness Classification

In the neatness classification experiments, we utilize SVM, DT, KNN, LR, and ResNet-18 models, each with three methods to deal with the data imbalance

Table 2. All results from neatness classification.

Model	Balance	5-fold	Precision	Recall	F1-Score
SVM	X	V	0.6637	0.6948	0.6787
SVM	Undersampling	V	0.4795	0.8291	0.6075
SVM	Oversampling	V	0.6624	0.6955	0.6782
DT	X	V	0.6927	0.7448	0.7174
DT	Undersampling	V	0.4744	0.8337	0.6046
DT	Oversampling	V	0.7378	0.7314	0.7346
KNN	X	V	0.6138	0.7683	0.6823
KNN	Undersampling	V	0.3212	0.8847	0.4710
KNN	Oversampling	V	0.3574	0.8621	0.5050
LR	X	V	0.8438	0.6714	0.7476
LR	Undersampling	V	0.5333	0.8153	0.6447
LR	Oversampling	V	0.6425	0.7689	0.6998
ResNet-18	X	V	0.8544	0.7309	0.7804
ResNet-18	Undersampling	V	0.7798	0.7955	0.7793
ResNet-18	Oversampling	V	**0.8281**	**0.7799**	**0.7997**

problem: undersampling, oversampling, and X which denotes that neither under-sampling nor oversampling techniques are applied. All results are presented in Table 2. Observing Table 2, the undersampling practice increases the proportion of the data with "Neatness = 0" in the training set, making the model more inclined to predict "Neatness = 0." This leads to an improvement in the Recall compared to the data imbalance case. However, the Precision experiences a significant drop due to the bolder prediction of "Neatness = 0." On the other hand, oversampling exhibits higher stability in the Precision than undersampling. Table 2 reveals that ResNet-18 with oversampling yields the best performance in this task.

Based on the experiment results presented in Table 2, it is evident that the ResNet-18 model with oversampling performs the best among all models and methods, achieving an F1-score of 0.7997. This result indicates that the ResNet-18 model, when trained with oversampling to address data imbalance, excels in both precision and recall compared to other models and methods tested. Overall, the results suggested that the ResNet-18 model with oversampling is most suitable for classifying neatness in handwriting. More efforts are needed to further improve the performance of the model for a practical use.

5.2   ASD/TD Classification

We introduce the notations of Ch_All and Ch_Neat where Ch_all is the training set in Chinese character-only dataset and Ch_Neat is the neat training set in Chinese character-only dataset. In the subsequent experiments, we employ

Table 3. The results of ASD/TD classification using Ch_ALL.

Balance	5-fold	Approach	Precision	Recall	F1-Score
X	V	Ours	0.9780	0.9224	0.9490
Undersampling	V	Ours	0.9388	0.9891	0.9629
Oversampling	V	Ours	**0.9807**	**0.9637**	**0.9720**
X	V	Yen et al.	0.982	0.762	0.856
Undersampling	V	Yen et al.	0.956	0.911	0.932
Oversampling	V	Yen et al.	0.934	0.975	0.954

ResNet-18 as the model. Under different training sets Ch_All and Ch_Neat, the corresponding testing sets are obtained from the respective datasets. Table 3 presents the results of ASD/TD classification using the Chinese character-only dataset, along with the results from Yen et al. [14]. The three employed methods for dealing with imbalance data were X, undersampling, and oversampling. Notably, the undersampling without 5-fold approach achieves the highest F1-score in Yen et al. [14]. From Table 3, it is evident that both our X and undersampling approaches surpass the results of Yen et al. [14]. Additionally, the oversampling approach outperforms the best performance reported in Yen et al. [14]. Consequently, we use the oversampling approach in our implementation for the following analyses of the handwriting characteristics.

Unlike Ch_All, Ch_Neat excludes data with "Neatness = 0," retaining only neat Chinese characters. The results of using the Ch_Neat training set are displayed in the lower three rows in Table 4, while the upper three rows present the results of using the Ch_All training set for a comparison. Using Ch_Neat falls short of surpassing Ch_All in X, undersampling, and oversampling. This is because Ch_Neat excludes data with "Neatness = 0." Despite the reduced amount of data, Ch_Neat still achieves similar results compared to Ch_All in oversampling. This demonstrates the feasibility of classifying ASD/TD using only neat Chinese characters. Additionally, the F1-Score for the Chinese character-only dataset is 0.9814.

Table 4. The results of ASD/TD classification using Ch_All and Ch_Neat.

	Balance	5-fold	Precision	Recall	F1-Score
Ch_All	X	V	0.9780	0.9224	0.9490
Ch_All	Undersampling	V	0.9388	0.9891	0.9629
Ch_All	Oversampling	V	**0.9807**	**0.9637**	**0.9720**
Ch_Neat	X	V	0.9804	0.8203	0.8881
Ch_Neat	Undersampling	V	0.9638	0.9281	0.9447
Ch_Neat	Oversampling	V	0.9739	0.9584	0.9658

5.3   Identifying Handwriting Characteristics

In this subsection, we aim to explore the differences between the performance using the Ch_All and Ch_Neat training sets. The performance of these two training sets is detailed in Table 5, and following the trend observed in ASD/TD classification using Ch_Neat, using Ch_All slightly outperforms Ch_Neat. Our testing set comprises a total of 3590 images. For each image, there are four possible prediction results: both correct, both wrong, Ch_All correct and Ch_Neat wrong, and Ch_All wrong and Ch_Neat correct. Out of the 3590 images, 3545 are both correctly predicted, 19 are both incorrectly predicted, and only 26 show different prediction results (17 for Ch_All correct and Ch_Neat wrong, 9 for Ch_All wrong and Ch_Neat correct). Despite having different training sets, the prediction results are very similar. In the subsequent analysis, we examine their differences from a CAM perspective.

Table 5. Results of using Ch_All and Ch_Neat without 5-fold verification.

Training set	Balance	5-fold	Precision	Recall	F1-Score
Ch_All	Oversampling	X	0.9789	0.9841	0.9815
Ch_Neat	Oversampling	X	0.9865	0.9656	0.9759

In order to solve the time-consuming and the subjective problems of a manual observation of the CAM results, we process these results in two steps. The first step is to specify a color area to divide a CAM result into two parts, as shown in Fig. 6. We choose the red and orange regions (the two most focused regions) in the CAM result to form this color area. The second step is to use formula (1) to encode the result of the first step, as shown in Fig. 7. In formula (1), each image (224*224 pixels per image) is divided into 16 blocks (56*56 pixels per block). If the block is conformed to formula (1), it is coded as 1, otherwise it is 0. The significance of this step is to encode the model focus since it likely refers to the area or features of the input data the model pays much attention to when making a prediction. Encoding the model focus involves converting the model focus into a format that can be visualized and analyzed.

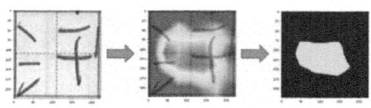

Fig. 6. First step of processing the CAM results.

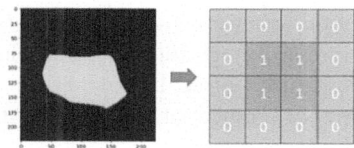

Fig. 7. Second step of processing the CAM results.

$$\frac{\text{\# of yellow pixels in the block}}{56 \times 56} \geq \frac{\text{\# of yellow pixels in the image}}{224 \times 224} \tag{1}$$

We define a *not-centered image* as one where the surrounding 12 blocks are marked 1 or 0 and the middle four blocks are marked 0, as illustrated in Fig. 8(a). The number of not-centered images is presented in Table 6. Among the not-centered images, the majority of them come from the TD children. Examples of the TD not-centered images are provided in Fig. 8(b). This result is quite surprising. It is generally believed that Chinese characters written by TD children are of moderate size and centered, so the model's focus area would be more centralized [14]. On the other hand, Chinese characters written by ASD children tend to have offsets or be written too large, potentially leading the model's focus area to be on the periphery. However, the results show that for Chinese characters written by TD children, the model's focus area is not in the center. This indicates that the model indeed uses unexpected features to do the prediction.

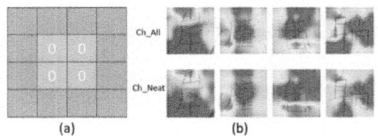

Fig. 8. Illustration (a) and examples (b) of not-centered images.

Table 6. Number of not-centered images.

Training set	Total	TD	ASD
Ch_All	2,273	2,200	73
Ch_Neat	2,386	2,278	108

We define a *not-peripheral* image as one where the surrounding 12 blocks are marked 0 and the middle four blocks are marked 1 or 0, as illustrated in Fig. 9(a).

The number of not-peripheral images is outlined in Table 7. Although there are not many not-peripheral images, a significant proportion of these images come from the ASD children. Examples of ASD not-peripheral images are provided in Fig. 9(b). This result further confirms that the features the model focuses on may differ from what we have expected.

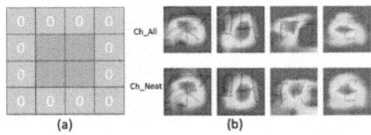

Fig. 9. Illustration (a) and examples (b) of not-peripheral images.

Table 7. Number of not-peripheral images

Training set	Total	TD	ASD
Ch_All	44	4	40
Ch_Neat	55	3	52

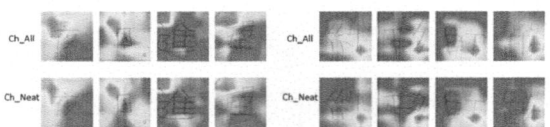

Fig. 10. Examples of TD upper left corner images and examples of ASD lower right corner images.

Handwriting characteristics of corner images. We further analyze the model focus by dividing an image into four corners. We define a *corner image* as one where the upper left (UL) corner or upper right (UR) corner or lower left corner (LL) or lower right (LR) corner is marked 1 or 0 while the other corners are marked 0. The number of corner images is outlined in Table 8. Observing the outcomes, using both training datasets exhibit similar tendencies. Specifically, the focus of the CAM on the images from the TD children is on UL while for ASD children, it tends to be on LR. Given the traditional habit of writing Chinese characters from top to bottom and from left to right, this pattern suggests that if a TD child writes a Chinese character, the CAM focuses on the starting stroke in the UL corner. Conversely, if the Chinese characters are written by an ASD child, the CAM tends to focus on the end stroke in the LR corner. This also resonates with the challenge faced by autistic children in initiating actions that result in subsequent movements or ultimate goals [6]. Examples of corner images are provided in Fig. 10 for TD and ASD children, respectively.

Table 8. Number of not-peripheral images.

Training set	UL (TD/ASD)	UR (TD/ASD)	LL (TD/ASD)	LR (TD/ASD)
Ch_All	524/7	78/5	61/12	79/66
Ch_Neat	393/4	39/11	192/12	87/70

6   Conclusion

In this paper, we introduce the concept of handwriting neatness and define its criteria for studying the handwriting characteristics of ASD children. By employing oversampling technique for data balancing, our approach surpasses the performance of the previous study on ASD/TD classification to achieve an F1-score of 0.9720 using the Ch_All training set. When using only neatly written Chinese characters, the F1-score is 0.9658. This demonstrates the model's capability to classify whether the Chinese characters are handwritten by ASD or TD children under neatly writing conditions. Finally, we encode the CAM results to address the shortcomings associated with the manual observation which is time-consuming and subjective. It reveals that the prediction results of the two training sets, Ch_All and Ch_Neat, are very similar. Moreover, the CAM results highlight the differences of the model focus for the data from the TD and ASD children.

Acknowledgments. This research was supported by National Science and Technology Council under Grant numbers NSTC 112-2221-E-468-009, NSTC 113- 2410-H-007-002, and NSTC 112-2410-H-007-003-SS2, and Ministry of Education's Yushan Young Fellows Program, which enabled the successful completion of this research.

Disclosure of Interests. The authors declare that they have no conflict of interest.

Ethical Approval. The Central Regional Research Ethics Committee at China Medical University reviewed and approved the research procedure [CRREC-112-002]. All research participants were fully informed about the study's methods and objectives, and they provided consent to participate in the research.

References

1. Association, A.: Diagnostic and Statistical Manual of Mental Disorders (2013)
2. Lord, C., Risi, S., DiLavore, P., Shulman, C., Thurm, A., Pickles, A.: Autism from 2 to 9 years of age. Arch. Gen. Psychiatry **63**, 694 (2006)
3. Hyman, S., et al.: Identification, evaluation, and management of children with autism spectrum disorder. Pediatrics **145** (2020)
4. Downey, R., Rapport, M.: Motor activity in children with autism: a review of current literature. Pediatr. Phys. Ther. **24**, 2–20 (2012)
5. Fournier, K., Amano, S., Radonovich, K., Bleser, T., Hass, C.: Decreased dynamical complexity during quiet stance in children with autism spectrum disorders. Gait Posture **39**, 420–423 (2014)
6. Chen, L., et al.: Postural control and interceptive skills in children with autism spectrum disorder. Phys. Ther. **99**, 1231–1241 (2019)
7. Sacrey, L., Germani, T., Bryson, S., Zwaigenbaum, L.: Reaching and grasping in autism spectrum disorder: a review of recent literature. Front. Neurol. **5** (2014)
8. Fabbri-Destro, M., Cattaneo, L., Boria, S., Rizzolatti, G.: Planning actions in autism. Exp. Brain Res. **192**, 521–525 (2008)
9. Mayes, S., Calhoun, S.: Learning, attention, writing, and processing speed in typical children and children with ADHD, autism, anxiety, depression, and oppositional-defiant disorder. Child Neuropsychol. **13**, 469–493 (2007)
10. Fuentes, C., Mostofsky, S., Bastian, A.: Children with autism show specific handwriting impairments. Neurology **73**, 1532–1537 (2009)
11. Beversdorf, D., et al.: Brief report: macrographia in high-functioning adults with autism spectrum disorder. J. Autism Dev. Disord. **31**, 97–101 (2001)
12. Johnson, B., Papadopoulos, N., Fielding, J., Tonge, B., Phillips, J., Rinehart, N.: A quantitative comparison of handwriting in children with high-functioning autism and attention deficit hyperactivity disorder. Res. Autism Spectrum Disord. **7**, 1638–1646 (2013)
13. Peebles, D.: SCML: A Structural Representation for Chinese Characters. Dartmouth College Undergraduate Theses (2007). https://digitalcommons.dartmouth.edu/senior%5Ftheses/51
14. Yen, L., Wong, J., Chen, A.: Identifying chinese handwriting characteristics for detecting children with autism. In: 39th ACM/SIGAPP Symposium on Applied Computing, pp. 856–865 (2024)
15. Selvaraju, R., Cogswell, M., Das, A., Vedantam, R., Parikh, D., Batra, D.: Grad-CAM: visual explanations from deep networks via gradient-based localization. In: 2017 IEEE International Conference on Computer Vision (ICCV) (2017)
16. Lakshmi Praveena, T., Muthu Lakshmi, N.: A methodology for detecting ASD from facial images efficiently using artificial neural networks. In: Advances in Computational and Bio-Engineering, pp. 365–373 (2020)
17. Sewani, H., Kashef, R.: An autoencoder-based deep learning classifier for efficient diagnosis of autism. Children **7**, 182 (2020)
18. Kong, Y., Gao, J., Xu, Y., Pan, Y., Wang, J., Liu, J.: Classification of autism spectrum disorder by combining brain connectivity and deep neural network classifier. Neurocomputing **324**, 63–68 (2019)
19. Mujeeb Rahman, K., Subashini, M.: Identification of autism in children using static facial features and deep neural networks. Brain Sci. **12**, 94 (2022)
20. Zhou, T., Zou, X., Li, M.: An automated assessment framework for speech abnormalities related to autism spectrum disorder. In: 3rd International Workshop on Affective Social Multimedia Computing (ASMMC) (2017)

21. Chojnicka, I., Wawer, A.: Social language in autism spectrum disorder: a computational analysis of sentiment and linguistic abstraction. PLoS ONE **15**, e0229985 (2020)
22. Li, Y., Jiang, J., Deng, H.: Understanding, description, calculation implementation and application introduction of chinese handwriting characteristics. E-Educ. Res. **36**, 62–69 (2015)
23. Fleiss, J.: Measuring nominal scale agreement among many eaters. Psychol. Bull. **76**, 378–382 (1971)
24. Landis, J., Koch, G.: The measurement of observer agreement for categorical data. Biometrics **33**, 159 (1977)
25. He, K., Zhang, X., Ren, S., Sun, J.: Deep residual learning for image recognition. In: 2016 IEEE Conference on Computer Vision and Pattern Recognition (CVPR) (2016)
26. Deng, J., Dong, W., Socher, R., Li, L., Li, K., Fei-Fei, L.: ImageNet: a large-scale hierarchical image database. In: 2009 IEEE Conference on Computer Vision and Pattern Recognition (2009)
27. Zhou, B., Khosla, A., Lapedriza, A., Oliva, A., Torralba, A.: Learning deep features for discriminative localization. In: 2016 IEEE Conference on Computer Vision and Pattern Recognition (CVPR) (2016)
28. Zhuang, Z.: Handwritten Chinese character recognition and aesthetic grading based on deep learning. Master thesis, Beijing University of Posts and Telecommunications (2019)

FNoDe: Faulty Node Detection in Microservices Architecture

Harsh Borse[(✉)], Utkalika Satpathy, Mainack Mondal, and Bivas Mitra

IIT Kharagpur, Kharagpur, India
harshzf2@gmail.com, utkalika.satapathy01@gmail.com,
{mainack.mondal,bivasmitra}@cse.iitkgp.ac.in

Abstract. As cloud services shift from monolithic architectures to microservices, post-failure fault and anomaly detection becomes increasingly challenging due to cascading effects across interdependent services and the overwhelming volume of heterogeneous logs and metrics. We propose **FNoDe** (Faulty Node Detection), a framework that integrates application logs, performance metrics, and distributed traces into a unified graph structure to detect both the root cause and type of anomaly. By leveraging a graph convolutional network (GCN), FNoDe learns system representations under normal and anomalous states from historical microservice data and uses these embeddings to classify new system states. Evaluated on five public benchmarks and two in-house microservice systems, FNoDe outperforms traditional methods by 20–30% in accuracy and maintains competitive performance with state-of-the-art frameworks, while also offering interpretability through XAI techniques.

1 Introduction

Microservices architectures (MSA) are increasingly adopted in IoT, mobile, cloud, and industrial applications due to their resilience, robustness, and adaptability [20]. MSA decomposes applications into independently deployable services with lightweight intercommunication, enabling flexible system development. However, their complexity and scale introduce fragility, making failures inevitable [13]. Rapid failure diagnosis is crucial for mitigating anomalies by identifying the faulty microservice and the type of fault [28]. Extensive research has focused on diagnosing failures, locating root-cause services, and categorizing anomalies.

Traditional root cause detection methods in microservices architectures have limitations that hinder failure analysis. These include: **(i)** A predominant focus on non-functional anomalies (e.g., CPU or memory shortages under high load), despite functional anomalies (unintended deviations in application logic or misconfiguration) being more prevalent in industry applications. Existing frameworks typically address either functional or non-functional anomalies, rather than integrating both [11,18,23]. **(ii)** Existing solutions typically rely on either deploying monitoring agents alongside applications to capture performance metrics [26,30], or on instrumenting distributed tracing mechanisms.

© The Author(s), under exclusive license to Springer Nature Switzerland AG 2026
C. K. Leung et al. (Eds.): DaWaK 2025, LNCS 16048, pp. 86–101, 2026.
https://doi.org/10.1007/978-3-032-02215-8_6

These approaches analyze the collected data using visualization tools or statistical techniques, and in some cases, construct topology or causality graphs to enhance anomaly detection. However, as distributed systems become increasingly complex, manually correlating signals across different observability dimensions becomes infeasible. This challenge is further exacerbated in microservice architectures due to their dynamic nature—characterized by frequent updates, redeployments, and elastic scaling—which demands adaptive and automated monitoring and diagnosis methods. [25,30]. **(iii)** Latency, extracted from trace data, is often regarded as a key indicator for anomaly detection. However, relying exclusively on latency can be misleading, as it may obscure the presence of multiple underlying anomaly sources within a microservice. For instance, elevated latency may arise from (i) an error caused by invoking a faulty downstream microservice, or (ii) a functional issue in the local code path, either preceding or following a successful remote invocation. Disambiguating these scenarios based on latency alone is inherently difficult.

Single-modality approaches often fail to capture the complex, interconnected behavior of microservices, where anomalies may manifest differently across monitoring dimensions. For instance, elevated latency could stem from high CPU usage, memory exhaustion, or application-level failures—ambiguities that are difficult to resolve from latency data alone. A holistic observability strategy incorporates traces, performance metrics, and application logs, each offering complementary perspectives: traces expose inter-service dependencies, metrics quantify resource usage, and logs provide semantic context. This paper addresses the challenge of robust root cause diagnosis by integrating these heterogeneous sources to analyze distributed system behavior more comprehensively.

Detecting root causes of anomalies in distributed microservices requires integrating diverse modalities—logs, performance metrics, and traces—while preserving their representational power. A graph-based representation effectively models microservices and hosts as nodes, with edges capturing dependencies. System and application health are embedded as node features using performance metrics and logs.

This paper introduces **FNoDe** (Faulty Node Detection), a novel framework for identifying functional and non-functional anomalies in microservices architectures. FNoDe unifies heterogeneous logs into a graph structure per user request, uncovering hidden dependencies [30,36]. *FNoDe* collects three heterogeneous observability modalities— performance metrics, traces and application logs corresponding to user requests under both normal and anomalous conditions. These modalities are integrated into a unified structure known as the Feature Graph (FG), where nodes represent individual microservices and edges capture inter-service interactions. The graph is then enriched with features derived from the collected modalities, assigning them as node and edge attributes to enable comprehensive analysis. Enabling broad anomaly detection without requiring separate models for different modalities. Labelled graphs are processed using a combination of Graph Convolutional Network (GCN) and a classifier to learn complex relationships across resources and metrics [19]. By leveraging historical anomalous requests, FNoDe classifies the faulty microservice node, providing a robust solution for root cause detection.

To assess the effectiveness of **FNoDe**, we conducted experiments on the open-source microservices benchmark [27] and two in-house systems, *Library* and *Ecom*, which simulate realistic user interactions and inject both functional and non-functional anomalies [30, 36]. Compared to traditional baseline frameworks, FNoDe achieved a **20%–30%** improvement in accurately identifying the root causes of anomalies, demonstrating its robustness and effectiveness in microservices-based architectures.

2   Related Work

Root cause analysis (RCA) has been widely explored in distributed, cloud, and microservices architectures. Existing methods fall into three categories based on the data source: logs, performance metrics (KPIs), or traces.

2.1   Log-Based Approach

Log-based RCA leverages native logs, avoiding extra instrumentation [24]. These methods build a *causality graph*, where nodes represent services and directed edges capture potential anomaly propagation. Traversing this graph helps localize root causes in multi-service applications [24]. Aggarwal et al. [10] model service logs as multivariate time series and apply Granger causality to infer dependencies among error events from frontend and backend logs.

2.2   Tracing-Based Approach

Tracing methods detect anomalies by analyzing execution paths and latency deviations. Some rely on visual comparison of traces [16, 36], while others apply statistical techniques [22] or extract topologies to guide RCA [21]. CloudDiag and TraceAnomaly [22] analyze trace-level service response times to detect interactions impacting frontend performance.

2.3   Monitoring-Based Approach

These approaches use infrastructure metrics to infer causal relationships. Statistical techniques over KPIs help identify suspect services [28, 31], while dependency graphs capture service interactions for tracing faults [32, 33]. Systems like Seer and MicroRCA [13, 32] employ deep learning on large-scale telemetry to detect faulty services and resource bottlenecks such as CPU contention, enabling efficient RCA.

3   Dataset and Problem Definition

To evaluate diverse anomalies and edge cases, we develop and deploy interactive, end-to-end microservices-based applications—*Library* and *Ecom*— which directly engage with users and simulate realistic request flows. In addition to these in-house systems, we utilize five publicly available benchmark microservices applications and datasets for comprehensive analysis [13,27].

Applications and Setup: We develop two microservices-based applications— *Library* (5 microservices) and *Ecom* (12 microservices)—to capture diverse interaction and data complexity. In *Library*, user requests traverse at least three services, including *Frontend, Recommend, Search, CRUD*, and *Database Management*, maintaining logical separation and minimizing redundant calls. In contrast, *Ecom* supports more complex workflows spanning 3 to 10 services per request, handling functionalities such as *Recommendation, Ads, Search*, and *Payment*. Figures 1 and 2 illustrate the system architecture and a representative request (r_i) flow. Both applications are deployed on five dedicated hosts using Docker Swarm, with services distributed across nodes for optimal performance. Hosts synchronize via a global clock and communicate over HTTP REST, while locally logging system metrics, application logs, and distributed traces, which are available for querying or centralized analysis.

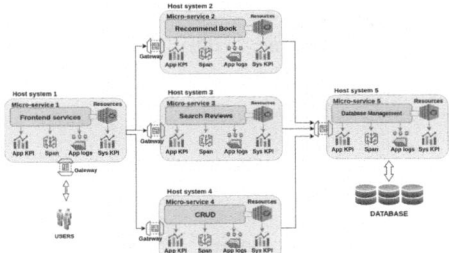

Fig. 1.   Fig. *library* system design (detailed)

Table 1. Faults definitions and number of requests generated

Faults	Description	#req.
F0	Normal requests where no faults are introduced in any microservices	66,332
F1	Microservices m increased the compute power consumption internally	40,100
F2	Zombie process eating unnecessary resources in host system m	56,420
F3	Error in internal code: for microservices m, the code for establishing communication with other microservices had a fault, resulting in a failed connection	46,300
F4	Algorithmic complexity increased in microservice m before sending a request to another microservice	50,800
F5	Algorithmic complexity increased in microservice m after receiving the response from another microservice	51,200
F6	Executing unnecessary heavy queries in microservices m with a relational database	48,500

3.1   Workload Generator

User Requests: We simulate realistic workloads via a mix of sequential and parallel user requests. *Sequential requests* mimic user navigation with human-like delays [17,27], while *parallel requests* vary concurrent user loads across endpoints, incorporating dynamic request patterns, seasonal spikes, and random user behavior.

Faults: We inject *non-functional* and *functional* anomalies across microservices (Table 1). Non-functional faults (F_1, F_2) affect system resources, while functional faults (F_3–F_6) stem from code errors or computational complexity. For instance,

F_4 delays interactions due to poor REST handling, and F_5 arises from complex data processing logic.

Fault Annotations: Notably, the attributes extracted from an individual microservice (such as high CPU utilization, high latency etc.) carries the signatures of health and fault observed in that microservice. Hence, we suitably annotate the health status of the microservices with the attributes obtained from multiple modalities triggered from each user request r. Labels are created as tuples (m_i, k_i) consisting of microservices $m_i \in V$ and the affected attributes $k_i \in K$ from different modalities. Since a fault can impact multiple attributes across microservices, the label for each request r_i represents affected microservices and their faulty attributes. For example, if fault $F1$ affects the compute power and memory usage of microservice m_x during request r_i, the label would be $R_i = \{(m_x, \mathrm{CPU}), (m_x, \mathrm{MEM})\}$.

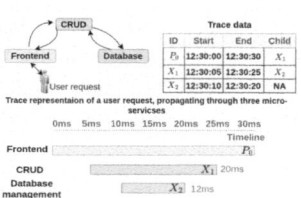

Fig. 2. Example trace showing span relationships for an insert operation in the *Library* system.

Fig. 3. Sample of application. (Blue extracted info.) (Color figure online)

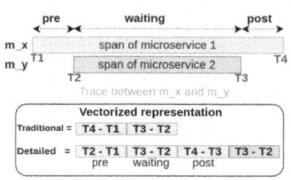

Fig. 4. Breakdown of Latency into pre-post-wait time.

3.2 Data Collection

We collect three key data types for monitoring microservices: *Traces*, *Performance Metrics*, and *Application Logs*.

Traces: A trace captures the complete execution flow of a user request across microservices as a service invocation chain [9,29]. It comprises multiple spans, each representing the execution time of an operation in a specific microservice, organized hierarchically and linked via a trace ID (Fig. 2). Traces provide latency and microservice interaction sequences.

Performance Metrics: Metrics assess system health and efficiency [12,13]. Key metrics include: (i) CPU utilization (application CPU time), (ii) memory utilization (used primary memory), (iii) resident set size (RSS: allocated physical memory), (iv) wait time (CPU idle awaiting I/O) and many more metrics.

Application Logs: Logs record system operations and user actions with fixed and variable fields, including host IP, PID/TID, and status codes. Extracted attributes include categorical fields (host IP, connection IP, request type, span ID, response code) and numerical ones (e.g., content length). Logs are generated per microservice, stored locally, and used for analysis (Sample log Fig. 3).

3.3   Problem Definition

Consider an application composed of $|V|$ microservices, where $V = \{m_1, m_2, m_3, \ldots, m_v\}$. Each microservice m_x is characterized by n attributes represented as $K = \{k_1, k_2, \ldots, k_n\}$, derived from multiple monitoring modalities. Workloads r_i are generated following both sequential and parallel execution strategies. The application may encounter anomalies arising from faults within one or more microservices. The objective of this paper is to propose a framework capable of detecting the root cause microservice $m_x \in V$ along with the specific faulty attributes $K_i \subset K$, and outputting the corresponding tuple (m_x, K_i).

4   Motivation

Traditional root cause detection frameworks in microservices architectures have several limitations, which we explore and address.

4.1   Limitations

(i) **Adaptability to Heterogeneous Modalities:** Microservices generate diverse monitoring data—traces, metrics, and logs. Relying on limited modalities often leads to incomplete or incorrect anomaly detection. To demonstrate this, we inject two faults into the *TrainTicket* application: (A) *CPU resource unavailability* and (B) a *code-level data-handling error* (F5, Table 1). For each user request, we extract: (i) *latency* from traces, (ii) *CPU usage* from metrics, and (iii) *content length* from logs. Figure 5 illustrates that focusing only on latency (top plot) misses the CPU spike in fault-A (middle plot), while memory and latency alone in fault-B miss the failed database responses, evident from zero content length (bottom plot). This emphasizes the need for multi-modal analysis.

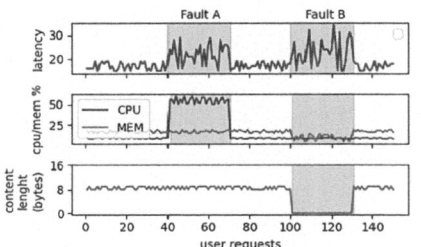
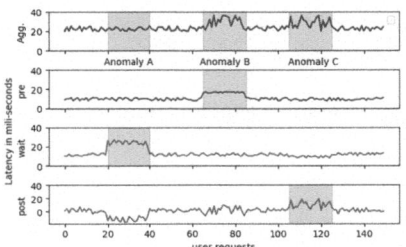

Fig. 5. Attribute changes across modalities for two anomalies (red indicates anomalies). (Color figure online)

Fig. 6. Aggregated and Detailed Breakdown of Trace.

(ii) Focus on Non-functional Anomalies: Most existing works focus on non-functional issues (e.g., high resource usage) and overlook functional anomalies [11, 18, 23]—errors like faulty database queries or increased logic complexity. These issues often resemble non-functional ones in behavior, making them harder to isolate with conventional clustering.

4.2 Opportunity

Detailed Span Decomposition in Traces: Relying solely on aggregated latency from traces can obscure critical insights into microservice interactions. A more granular breakdown of trace spans reveals detailed request flow across services. For example, a user request r that starts in microservice m_x, invokes m_y, and then continues in m_x, can have its span decomposed into: (i) pre-time—processing before calling m_y, (ii) wait time—idle while waiting for m_y's response, and (iii) post-time—processing after receiving data from m_y (Fig. 4). To demonstrate the utility of this breakdown, we inject three functional faults into m_x in the *TrainTicket* benchmark [27], which communicates with a database microservice m_y. In fault A (F5 in Table 1), a database disconnection error is introduced, resulting in increased wait time and reduced post-time due to retries and empty responses (Fig. 6); however, this is not reflected in the aggregated latency. In faults B and C, we inject algorithmic slowdowns before and after the invocation of m_y, respectively. While the aggregated latency increases in both cases, it fails to distinguish between them. By analyzing span breakdowns, fault B is indicated by elevated pre-time, and fault C by increased post-time in m_x, demonstrating the diagnostic power of fine-grained trace decomposition.

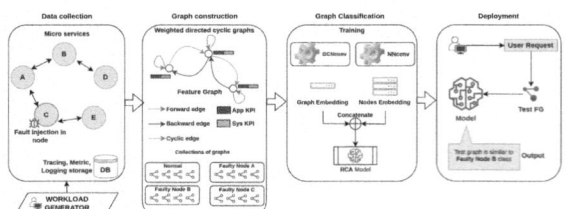

Fig. 7. FNoDe Pipeline and graph formation from trace.

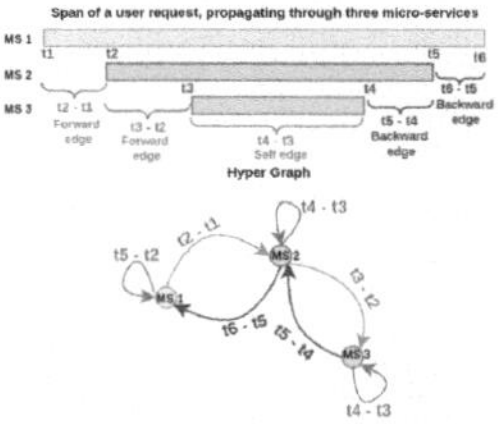

Fig. 8. Edge Formation

5 Methodology

FNoDe detects root causes in microservices by converting heterogeneous data into a unified graph. It uses a Graph Convolutional Network (GCN) to model dependencies and classify system states, improving anomaly detection through cross-modality learning.

5.1 Graph Construction

We construct a *Feature Graph (FG)* for each user request r_i by unifying heterogeneous microservice logs into a graph-based representation. Nodes represent microservices, enriched with features from application logs and performance metrics as features, while directed edges capture inter-service dependencies. FG construction details are provided below.

Table 2. Comparison of representation approaches

Representation	Information Preservation	Dependency Modeling	Multi-Modal Integration
Time Series	Temporal only	Indirect correlation	Separate models
Vector Space	No structure	Statistical correlation	Concatenation
Tree	Hierarchical only	Parent-child only	Node-level only
Graph	Complete topology	Direct representation	Natural integration

Table 3. Edge definitions with illustrative examples (refer to Fig. 8)

Edge Type	Definition	Example
Forward Edge	For a request r_i, microservice MS_x invokes MS_y. This edge represents the time spent by MS_x on r_i before calling MS_y. $(t_r - t_p)$	(t2 - t1), (t3 - t2)
Backward Edge	For a request r_i, MS_y returns a response to MS_x. This edge captures the time spent by MS_x on r_i after receiving the response. $(t_q - t_s)$	(t6 - t5), (t5 - t4)
Wait Edge (self-edge)	Represents the waiting period for MS_x between sending a request to and receiving a response from MS_y during r_i. $(t_s - t_r)$	(t5 - t2), (t4 - t3)
Execution Edge (self-edge)	For a terminal service MS_z in r_i, this edge denotes the final execution duration. $(t_s - t_r)$	(t4 - t3)

Node-Edge Formation. Each user request r_i is modeled as a graph $G_i = \langle V_i, E_i \rangle$, where V_i denotes the set of microservices traversed by r_i, and E_i represents the directed edges capturing inter-service interactions, as inferred from distributed trace data. The request execution path is decomposed into segments representing caller-callee transitions and corresponding responses. Edges are categorized based on temporal and directional flow (as shown in Table 3): *forward*, *backward*, and *self* edges. The resulting FG is a directed cyclic graph where: - Forward edges represent request propagation, - Backward edges represent response return paths, and - Self-edges capture intra-service execution or wait times.

Consider a request r_i traveling from microservice m_x to m_y. Let m_x execute r_i over $[t_p, t_q]$ and invoke m_y at t_r. Meanwhile, m_y executes r_i over $[t_r, t_s]$ and responds to m_x at t_s. The corresponding edges are: - Forward edge: (m_x, m_y) - Backward edge: (m_y, m_x) - Self-edges: $(m_x, m_x), (m_y, m_y)$ Thus, the complete edge set is $E_i = \{(m_x, m_y), (m_x, m_x), (m_y, m_y), (m_y, m_x)\}$ with node set $V_i = \{m_x, m_y\}$, forming a directed graph for request r_i.

Edge Feature Extraction. Each edge in E_i is annotated with a single feature: *latency*, derived from trace data. *Latency* is computed as $|t_r - t_p|$ for forward edges (m_x, m_y), $|t_q - t_s|$ for backward edges (m_y, m_x), and $|t_s - t_r|$ for self-edges (m_x, m_x) or (m_y, m_y).

Thus, the feature of an edge (m_x, m_y) is represented by a scalar value indicating the time delay associated with that interaction. These latency values capture the temporal dynamics of the request flow through different microservices. All edge latencies in E_i are aggregated into the feature set $\mathcal{E}_i$, resulting in the enriched graph representation $G_i = \langle V_i, E_i, \mathcal{E}_i \rangle$ for request r_i.

Node Feature Extraction. We leverage performance metrics and application logs to compute node features for each microservice m_x involved in processing a user request r_i. Numerical fields extracted from application logs are used directly, while categorical fields are encoded using one-hot encoding to ensure effective representation.

For each request r_i, we construct a log-based feature vector $L^i_{x \in V}$ of dimension $1 \times |L|$, where $|L|$ represents the total number of numerical and one-hot encoded categorical log fields specific to microservice $m_x \in V_i$.

In parallel, we extract performance metrics from A distinct numerical attributes (e.g., CPU usage, memory consumption), forming a vector P^i_x of dimension $1 \times |A|$ for each microservice $m_x \in V_i$. In our case, $|A| = 70$.

The final node feature vector is formed by concatenating the log vector L^i_x with the performance metric vector P^i_x, resulting in $\mathcal{V}_{x,i} = (L^i_x \oplus P^i_x)$ of dimension $1 \times (|L| + |A|)$ for each $m_x \in V_i$.

All such vectors are collected and stacked to form the node feature matrix $\mathcal{V}_i = \cup_{x \in V_i} \mathcal{V}_{x,i}$, with dimensions $|V| \times (|L| + |A|)$. This matrix $\mathcal{V}_i$ captures the node-level feature representation for the graph corresponding to request r_i, leading to the final feature-enriched graph $G_i = \langle V_i, E_i, \mathcal{E}_i, \mathcal{V}_i \rangle$.

5.2   Graph and Node Embedding

The Feature Graph $G = \langle V, E, \mathcal{E}, \mathcal{V} \rangle$ encodes system health and microservice status for each request r_i. To extract meaningful representations, we employ graph neural networks, obtaining both graph-level and node-level embeddings. Each node embedding is labeled as (M_v, F_i), indicating whether the microservice was healthy or anomalous.

(i) Graph Embedding: We use a Graph Convolutional Network (GCNconv) [19] to learn vector representations of Feature Graphs. Global average pooling is applied to obtain a holistic view of the system's state. The layer-wise propagation rule is: $H^{(l+1)} = \sigma(D^{-1/2} \mathcal{E} D^{-1/2} H^{(l)} W^{(l)})$ where D is the node degree matrix constructed from Edge connections E matrix, $\mathcal{E}$ represents weighted edge attributes, $H^0 = \mathcal{V}$ is initial node features. and $W^{(l)}$ is a trainable weight matrix. The final graph embedding $\mathbb{E}^{graph}_i$ has a dimension of $1 \times |D_g|$, where D_g is the embedding size.

(ii) Node Embedding: To capture fine-grained microservice behaviors, we employ NNconv [14]. The update rule for node embeddings in the NNConv layer is defined as: $\mathbf{h}^{(l+1)}_x = \sigma \left(\sum_{y \in V} f_\theta(\mathcal{E}_{xy}) \cdot \mathbf{h}^{(l)}_y \right)$, where $h^0_x = \mathcal{V}_x$ and $\mathcal{E}_{xy}$ is edge feature between edge of microservices (m_x, m_y). The resulting node embeddings $\mathbb{E}^{node_v}_i$ have dimensions $1 \times |D_n|$, where D_n is the output size of NNconv. These embeddings represent the operational state of each microservice during request.

5.3   Root Cause Detection

We utilize both graph and node embeddings to detect root causes of anomalies. Since microservices may exhibit similar embeddings due to operational factors (e.g., high load, user spikes), incorporating graph embeddings enhances system-wide anomaly assessment.

For effective classification, we enhance node with graph embeddings: $\mathbb{E}_i^{nodes_v} = \mathbb{E}_i^{nodes_v} \oplus \mathbb{E}_i^{graph}$ where $\mathbb{E}_i^{nodes_v}$ represents the node embedding of each microservice, and $\mathbb{E}_i^{graph}$ captures the overall system state. These vectors are input into a deep neural network (DNN) for classification. The input layer size is: $|D_g| + |D_n|$ where D_g and D_n are the dimensions of the graph and node embeddings, respectively. The output layer consists of $R = |V| \times |F|$ neurons, representing all possible microservice-fault pairs (M_v, F_i). The whole model (GCN+DNN) is trained on labeled data, where ground truth labels indicate the faulty microservice and anomaly type.

The core idea is as follows. At runtime, the root cause detection module constructs the Feature Graph (FG) using data samples collected from multiple microservices. This graph G_i is then passed through a pre-trained model combining Graph Convolutional Networks (GCN) and a Deep Neural Network (DNN), which produces a probability distribution of dimension $|R|$. Each entry in the output corresponds to a label (M_v, F_i), indicating the likelihood of fault F_i occurring in microservice M_v. To identify the root cause, FNoDe ranks the predicted probabilities and selects the top-k candidate labels as the most probable root causes.

6   Theoretical Foundation for Graph Representations

This section provides formal justification for graph-based representations in microservices fault detection.

6.1   Formal Problem Definition

A microservices system $S = \{s_1, s_2, ..., s_n\}$ processes requests $R = \{r_1, r_2, ..., r_m\}$ and may experience faults $F = \{f_1, f_2, ..., f_k\}$. Given observable states $O = \{o_1, o_2, ..., o_t\}$ from traces, metrics, and logs, we seek a mapping $\phi : O \to S \times F$ identifying both the faulty service and fault type.

6.2   Information Preservation Properties

Table 2 compares representation techniques. We provide theoretical explanation of graph-based representations [19,34].

Theorem 1 (Structural Completeness). *A directed graph $G = (V, E)$ with V representing microservices and E representing service interactions preserves all structural dependencies present in a microservices architecture.*

Proof. For any pair of microservices (s_i, s_j) with a dependency relationship, there exists a corresponding edge $e_{ij} \in E$. The graph structure maintains complete topology, unlike vector or time-series representations that lose structural information through flattening or aggregation.

Theorem 2 (Locality of Fault Propagation). *Faults in microservices exhibit locality, where anomaly effects propagate through adjacent nodes in the dependency graph.*

Proof. Consider a fault in service s_i. Primary effects are observable in s_i's metrics and logs. Secondary effects propagate to dependent services s_j where $(s_i, s_j) \in E$ or $(s_j, s_i) \in E$, manifesting as increased latency or error rates. This propagation follows the graph's edge structure, creating a neighborhood effect that diminishes with graph distance—naturally captured by GCN's message-passing operations.

7   Experimental Setup

This section outlines the infrastructure, datasets, fault injection strategies, baselines, and evaluation metrics used in our study.

7.1   Setup and Dataset Details

We use a custom microservices setup along with public benchmark datasets.
Microservices Applications: Each microservice runs on a dedicated host (8GB RAM, Intel i5, HDD) using Python 3.8 with Flask [15]. Workloads are generated via Locust 4, simulating 50–300 users and 20–600 requests/sec [17]. Tracing and metrics are collected using Jaeger, Prometheus, PIDstat, node-exporter, and SARstat [1–5].
Fault Injection: We introduce functional and non-functional anomalies. Non-functional faults (F_1, F_2): High resource use and slow responses via stress-ng, cpu-load-gen, and Traffic Control [6–8]. Functional faults (F_3–F_6): Code-level issues increasing algorithmic complexity and inter-service delays.
Public Datasets: We also use five public microservice benchmarks—*hotel-reservation, media-service, social-network, ticket-booking,* and *sock-shop* [27]. These include traces with injected anomalies (e.g., latency, overuse), span 5–32 services, 5–12 fault types, and cover up to 90K requests or 8 h of logs.

7.2   Baseline Methods

We compare **FNoDe** with three baseline methods:

FIRM [27] Identifies target microservices using per-instance performance variability and detects performance violations using SVM.

RCD [17] Establishes relationships between faulty node metrics and anomalies to identify the root cause, narrowing search via hierarchical learning.

PUTraceAD [35] Represents traces as a call graph with node features such as operation name, response code, and duration. A GNN and PU-learning model are used for root cause detection.

7.3   Evaluation Metric

The task is framed as a multi-class classification problem, evaluated using accuracy: $Accuracy = TP + TN/TP + TN + FP + FN$ where TP, TN, FP, and FN denote true positives, true negatives, false positives, and false negatives, respectively. A true positive occurs when the ground truth label (faulty microservice and anomaly type) is within the top-k predictions.

8   Evaluation

In this section, we conduct a rigorous performance evaluation of FNoDe.

8.1   Overall Performance

(i) Table 4 presents the anomaly-wise performance of FNoDe for the *Library* and *Ecom* applications. The results indicate consistent accuracy across different anomaly types (top-k, $k = 3$), with minimal variations. This is attributed to FNoDe's ability to leverage heterogeneous microservices logs, generating distinct embeddings, as evidenced by clustering results in Fig. 11(d). To validate the impact of multiple modalities, we conducted experiments using subsets of the modalities, presented in Fig. 10(b).

Table 4. FNoDe anomaly-wise performance for *Library* and *Ecom*.

Anomalies	Library	Ecom
F_0	99.8%	99.1%
F_1	99.4%	95.2%
F_2	99.5%	91.8%
F_3	99.1%	90.5%
F_4	99.5%	94.6%
F_5	99.6%	92.4%
F_6	99.3%	95.3%
Avg.	**99.4%**	**94.1%**

Table 5. Microservices-wise performance for *Ecom*.

Service Type	#Services	Accuracy
User-facing	4	96%
Intermediate	6	95%
Data management	2	91%

Table 6. Baseline comparison on *Library* and *Ecom* applications.

Framework	Library	Ecom
$FIRM$	77.5%	75.8%
RCD	65.2%	60.4%
$PUtraceAD$	75.9%	69.5%
$FNoDe$	99.4%	94.1%

(ii) Table 5 presents root cause detection accuracy across different microservice categories in *Ecom*. User-interactive and intermediate services exhibit high accuracy, as their behavior varies by operation and call paths. In contrast, accuracy declines for data management services, which are invoked by multiple services simultaneously, making health status generalization challenging.

8.2   Baseline Comparison with Multiple Datasets

We evaluate FNoDe against baseline algorithms using publicly available microservices datasets, selected for diversity in data source, information richness, and scale. The microservices architectures in these datasets range from 5 to 30 services.
(i) Table 6 compares FNoDe with baseline algorithms on the *Library* and *Ecom* datasets. The baseline frameworks struggle with heterogeneous microservices data, where anomaly signatures manifest across multiple modalities.

FIRM extracts a "critical path" from tracing and telemetry data but may miss branches in the call graph. Additionally, SVM-based detection cannot capture hidden relationships between different modalities. *PUtraceAD* constructs an undirected microservices graph, relying only on logs and traces, limiting its ability to correlate latency and other metrics. *RCD* models failures using a probability-based causation approach but only considers latency variations, making it challenging to learn dependencies across multiple data sources.

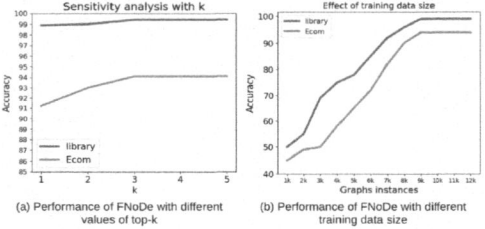

(a) Performance of FNoDe with different values of top-k

(b) Performance of FNoDe with different training data size

Fig. 9. Performance of FNoDe with different values of top-k and varying training sizes.

Table 7. Baseline evaluation on multiple datasets.

Framework	Hotel	Media	Social	Ticket	Socks
FIRM	95%	96%	95%	94%	96%
RCD	88%	87%	85%	87%	88%
PUtraceAD	90%	91%	90%	89%	93%
FNoDe	95%	95%	96%	96%	95%

(ii) Table 7 presents results across benchmark datasets. FNoDe achieves comparable performance to baseline algorithms, as these datasets primarily focus on non-functional anomalies (e.g., resource usage, latency) that traditional methods are designed to handle. Moreover, these datasets lack the diverse modalities FNoDe utilizes, demonstrating its capability to operate effectively with partial modality information.

8.3   Breakdown of FNoDe

We analyze the performance and interpretability of FNoDe using ablation studies and GNNExplainer, and highlight its strengths and limitations.
(i) Sensitivity to top-k predictions: In Fig. 9(a), accuracy remains consistent across varying values of top-k, indicating the true faulty microservice is usually ranked at the top. However, higher k values increase false positives, suggesting smaller k values are preferable for operational efficiency.
(ii) Training volume impact: FNoDe performs well even with limited training data, showing its ability to learn distinct anomaly signatures from heterogeneous

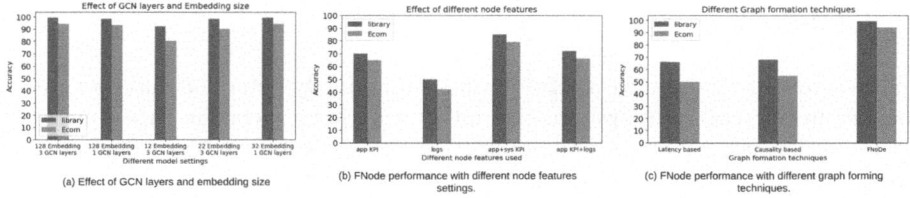

Fig. 10. FNoDe performance across different model settings.

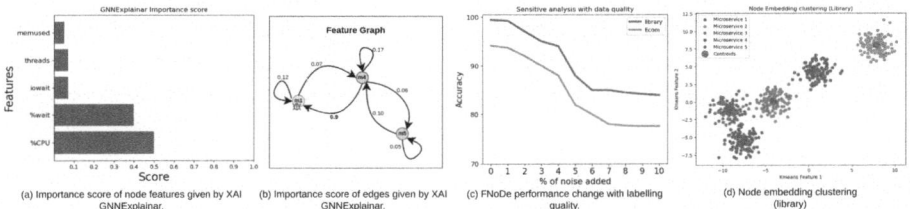

Fig. 11. Breakdown of FNoDe.

sources. Figure 9(b) demonstrates that FNoDe achieves optimal performance with a small training volume, validating its ability to learn hidden signatures within heterogeneous data efficiently

(iii) Embedding size and GCN depth: Optimal performance is achieved with intermediate embedding sizes and shallow GCN layers as shown in Fig. 10(a). Larger embeddings risk overfitting, and deeper GCNs suffer from over-smoothing. Combining node- and graph-level embeddings allows FNoDe to capture both localized and global system behavior without deep architectures.

(iv) Graph formation techniques: Feature graphs used by FNoDe outperform traditional Causality-based and latency-based approaches by preserving inter-service dependencies and reducing ambiguity. Figure 10(c) shows that capturing call structure explicitly improves detection accuracy. Table 5 also reveals that services with more stable call patterns yield better results, while shared services like data nodes show slightly reduced accuracy due to their variable behavior.

(v) Explainability Analysis: Using GNNExplainer in PyTorch, we analyze FNoDe's learned representations. Figure 11(a) shows that $\%CPU$ and $\%wait$ were most influential for anomaly F_1, aligning with ground truth. Figure 11(b) indicates that edges from m_4 to m_1 were crucial in detecting anomaly F_5. These results confirm that FNoDe effectively learns and predicts anomalies without overfitting.

(vi) Sensitivity to Noisy Labels: Fig. 11(c) demonstrates that FNoDe's performance degrades significantly with noisy labels, highlighting its sensitivity due to the adaptability of GCN layers and the supervised training approach.

9   Conclusion

FNoDe proves to be an effective framework for anomaly detection and root cause analysis in microservices systems. By integrating heterogeneous data sources—traces, performance metrics, and application logs—FNoDe enables a comprehensive analysis of system behavior and anomaly identification.

Its graph-based approach effectively captures complex dependencies between microservices, facilitating accurate detection of anomalies and their root causes. Experimental results demonstrate that FNoDe consistently outperforms baseline methods, particularly in detecting functional anomalies. Furthermore, its ability to handle both functional and non-functional anomalies across multiple heterogeneous modalities makes it a robust solution for diverse microservices environments.

Acknowledgments. This work has been partially supported by the DST-SERB funded project titled "SysFix: Building an End-to-end Troubleshooting Solution for Large-scale Systems with Automated Failure diagnosis, Root cause analysis and Remediation".

References

1. Jaeger: open source, end-to-end distributed tracing. https://www.jaegertracing.io
2. node_exporter. https://github.com/prometheus/node_exporter
3. pidstat. https://man7.org/linux/man-pages/man1/pidstat.1.html
4. Prometheus. https://prometheus.io
5. sar. https://man7.org/linux/man-pages/man1/sar.html
6. sleep. https://linuxhint.com/sleep_command_linux
7. stress-ng. https://wiki.ubuntu.com/Kernel/Reference/stress-ng
8. tc. https://linux.die.net/man/8/tc
9. Zipkin. http://zipkin.io
10. Aggarwal, P., et al.: Localization of operational faults in cloud applications by mining causal dependencies in logs using golden signals. In: Hacid, H., et al. (eds.) Service-Oriented Computing – ICSOC 2020 Workshops: AIOps, CFTIC, STRAPS, AI-PA, AI-IOTS, and Satellite Events, Dubai, United Arab Emirates, December 14–17, 2020, Proceedings, pp. 137–149. Springer International Publishing, Cham (2021). https://doi.org/10.1007/978-3-030-76352-7_17
11. Chen, Y., Jin, Y., Tao, B., Qiu, M.: MFRL-CA: microservice fault root cause location based on correlation analysis. In: ICBASE, pp. 85–91 (2021)
12. Gan, Y., Zhang, Y., Cheng, D., et al.: Sage: practical and scalable ML-driven performance debugging in microservices. In: ASPLOS, pp. 135–151 (2021)
13. Gan, Y., et al.: Seer: leveraging big data to navigate the complexity of performance debugging in cloud microservices. In: ASPLOS, pp. 19–33 (2019)
14. Gilmer, J., Schoenholz, S., Riley, P., Vinyals, O., Dahl, G.: Neural message passing for quantum chemistry. arXiv preprint arXiv:1704.01212 (2017)
15. Grinberg, M.: Flask web development: developing web applications with python. O'Reilly (2018)

16. Guo, X., et al.: Graph-based trace analysis for microservice architecture under-standing. In: ESEC/FSE, pp. 1387–1397 (2020)
17. Ikram, A., Chakraborty, S., Shah, S., Co, R., Baset, S.: Root cause detection in microservice architecture: a hierarchical approach. In: NeurIPS (2022)
18. Kalinagac, O., Aral, A., Ovatman, T.: Graph based liability analysis for the microservice architecture. In: CNSM, pp. 314–322 (2022)
19. Kipf, T., Welling, M.: Semi-supervised classification with graph convolutional networks. In: ICLR (2017)
20. Kratzke, N., Quint, P.: Understanding cloud-native applications after 10 years of cloud computing. J. Syst. Softw. **126**, 1–16 (2017)
21. Liu, D., He, C., Peng, X., et al.: MicroHECL: high-efficient root cause localization in large-scale microservice systems. In: ICSE-SEIP, pp. 338–347 (2021)
22. Liu, P., Xu, H., Ouyang, Q., et al.: Unsupervised detection of microservice trace anomalies through service-level deep Bayesian networks. In: ISSRE, pp. 48–58 (2020)
23. Ma, M., Xu, W., Luo, L., Chen, Y., Scott, S.: ServiceRank: root cause identification of anomaly in large-scale microservice architectures. IEEE Trans. Dependable Secure Comput. **19**(5), 3087–3100 (2022)
24. Nandi, A., Mandal, A., Atreja, S., Dasgupta, G., Bhattacharya, S.: Anomaly detection using program control flow graph mining from execution logs. In: KDD, pp. 215–224 (2016)
25. Nguyen, H., Dang, K., Tran, D., Vuong, Q., Do, T.: A survey on graph neural networks for microservice-based cloud applications. Sensors **22**(23), 9492 (2022)
26. Qiu, H., Banerjee, S., Jha, S., Kalbarczyk, Z., Iyer, R.: FIRM: an intelligent fine-grained resource management framework for SLO-oriented microservices. In: OSDI, pp. 805–825 (2020)
27. Qiu, J., Du, Q., Yin, K., Zhang, S., Qian, C.: A causality mining and knowledge graph based method of root cause diagnosis. Appl. Sci. **10**(6), 2166 (2020)
28. Shan, H., et al.: μ-Diagnosis: unsupervised and real-time diagnosis of small-window long-tail latency. In: SOSP (2019)
29. Sigelman, B., et al.: Dapper, a large-scale distributed systems tracing infrastructure (2010), google Technical Report
30. Soldani, J., Tamburri, D., Van Den Heuvel, W.: Anomaly detection and failure root cause analysis in (micro) service-based cloud applications. ACM Comput. Surv. **55**(3), 1–39 (2022)
31. Wang, L., Chen, N., Li, P., et al.: Root-cause metric location for microservice systems via log anomaly detection. In: ICWS, pp. 142–150 (2020)
32. Wu, L., Bogatinovski, J., Nedelkoski, S., Tordsson, J., Kao, O.: MicroRCA: root cause localization of performance issues in microservices. In: NOMS, pp. 1–9 (2020)
33. Wu, L., Bogatinovski, J., Nedelkoski, S., Tordsson, J., Kao, O.: Performance diagnosis in cloud microservices using deep learning. In: ICSOC Workshops, pp. 85–96 (2021)
34. Xu, K., Hu, W., Leskovec, J., Jegelka, S.: How powerful are graph neural networks? In: ICLR (2019)
35. Zhang, K., Xu, J., Li, M., Liu, Y., Xie, T., Mei, H.: PUTraceAD: trace anomaly detection with partial labels based on GNN and PU learning. In: ISSRE, pp. 345–356 (2022)
36. Zhou, X., et al.: Fault analysis and debugging of microservice systems. IEEE Trans. Softw. Eng. **47**(2), 243–260 (2021)

An Enhanced FP-Growth Algorithm with Hybrid Adaptive Support Threshold for Association Rule Mining

Kanda Runapongsa Saikaew[1] , Carson K. Leung[2] ,
and Kritbodin Phiwhorm[3(✉)]

[1] Khon Kaen University, Khon Kaen, Thailand
krunapon@kku.ac.th
[2] University of Manitoba, Winnipeg, MB, Canada
Carson.Leung@UManitoba.ca
[3] Sisaket Rajabhat University, Sisaket, Thailand
kritbodin.p@sskru.ac.th

Abstract. Finding frequent itemsets remains challenging due to manual threshold specification requirements in existing algorithms. This paper presents an Enhanced FP-Growth algorithm incorporating a hybrid adaptive support threshold that combines statistical variance analysis, frequency distribution patterns, and transaction density metrics. The algorithm automatically adjusts support levels based on dataset characteristics, eliminating manual threshold tuning. Experimental evaluation on five benchmark datasets against Aprior, FP-growth, and FP-Max shows our Enhanced FP-Growth consistently achieves superior execution time and improved memory efficiency. The hybrid threshold mechanism dynamically calibrates according to dataset characteristics, offering substantial efficiency gains across diverse data types.

Keywords: Frequent Itemset Mining · Enhanced FP-Growth · Hybrid Threshold · Association Rules · Data Mining · Big Data Analytics · Knowledge Discovery

1 Introduction

Association rule mining [1] enables discovery of relationships within transaction databases, with applications spanning retail analysis, healthcare, and web usage mining [2–7]. The core challenge lies in efficiently finding frequent itemsets, which forms the foundation for generating meaningful association rules [8].

While early algorithms like Apriori suffered from candidate generation overhead and FP-growth offered a more efficient tree-based approach, both share a critical drawback: manual specification of minimum support thresholds. Traditional methods require domain expertise or trial-and-error iterations [9]. Inappropriate thresholds cause either pattern explosion (too low) or valuable pattern loss (too high).

© The Author(s), under exclusive license to Springer Nature Switzerland AG 2026
C. K. Leung et al. (Eds.): DaWaK 2025, LNCS 16048, pp. 102–108, 2026.
https://doi.org/10.1007/978-3-032-02215-8_7

Recent adaptive approaches focus on statistical measures [9] or data density patterns [10], but most address single aspects rather than comprehensive dataset characteristics.

Our *key contributions* of this paper include our Enhanced FP-Growth with a hybrid threshold mechanism. It (a) combines statistical variance, frequency distribution, and transaction density metrics; (b) eliminates manual threshold tuning through automatic calibration; and (c) improves computational efficiency. We evaluate Enhanced FP-Growth against Apriori, FP-growth, and FP-Max using five benchmark datasets, demonstrating significant improvements in execution time and memory utilization.

The remainder of this paper is organized as follows: The next section reviews related work. Section 3 presents our Enhanced FP-Growth algorithm. Section 4 describes experimental evaluation. Section 5 draws conclusions and outlines future directions.

2   Related Work

The Apriori algorithm [1] uses candidate generation with multiple database scans, causing performance bottlenecks. The FP-growth algorithm [8] improves efficiency through compressed FP-tree structures and divide-and-conquer strategies. FP-Max [11] extends FP-growth for maximal frequent itemset mining. However, all these algorithms depend significantly on manual minimum support threshold selection.

Traditional static thresholds require domain expertise and trial-and-error, potentially causing pattern explosion or information loss. Recent adaptive approaches include statistical and density measures [9] or focus on data density patterns alone [10], but most address single aspects rather than the comprehensive dataset characteristics targeted by our hybrid approach.

3   Enhanced FP-Growth

3.1   Hybrid Threshold Mechanism

Our hybrid threshold (HT) mechanism combines statistical, distribution-based, and density-based measures to compute an optimal threshold value. More specifically, HT integrates three components: Statistical threshold (ST), distribution threshold (DT), and density threshold (NT). Each component captures different dataset characteristics, with relative importance determined by weights w_1, w_2, and w_3. Consequently, the hybrid threshold (HT) is calculated as:

$$HT = w_1 ST + w_2 DT + w_3 NT \tag{1}$$

where weights $w_1 + w_2 + w_3 = 1$ and $w_i \in [0, 1]$.

To elaborate, the three components are calculated as follows. Statistical threshold (ST) uses average support:

$$ST = \frac{\sum_{i=1}^{m} \text{support}(item_i)}{m} \tag{2}$$

where m is total unique items. Distribution threshold (DT) employs median support for outlier robustness [9]:

$$DT = \text{median}(\text{support}) \tag{3}$$

Density threshold (NT) evaluates transaction density:

$$NT = \frac{\sum_{i=1}^{n} |t_i|}{n \times m} \tag{4}$$

where $|t_i|$ is length of transaction i and n is number of transactions.

To optimize ensemble weights, we follow established methodologies in ensemble learning research [12,13]. We determine optimal weights through systematic multi-criteria evaluation. Our weight configuration ($w_1 = 0.5$, $w_2 = 0.3$, $w_3 = 0.2$) was determined through:

1. *Systematic evaluation:* Comprehensive assessment of weight combinations with 0.1 intervals across five benchmark datasets.
2. *Performance assessment:* Combined evaluation of execution time and memory utilization.
3. *Component-wise validation:* Evaluation of statistical, distribution, and density threshold contributions.

Statistical threshold receives primary weight ($w_1 = 0.5$) as it captures fundamental frequency distribution patterns across diverse datasets. Distribution threshold receives secondary weight ($w_2 = 0.3$) for providing robustness against outliers, while density threshold receives tertiary weight ($w_3 = 0.2$) as a contextual scaling factor. This configuration ensures optimal threshold calculation while maintaining computational efficiency.

3.2   Enhanced FP-Growth for Frequent Itemset Mining

Enhanced FP-Growth extends traditional FP-growth by incorporating hybrid threshold mechanism for automatic pruning. While traditional FP-growth stores transactions with common prefixes, our Enhanced FP-Growth dynamically determines which items to include based on dataset characteristics, eliminating manual threshold tuning.

Algorithm 1 outlines the Enhanced FP-Growth construction process. The critical improvement appears in line 3, where CalculateHybridThreshold function automatically determines frequent items.

Enhanced FP-Growth follows divide-and-conquer strategy with pre-filtered tree representation. Unlike traditional methods, it combines efficiency from adaptive threshold and optimized tree structure through: (1) streamlines pattern verification against hybrid threshold; and (2) reduces overhead by building conditional trees only for promising patterns.

Algorithm 1. Enhanced FP-Growth Construction

1: **Input:** Transaction database D, Items I
2: **Output:** Enhanced FP-Growth
3: $HT \leftarrow$ CALCULATEHYBRIDTHRESHOLD(D, I)
4: $F \leftarrow \{i \in I \mid \text{support}(i) \geq HT\}$
5: HeaderTable $\leftarrow$ SORTBYFREQUENCY(F)
6: Root $\leftarrow$ CREATENODE(null)
7: **for** each transaction $t \in D$ **do**
8:     $t' \leftarrow$ FILTERANDSORT(t, HeaderTable)
9:     INSERTTRANSACTION(Root, t')
10: **end for**
11: **return** Root, HeaderTable
---

4   Evaluation

4.1   Performance Metrics and Datasets

Time efficiency combines threshold computation and mining execution time. Memory utilization captures maximum memory across construction and generation phases.

We used five benchmark datasets for evaluation: Mushroom (dense categorical), Retail (real-world sparse), Accidents (highly correlated), T25I10D10K and T10I4D100K (synthetic controlled). Table 1 summarizes their characteristics.

Table 1. Dataset characteristics for association rule mining.

Dataset name	#transactions	#items	Source
Mushroom	8,124	119	UCI [14]
Retail	88,162	16,470	BRS [2]
Accidents	340,183	468	FIMI [15]
T25I10D10K	9,976	929	FIMI [15]
T10I4D100K	100,000	870	IBM [1]

4.2   Performance Results

Execution Time. As shown in Fig. 1, our Enhanced FP-Growth demonstrates consistently superior execution time, with the most significant improvement seen on the T10I4D100K dataset, where it was 81.5% faster than Apriori and 47.4% faster than FP-growth. The hybrid threshold enables early elimination of infrequent items, resulting in streamlined tree construction and faster pattern extraction. For the Accidents dataset, both FP-growth and our Enhanced FP-Growth perform significantly faster than others. Between the two, FP-growth performs just slightly faster than Enhanced FP-Growth (9.92 s vs. 11.39 s), attributed to uniform item distribution where hybrid threshold overhead incurs additional time.

Memory Usage. As shown in Fig. 2, our Enhanced FP-Growth demonstrates notable reduction in memory consumption: Mushroom (2.7% reduction: 251.61 MB vs. 258.55-258.64 MB) and T25I10D10K (2.7% reduction). For Accidents, both FP-Max and our Enhanced FP-Growth achieve 25% memory reduction versus Apriori. Reduction in memory consumption stems from the hybrid threshold identifying and eliminating low-utility patterns early.

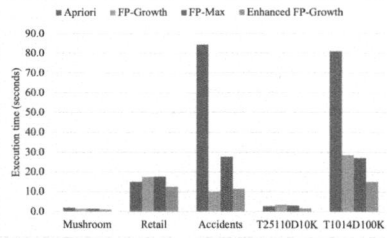

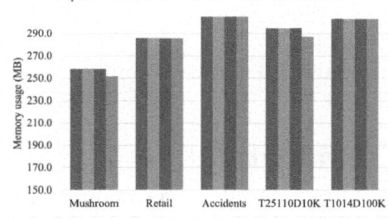

Fig. 1. Execution time comparison.   **Fig. 2.** Memory usage comparison.

Algorithm Completeness. Our Enhanced FP-Growth generates the same frequent itemsets as traditional FP-growth when using equivalent threshold values, confirming that pattern completeness is preserved. This study focused on the performance and completeness of frequent itemset mining; an analysis of the quality of the resulting association rules is noted as an area for future work.

Weight Sensitivity Analysis. The optimized weight configuration (w_1=0.5, w_2=0.3, w_3=0.2) demonstrates consistent performance across all tested datasets. The statistical component shows most influential for threshold calculation, with weights determined through systematic evaluation across benchmark datasets. Deviating significantly from these weights led to performance degradation. For instance, shifting to a (0.7, 0.2, 0.1) configuration or an inverted (0.2, 0.3, 0.5) configuration resulted in suboptimal threshold calculation. The selected weights provide the most robust balance between computational efficiency and algorithmic effectiveness across all tested datasets, while distribution and density components provide essential refinement for different data characteristics.

5   Conclusions

In this paper, we presented Enhanced FP-Growth with hybrid adaptive support threshold, which successfully addresses manual threshold specification challenges in frequent itemset mining. Our hybrid mechanism automatically calibrates based on statistical variance, frequency distribution, and transaction density. Experimental evaluation demonstrated consistent performance improvements in both execution time and memory efficiency across benchmark datasets while preserving pattern completeness. The empirically validated weight configuration shows robustness across diverse data characteristics.

For *ongoing and future research*, we focus on algorithm integration with other mining approaches, advanced optimization techniques, and extension to multi-level pattern mining. To further validate generalization, evaluation on larger-scale datasets is necessary. Additionally, a detailed analysis of the quality of the generated association rules using metrics like confidence and lift represents a promising direction.

Acknowledgments. This work is partially supported by Faculty of Engineering at Khon Kaen University (Thailand), NSERC (Canada) and University of Manitoba.

Disclosure of Interests. The authors have no competing interests to declare. We acknowledge the use of AI-assisted tools (Claude and Gemini) for grammar checking, proofreading, and language refinement. However, all scientific content, analyses, and conclusions presented in this paper were verified and approved by the authors.

References

1. Agrawal, R., Srikant, R.: Fast algorithms for mining association rules. In: VLDB 1994, pp. 487–499 (1994)
2. Brijs, T., et al.: Using association rules for product assortment decisions: a case study. In: ACM KDD 1999, pp. 254–260 (1999)
3. Capillar, E., et al.: Bitwise vertical mining of minimal rare patterns. In: DaWaK 2023. LNCS, vol. 14148, pp. 135–141 (2023)
4. Czubryt, T.J., et al.: Q-VIPER: quantitative vertical bitwise algorithm to mine frequent patterns. In: DaWaK 2022. LNCS, vol. 13428, pp. 219–233 (2022)
5. Leung, C.K.-S., Hayduk, Y.: Mining frequent patterns from uncertain data with MapReduce for big data analytics. In: Meng, W., Feng, L., Bressan, S., Winiwarter, W., Song, W. (eds.) DASFAA 2013. LNCS, vol. 7825, pp. 440–455. Springer, Heidelberg (2013). https://doi.org/10.1007/978-3-642-37487-6_33
6. Madill, E.W.R., et al.: Enhanced sliding window-based periodic pattern mining from dynamic streams. In: DaWaK 2022. LNCS, vol. 13428, pp. 234–240 (2022)
7. Rizvee, R.A., et al.: A tree-based framework to mine top-k closed sequential patterns. Appl. Intell. **55**(3), 221 (2025)
8. Han, J., et al.: Mining frequent patterns without candidate generation. In: ACM SIGMOD 2000, pp. 1–12 (2000)
9. Hikmawati, E., Maulidevi, N.U., Surendro, K.: Minimum threshold determination method based on dataset characteristics in association rule mining. J. Big Data **8**(1), 1–17 (2021). https://doi.org/10.1186/s40537-021-00538-3
10. Ogedengbe, M., et al.: Adaptive minimum support threshold for association rule mining. Indonesian J. Data Sci. **5**(2), 101–108 (2024)
11. Grahne, G., Zhu, J.: High performance mining of maximal frequent itemsets. In: SIAM SDM 2003 Workshop on HPDM (2003)
12. Ganaie, M., et al.: Ensemble deep learning: a review. Eng. Appl. Artif. Intell. **115**, 105151 (2022)

13. Kaya, E., Guneyi, E.: Optimizing ensemble weights and hyperparameters of machine learning models for regression problems. Mach. Learn. Appl. **7**, 100251 (2022)
14. Kelly, M., et al.: The UCI machine learning repository (2025). https://archive.ics.uci.edu
15. Goethals, B., Zaki, M.: Proceedings of IEEE ICDM Workshop on FIMI 2003 (2003). https://ceur-ws.org/Vol-90/

Sequential Data Analytics
and Recommendation Systems

Entity Resolution for Streaming Data with Embeddings

Zhongwei Ma[1,2]($\boxtimes$)(iD), Philippe Roose[2](iD), and Jiefu Song[3](iD)

[1] Technopôle Domolandes, Saint Geours de Maremne, France
[2] Universite de Pau et des Pays de l'Adour, E2S UPPA, LIUPPA, Anglet, France
`{zhongwei.ma,philippe.roose}@univ-pau.fr`
[3] Institut de Recherche en Informatique de Toulouse - Université Toulouse Capitole,
31000 Toulouse, France
`jiefu.song@ut-capitole.fr`

Abstract. Most data streaming systems collect and process real-time data from various sources. However, many decisions require both streaming and historical data to obtain a comprehensive view. Entity resolution (ER) determines whether two records refer to the same real-world entity in the absence of a shared identifier. Therefore, it is crucial to integrate stream data and stored data together through entity resolution and to ensure the accessibility and usability of data from different sources. Existing ER methods often fail to support incremental stream processing while efficiently handling complex data like text. Given the strength of embedding techniques in capturing semantic and syntactic information, we aim to adapt embedding-based ER to streaming data for integrating incoming and existing records. We propose a dynamic graph embedding suitable for stream processing to perform entity resolution within relational tables.

Keywords: entity resolution · data stream · embedding

1 Introduction

Entity resolution, also called record linkage or data matching, aims to find different descriptions that refer to the same entity in the real world appearing either within or across data sources, when there are no unique entity identifiers, ensuring the accuracy and consistency of integrated datasets [10]. Assuming that we have real-time sales records across different retailers' platforms, if these records are integrated with historical records in our database, we can gain deeper insights into popular products and consumer purchasing trends. This is a crucial step when we perform data integration and want to offer a uniform view of autonomous and heterogeneous data sources, facilitating further processing [20].

In the big data era, streaming data—continuous and real-time—is increasingly common [3], originating from sources like social media and IoT devices. The requirements for data stream analytics come from two aspects: real-time

© The Author(s), under exclusive license to Springer Nature Switzerland AG 2026
C. K. Leung et al. (Eds.): DaWaK 2025, LNCS 16048, pp. 111–125, 2026.
https://doi.org/10.1007/978-3-032-02215-8_8

processing to support immediate decision-making with low latency, and storage of data for subsequent queries or more complex and time-consuming analysis. For instance, in the IoT industry, millions of devices generate continuous data streams and report abnormal or significant events in real time, optimizing online production. By integrating and analyzing this data with historical consumer trends, companies can unlock greater value, enhance consumer insights, and maximize profitability. Therefore, modern data stream processing systems need to incorporate both stream and batch processing [17] within a unified framework, as outlined in the Lambda architecture originally introduced by Matz and Werren [22]. To enhance the efficiency of these systems, integrating stream data with historical data by entity resolution is essential.

The main challenge of stream data lies in its unbounded nature, which requires a dynamic framework capable of performing entity resolution incrementally. The lack of a complete dataset also makes it complicated because we can't determine similarity between two descriptions and identify the best match within a finite amount of data.

ER has been defined historically as an offline task performed during data integration to improve data quality in databases [10], and many attempts have been made to focus on batch processing using different methods [9,13,19,25,26]. However, these batch-oriented approaches designed for static databases are not well suited for streams. So far, two main approaches have been adapted for incremental data. The first is rule-based methods [15,30]. While effective in constrained domains, these approaches often rely on complex rules and domain-specific knowledge, which limits their generalizability. The second approach involves machine learning, particularly clustering techniques [11,31]. However, this approach faces challenges when applied to unstructured textual data, limiting its effectiveness in real-world environments.

To address the limitations in text analysis and the complexity of domain-specific algorithms, we turn our attention to embedding techniques to map text into low-dimensional vectors capturing semantic relationships [34]. For now, this technique is usually used for batch processing [19]. Additionally, significant research has been conducted on incremental embeddings [4], laying the foundation of entity resolution in streaming data.

In this work, we introduce a framework for data integration of streaming data by dynamic graph embedding. This framework includes a complete process from pre-trained embedding models by local datasets to accepting streaming data and gradually involves the added data and determining whether the added data belongs to any entity present in the original dataset. The construction of embedding represents well the semantic and syntactic information [34].

Our contributions is a dynamic graph embedding model for streaming entity resolution. Building on embedding techniques that capture relationships and semantic information, our model incrementally updates both the graph and its embeddings. Specifically, when a new record arrives, we add a corresponding node, create edges connecting it to relevant parts of the existing graph, and generate evolving random walks from the new node to assess similarity. This incremental approach avoids retraining from scratch, significantly reducing computational costs while enabling efficient and scalable resolution of entities in data streams.

The article is structured as follows. In Sect. 2, we list common methods used for Entity Resolution and analyze their characteristics. Section 3 introduces the conception of our proposal. In Sect. 4, we describe in detail the implementation of the protocol and present an analysis of the results. We conclude in the last section and outline directions for future work.

2   Related Work

We review the literature related to this paper in three areas. Section 2.1 summarizes the methods suitable for streaming data in order to present the current situation of incremental entity resolution. Section 2.2 focuses on embedding techniques and illustrates how they can be applied to achieve entity resolution. Finally, Sect. 2.3 explores dynamic embeddings, detailing how existing tools can adapt embedding methods in batch for real-time stream data processing.

2.1   Entity Resolution for Stream

Many studies mentioned in [5] have explored various approaches for entity resolution in order to face the challenge of incremental data. We categorize them into three main groups: Query-driven approaches, Rule-based approaches, and Learning-based approaches.

Query-driven approaches [2,32] store all incoming data and perform entity resolution at query time. However, these on-demand methods can take several minutes to wait for a result. Additionally, redundant records further increase storage demands, adding to resource inefficiency.

Rule-based approaches [14,15,30] are one of the most commonly used methods in entity resolution, which rely on deterministic and interpretable rules to compare record attributes. However, the realization of this approach can be very complex and requires sophisticated algorithms designed for streaming data.

Learning-based approaches have become increasingly prevalent due to advancements in machine learning techniques. Most of these approaches operate in batch mode [26] and aim to enhance accuracy and efficiency through methods such as graph matching and neural networks. However, training machine learning models presents significant challenges, particularly with noisy data and limited labeled resources, especially in streaming scenarios, where datasets are often incomplete. Incremental methods often use unsupervised techniques like clustering [31]. Instead of linking records individually, the objective of clustering is to group records into their true and often latent entities. Despite their advantages, the performance of clustering methods is limited when processing high-dimensional data, such as text, which is crucial in entity resolution tasks. Moreover, dirty data, such as spelling mistakes or data with missing values, can cause noise and false positives, which amplifies inaccuracies, especially in the context of continuous data streams [10].

Therefore, existing propositions of entity resolution for streaming satisfy some requirements in real-time situations, like incremental update. But they do not have excellent performance as for large-scale and complex data streams such as unstructured textual data. This kind of data can be analyzed by embedding models efficiently, which we will present in the next section.

2.2 Embeddings and Entity Resolution

Word embeddings are a key technique in machine learning, mapping high-dimensional data into fixed-length vectors. This representation captures semantic relationships between words, ensuring that similar words have similar vector representations [1].

Various methods can be used to obtain this type of representation. Prediction-based approaches (e.g., word2vec [23]) train neural networks by leveraging local data (e.g., a word's context) to either predict a word based on its context or infer the context from a given word. This process ensures that words with similar semantic meanings are mapped to similar vector representations. Count-based methods (e.g., GloVe [28]), on the other hand, build co-occurrence matrices to extract global statistical patterns, such as corpus-wide word frequencies, then reduce dimensions to obtain semantically meaningful vector representations. These powerful semantic understanding capabilities significantly improve the analysis of dirty data and reduce errors. Below are key applications of embedding techniques in the domain of entity resolution.

DeepER [12] was one of the first methods to use word embeddings (e.g. GloVe). It applies Long Short-Term Memory (LSTM) networks to learn relationships between attributes in the tuple by labeling data and converting them into a vector representation of fixed dimensions. Then similarity features are calculated and fed into a binary classifier to determine entity matches. DeepMatch [25] follows a similar approach but with more choices for attribute embedding and similarity representation. They demonstrate competitive performance when handling the datasets containing a moderate level of dirty data.

Ditto [19] uses pre-trained transformer-based models for feature extraction and fine-tunes them as a binary classifier for entity matching. It handles dirty data better through a structured serialization process and data augmentation techniques. [7] also evaluate transformer-based models and explore attention mechanisms for the entity resolution task. However, there are significant energy and computational costs associated with using large language models (LLM).

More recently, several frameworks have represented data or record pairs using embedding-based approaches such as multiEM [35], Unicorn [33] and FlexER [16]. These frameworks employ advanced techniques, including graph neural networks, to identify record similarities and enhance generalization capabilities. Nevertheless, they encode data as fixed, linear sequences based on predefined rules, which restricts their ability to interpret relational tables, for example, column relationships. To achieve a more comprehensive interpretation, Embdi [8] introduces graphical representations into entity resolution tasks to enrich the information captured by embeddings.

Despite these advancements, all the mentioned methods operate in batch mode, which limits their application to real-time scenarios. In order to deal with continuously arriving data, an executable method is required to incorporate the data progressively without manually restarting the program each time.

2.3 Dynamic Embeddings

Traditionally, word embeddings are trained in batch mode. To support incremental learning, several models have extended classical algorithms. Incremental Skip-Gram (ISG) [18] adapts Word2Vec to streaming data by updating word frequencies and noise distributions using the Misra-Gries algorithm [24] for fixed vocabulary management. The incremental GloVe model [27] reformulates the original loss function recursively, allowing continuous updates without retraining on the full dataset. The incremental word matrix approach [6] maintains a word-context matrix by replacing infrequent or outdated entries when memory is limited.

As far as the entity resolution problem is concerned, in addition to dealing with plain text itself, we can construct a graph to deal with the problem, such as Embdi. In contrast to limited evolution in text (e.g., adding words), graph evolution can take many forms, such as the addition of nodes or edges, the evolution of edge weights, and more. The field of dynamic graph embedding is well-researched, with various methods [4] depending on how graph embeddings are constructed and how the graph evolves over time. For instance, dynnode2vec [21] utilizes the dynamic Skip-Gram model to generate embedding vectors. It leverages the pre-trained Skip-Gram model as initial weights and updates the vocabulary based on new evolving walks, instead of retraining the entire graph.

Therefore, although most embedding methods for entity resolution still operate in batch mode, incremental language models suggest that existing batch processing techniques can be adapted for streaming processing.

In summary, while existing methods for incremental entity resolution provide valuable experience, they still lack generalizability across diverse data sources and scalability for continuous data streams. Besides, ensuring stability when handling dirty data remains a challenge. Embedding techniques offer a universal solution for capturing text semantics, with dynamic embeddings showing strong potential for streaming applications. Additionally, we propose a dynamic embedding-based approach for entity resolution tasks, detailed in the next section.

3   Proposal

3.1   Problem Statement

In this section, we clarify the problem and introduce key definitions that guide our approach.

To simulate the integration of streaming data into a local dataset, we assume two data sources: one is a static dataset in which each record corresponds to a different entity, and another consists of incoming streaming data. The goal of incremental entity resolution is to match an arriving record to its most similar entity in an evolving reference dataset. This reference dataset includes both the static dataset and previously observed streaming records, dynamically updating over time as new data arrives.

To formalize the problem, we introduce some main symbols and concepts used in incremental approaches in Table 1.

Table 1. Symbol Definitions

Symbol	Explanation
D	A static dataset, which can be empty
$\mathcal{D}$	A dataset of streaming data
ΔD	The increment of incremental dataset over a period of time
R_i	An arbitrary record
$\mathcal{C}_t$	A set of all processed data prior to time t, which serves as the input for the next round of processing
M_{t,R_i}	The best match for record R_i at time t
$\mathcal{M}_t$	The match list for all records at time t
$sim(\cdot,\cdot)$	A similarity function for matching records

Entity resolution in data streaming presents two main problems.

The first is from an incomplete dataset for streaming. At time t, we only have a partial view of $\mathcal{D}$, making it impossible to wait until the full dataset is available before performing the process. More specifically, we can represent the dataset $\mathcal{D}$ at one moment as follows,

$$\mathcal{D}_n = \bigsqcup_{i=1}^{n} \Delta D_i \qquad (1)$$

$\mathcal{D}_n$ represents a dataset formed by combining data increments $\Delta D_1, \ldots, \Delta D_n$, where each increment ΔD_i corresponds to new data arriving within a specific time interval $[t_{i-1}, t_i]$. We assume that these intervals are sequential and non-overlapping, meaning that each new time interval meets the previous one, expressed by $[t_{i-1}, t_i]$ meets $[t_i, t_{i+1}]$. Thus, the overall dataset $\mathcal{D}_n$ evolves over time by sequentially adding these increments.

Instead of comparing the whole dataset in batch mode, which is impossible because the data stream is generated continuously with no end point, the process should operate incrementally. A reasonable way is the execution of the entity resolution process each time ΔD_n arrives in order to not accumulate too much data. The timing of these increments can be determined by various factors, such as a predefined number of newly arrived records or fixed time intervals.

The main idea of incremental entity resolution is reusing the results of previous processing to avoid repetitive calculation. In this case, for incremental data ΔD_n et time t, the previously processed data is stored in a candidates set, represented by $C_t = D \cup D_{n-1}$. These data have been transformed from their original state to a form ready for comparison. At each incremental update, the candidate set is continuously integrated with new data, while the previously processed data remains unchanged and is only used for comparison in the similarity function. Figure 1 shows the data involved in this incremental process over time.

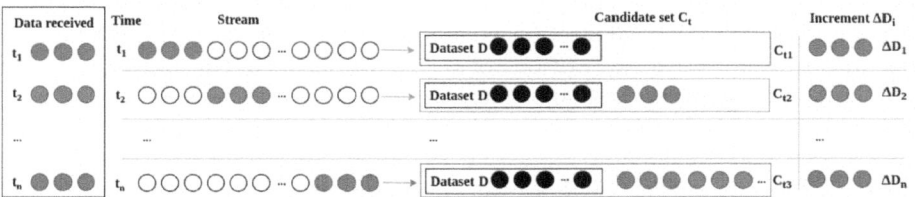

Fig. 1. Data involved in incremental process over time.

So we can define the first problem as follows:

Definition 1. *At any time t, incrementally update the candidate set of C_t and compare data pairs by similarity function $sim(\Delta D_n, C_t)$ to determine M_{t,R_i} for each element $R_i \in \Delta D_n$.*

The second problem is that matching is not a static process over time. Since candidate records change over time because of the continuous nature of the stream, previously resolved matches M_{t',R_i} (where $t' < t$) may need to be updated when new records arrive. This requires continuously maintaining consistency in match results. We define the match set at time t as $M_t = \{(R_i, M_{i,R_i}) \mid i \leq t\}$. With growth of M_t, storing all historical matches becomes infeasible. To ensure efficiency, we need a strategy to determine when and which matches should be retained or discarded. We resume as the following:

Definition 2. *Handle and maintain the dynamic matching results over time M_t, such that at any time t, the correct corresponding similar result is included in the list once it appears, and its similarity value remains relatively high compared to all potentially similar records in the list.*

Our proposed pipelines address these problems through an incremental resolution process that efficiently updates matches while minimizing storage overhead. The key feature of our mechanism is to maintain all relevant matching information in a single index, enabling uniform query access. The details of our solution are presented in the following section.

3.2 Proposition

Our framework[1] for incremental entity resolution consists of three main parts: data preparation, incremental embedding construction, and matching list con-

---

[1] https://github.com/Mzhongwei/er_embedding_streaming.

struction. In this work, we reuse the approach for building an embedding model in [8], which is suitable for extension for streaming data and represents rich semantic information in a lightweight language model. However, instead of constructing an embedding model based on two complete datasets at once, we introduce incremental embeddings that update the model over time to improve the scalability of the streaming application. To illustrate how these three parts fit together and the flow of data between them, we summarize the process as follows.

Data Preparation. Data preparation processes raw data increments ΔD_i from data sources. Metadata extraction is conducted at this stage to provide concise and representative descriptions of real-world entities. For instance, in the case of scientific articles, metadata may include attributes such as the author's name, title, and publication date. A relational table is then constructed based on the extracted metadata, providing a flexible foundation for various tasks across different domains. In this work, it is specifically used for entity resolution. This data preparation step transforms unstructured data into a structured format, effectively reducing data volume and standardizing it for downstream processing.

Incremental Embedding Construction. Based on this relational table of data increments, we construct an incremental embedding model. The model used in this process is pre-trained on a local dataset D, but pre-training can be ignored in case we integrate data streams from scratch. The training pipeline is structured as follows: each record in the relational table is first assigned a unique identifier. Every value, including ID numbers and column names, is treated as a node in an undirected heterogeneous graph. A random walk is then performed on this graph to generate contextual representations of these values and tokens in the form of sentences. These sentences are subsequently mapped into a fixed-length vector space, which forms the embedding representation for values [8].

Algorithm 1: Incremental Embedding Algorithm

    Input: data increment ΔD_i, embedding model $\mathcal{E}_{i-1}$, graph $\mathcal{G}_{i-1}$
    Output: updated embedding model $\mathcal{E}_i$, updated graph $\mathcal{G}_i$
1: Initialize evolving nodes list $\mathcal{L}$
2: Initialize graph $\mathcal{G}_i \leftarrow \mathcal{G}_{i-1}$
3: **foreach** $R \in \Delta D_i$ **do**
4:    $\mathcal{G}_i, \Delta N_1 \leftarrow \text{addNodesToGraph}(\mathcal{G}_i, R)$
5:    $\mathcal{G}_i, \Delta N_2 \leftarrow \text{addEdgesToGraph}(\mathcal{G}_i, R)$
6:    $\mathcal{L} \leftarrow \text{addEvolingNodeToList}(\mathcal{G}_i, \Delta N_1, \Delta N_2)$
7: Generate new random walks for $\mathcal{L}$: $\mathcal{W} = \text{GenerateRandomWalk}(\mathcal{G}, \mathcal{L})$;
8: Train model with walks $\mathcal{W}$: $\mathcal{E}_i = \text{UpdateEmbedding}(\mathcal{E}_{i-1}, \mathcal{W})$
9: **return** $\mathcal{E}_i,$, $\mathcal{G}_i$

The incremental pipeline follows the same structure but incorporates new data dynamically. The whole process is shown in Algorithm 1. When a data increment ΔD_i arrives at time t_i, we take the graph $\mathcal{G}_{i-1}$ and the embedding model $\mathcal{E}_{i-1}$ at time t_{i-1} as initial variables. The graph is updated by adding new nodes with values extracted from each record R of data incremental ΔD_i if the corresponding values do not already exist; then new edges are added to the graph in order to represent the relation with different records and different values within one record (cf. 3–5). In this step, all the nodes that have evolved, both new nodes ΔN_1 as well as nodes with changed edges ΔN_2, are recorded in an evolving nodes list $\mathcal{L}$ (cf. line 6). New random walk $\mathcal{W}$ then starts from nodes in this list and generates new sentences (cf. line 7). These sentences are used to refine the embedding model through incremental learning, ensuring that the representations remain up to date without retraining from the beginning (cf. line 8). The updated graph $\mathcal{G}_i$ and embedding model $\mathcal{L}_i$ are stored for the next iteration.

Matching List Construction. Matching list construction takes as input ID numbers of incremental records to represent these records and the embedding model updated by the previous step. As shown in Algorithm 2, for each record represented by one identifier, we can calculate similarity degrees between vectors of the ID numbers to identify the most similar records according to the vector representations in the embedding model (cf. line 2). Then we select the top k most similar matches (cf. line 3), store their relationships and their degree of similarity symmetrically in a list, where we retain only n most similar matches for each record and remove the matches sorted after n (cf. line 4–6).

Algorithm 2: Matching List Construction

Input: Incremental record IDs $\mathcal{ID} = \{id_1, id_2, \ldots, id_n\}$, embedding model $\mathcal{E}_i$, candidate set $\mathcal{C}$, number of most similar records k after calculation and n in matching list, previous matching list $\mathcal{M}_{i-1}$

Output: Updated matching list $\mathcal{M}_i$

1: **foreach** $id \in \mathcal{ID}$ **do**
2:      Execute similarity function: $\mathrm{Sim}(\mathcal{E}_i, \mathrm{id})$
3:      Select top-k most similar IDs with their similarity s:
      $\mathcal{N}_{id} = \{(j_1, s_1), (j_2, s_2), \ldots, (j_k, s_k) \mid j \in \mathcal{C}\}$
4:      $\mathcal{M}_i[id] \leftarrow$ top-n most similar elements of $\mathcal{M}_{i-1}[id] \cup \mathcal{N}_{id}$;
5:      **foreach** $(j, s) \in \mathcal{N}_{id}$ **do**
6:         $\mathcal{M}_i[j] \leftarrow$ top-n most similar elements of $\mathcal{M}_{i-1}[j] \cup \{(id, s)\}$;

7: **return** $\mathcal{M}_i$

For example, given an ID number 001, we compute its most similar ID 002, 003, 004 with similarity representatively 0.8, 0.9, 0.7. To ensure bidirectional consistency, we record both the forward and reverse association, such as $001 \rightarrow (002, 0.8), (003, 0.9), (004, 0.7)$, $002 \rightarrow (001, 0.8)$, $003 \rightarrow (001, 0.9)$, $004 \rightarrow (001, 0.7)$. Consider also that we keep only the 2 most similar values for each record, so the final result is like $001 \rightarrow (002, 0.8), (003, 0.9)$, $002 \rightarrow (001, 0.8)$, $003 \rightarrow (001, 0.9)$, $004 \rightarrow (001, 0.7)$. This list dynamically evolves as new comparisons are performed over time.

4   Experiments

4.1   Protocol

In this section, we present an experimental evaluation of our proposition. The main objective of our experiments is to evaluate the performance of our proposed real-time processing architecture and its capacity to handle incremental data. Our evaluation aims to answer the following research questions:

RQ1. What are the effectiveness and performance of our proposition ?

RQ2. How does the size of the incremental training data affect the experimental results of incremental learning?

Datasets. We use three open-source datasets[2] created from real-world data and are dedicated to the entity resolution task. Their information is described in Table 2.

Table 2. Overview of datasets used in the evaluation.

Name	Columns	Tuples - Source A	Tuples - Source B	Matches	Description
Amazon	Id, title, price, manufacturer	1363	3226	1167	Commodity list from Amazon
DBLP-ACM	Id, title, venue, year, author_1, author_2, author_3, author_4	2294	2617	2224	Scientific articles
Fordors-zagats	Id, name, addr, city, phone, type, class	533	331	110	Personal information

Evaluation Metrics. We use precision, recall and F1-score as primary metrics for performance. Since it is difficult to define the end of the task during stream processing, we calculate metrics based on the overall results after all Source B data have been processed. We sort the list in descending order of similarity and select $x(x \leq listLength)$ records with similarity ranked in the top k. These records are then grouped into candidate pairs. After removing duplicates, the remaining predicted pairs are compared with the ground truth, where records refer to the same entity. The number of correctly predicted pairs corresponds to the intersection between the predicted and ground truth pairs.

Implementation. To simulate the data streaming application, we take the tuples in Data Source A as the local dataset used for pre-training, and send the data from source B to kafka producer incrementally as a data stream at a regular and fixed frequency without concurrency. For each record of source B, we look for its similar values from Source A and construct a sequence of up to 10 in both

---

[2] https://zenodo.org/records/7930461.

directions (as mentioned in Sect. 3.2) to analyze how well the similarity-based ranking aligns with the correct pairing provided by the ground truth. In this article, we directly utilize relational data. Therefore, the Data Preparation will not be discussed here. The whole process is illustrated in Fig. 2.

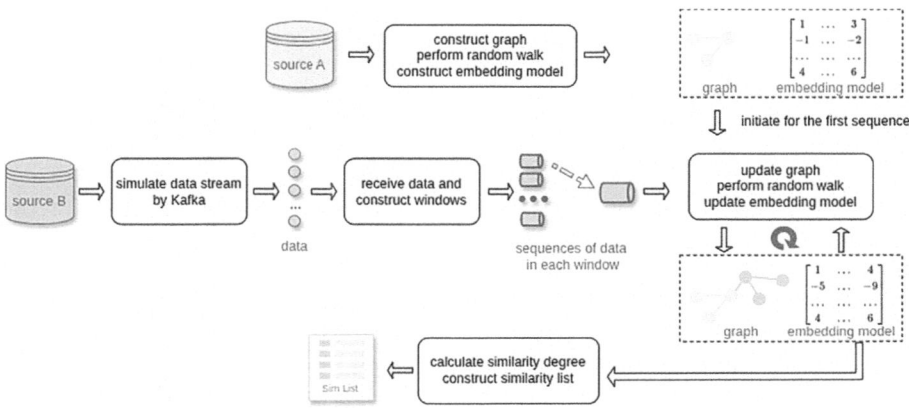

Fig. 2. Pipeline of implementation process

For the machine learning model, we use the Gensim [29] python library to execute word embedding tasks and support online learning. By default, we set configuration parameters as follows: for each token, a vector size of 300 and a context window of 3. During the random walk, we set the walk length to 60 and apply log smoothing to enhance the likelihood of selecting low-probability nodes, thereby expanding the coverage of nodes in the graph by the generated sentences. Considering online learning, we choose the skip-gram algorithm and word2vec model as the main training method. We calculate the inner product of the vectors and obtain the cosine similarity by normalizing the vectors, i.e., dividing their inner product by the product of their norms. Each similarity value is retained to 16 decimal places. Since each random walk introduces variability and generates different sentences for word embeddings, the results of each experiment were averaged across three independent runs to mitigate the influence of randomness on the outcomes.

Experiments have been conducted on a laptop with CPU $12 \times$ 12th Gen Intel(R) Core(TM) i5-1245U with 15.3 GB RAM.

4.2   Results Analysis

Experiments on Effectiveness. Figure 3 illustrates the effect of the most similar value ordering on recall, in order to demonstrate the effectiveness of our algorithm. The higher the recall, the more pairs of records are correctly identified. As the value of k increases, we can see that the incremental algorithm

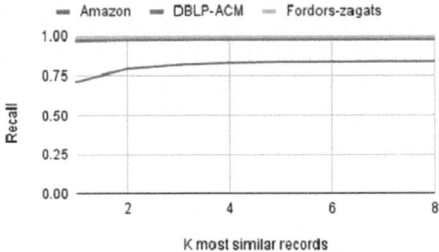

Fig. 3. The effect of the k most similar predicted pairs on the recall rate

successfully predicts most of the similar record pairs. Moreover, most of the correct predictions are concentrated in the records with relatively high similarity ranks.

Adjusting the k-value can effectively enhance the recall for entity resolution tasks. Furthermore, beyond a certain threshold, the length of the similarity list has minimal impact on the accuracy of predicted pairs. Most of the less similar pairs tend to be irrelevant, and focusing on the top candidates with high similarity can significantly improve both the efficiency and accuracy of the algorithm.

Effect of Increment Size on Results. In incremental learning, the size of the newly added data at each step impacts the model's training process. Frequent updates or introducing too much data at once can potentially destabilize the model. To analyze the effect of the incremental data size on the entity resolution task performance, we varied the proportion of the incremental data. The size of the incremental data is described as a percentage of the pre-trained data. For example, a 5% increment means that the amount of data trained in each incremental process is equivalent to 5% of the original pre-trained data volume.

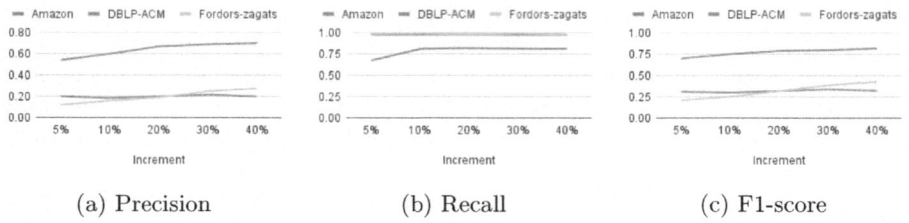

(a) Precision   (b) Recall   (c) F1-score

Fig. 4. Effect of increment size on results

Figure 4 illustrates the changes in precision, recall and f1 score as increment size increases. As shown in the figure, the recall improves as increment size increases. This suggests that for the ER task, updating the text corpus too frequently has a greater impact on the results and leads to a decrease in the model's ability to predict similar records. Moreover, different increment sizes influence the metrics in varying ways. These performance indicators tend to stabilize once the increment size surpasses a certain threshold.

The similarity list approach used here does not deeply evaluate dissimilarity, focusing mainly on whether two records are similar. As a result, accuracy may be suboptimal for certain datasets. However, a high recall indicates that the majority of correct pairings are identified, which is crucial in a context where large volumes of unbounded data are rapidly incoming. Reducing the range of potential pairs lays the foundation for subsequent stages that progressively enhance precision, for example, the transition from a streaming processing approach to a more efficient batch processing method for the next steps.

Additionally, the performance differences across datasets, as illustrated in the figures, highlight the significant impact of dataset structure on outcomes. In particular, variations in duplication rates (as shown in Table 2) and the presence of different types of data noise contribute to the variability in results. Noise types such as transposed characters, missing characters, or inconsistent abbreviations reduce textual similarity and increase the complexity of accurate record matching.

5   Conclusion

With the increasing demand for various streaming data analysis, it is essential to integrate streaming data with stored data efficiently. In this study, we addressed the challenge of entity resolution in data integration by adapting existing embedding techniques, traditionally used in batch processing, to a streaming context. We proposed a dynamic solution based on an incremental embedding model that continuously updates representations with the incoming stream of new data.

We construct a heterogeneous graph using relational data and incrementally update it using streaming data sequences. Using this evolving graph, we perform random walks on newly added nodes to generate sentences, which are then used to update the embedding model. Results vary widely across datasets, while our method achieves up to 91% precision, 97% recall, and 94% F1-score, demonstrating its effectiveness in identifying similar matches in streaming contexts.

Nevertheless, our approach focuses primarily on detecting potentially similar record pairs. As a next step, we aim to refine our results using similarity lists to identify mismatches in the list to determine more definitive matches. In addition, since the results vary across datasets, enhancing the adaptability of our method to different data characteristics is essential. We plan to incorporate more specific noise-handling strategies to improve robustness and generalization across diverse scenarios.

References

1. Almeida, F., Xexéo, G.: Word embeddings: a survey (2023)
2. Altwaijry, H., Mehrotra, S., Kalashnikov, D.V.: Query: a framework for integrating entity resolution with query processing. Proc. VLDB Endow. **9**(3), 120–131 (2015)
3. Bahri, M., Bifet, A., Gama, J., Gomes, H.M., Maniu, S.: Data stream analysis: foundations, major tasks and tools. WIREs Data Min. Knowl. Discov. **11**(3), e1405 (2021). https://doi.org/10.1002/widm.1405
4. Barros, C.D.T., Mendonça, M.R.F., Vieira, A.B., Ziviani, A.: A survey on embedding dynamic graphs. ACM Comput. Surv. **55**(1), 1–37 (2021)
5. Binette, O., Steorts, R.C.: (Almost) all of entity resolution. Sci. Adv. **8**(12), eabi8021 (2022)
6. Bravo-Marquez, F., Khanchandani, A., Pfahringer, B.: Incremental word vectors for time-evolving sentiment lexicon induction. Cogn. Comput. **14**(1), 425–441 (2021). https://doi.org/10.1007/s12559-021-09831-y
7. Brunner, U., Stockinger, K.: Entity matching with transformer architectures - a step forward in data integration. In: Proceedings of the 23rd International Conference on Extending Database Technology (EDBT) (2020)
8. Cappuzzo, R., Papotti, P., Thirumuruganathan, S.: Creating embeddings of heterogeneous relational datasets for data integration tasks. In: Proceedings of the 2020 ACM SIGMOD International Conference on Management of Data, Portland, OR, USA, pp. 1335–1349. ACM (2020)
9. Chen, R., Shen, Y., Zhang, D.: Gnem: a generic one-to-set neural entity matching framework. In: Proceedings of the Web Conference 2021, Ljubljana, Slovenia, pp. 1686–1694. ACM (2021). https://doi.org/10.1145/3442381.3450119
10. Christophides, V., Efthymiou, V., Palpanas, T., Papadakis, G., Stefanidis, K.: An overview of end-to-end entity resolution for big data. ACM Comput. Surv. (2020)
11. Do Nascimento, D.C., Santos Pires, C.E., Gomes Mestre, D.: Heuristic-based approaches for speeding up incremental record linkage. J. Syst. Softw. **137**, 335–354 (2018)
12. Ebraheem, M., Thirumuruganathan, S., Joty, S., Ouzzani, M., Tang, N.: Distributed representations of tuples for entity resolution. Proc. VLDB Endow. **11**(11), 1454–1467 (2018)
13. Efthymiou, V., Papadakis, G., Stefanidis, K., Christophides, V.: Minoaner: schema-agnostic, non-iterative, massively parallel resolution of web entities. In: Proceedings of the 22nd International Conference on Extending Database Technology (EDBT) (2019)
14. Gazzarri, L., Herschel, M.: End-to-end task based parallelization for entity resolution on dynamic data. In: 2021 IEEE 37th International Conference on Data Engineering (ICDE), Chania, Greece, pp. 1248–1259. IEEE (2021)
15. Gazzarri, L., Herschel, M.: Progressive entity resolution over incremental data. In: Proceedings of the 26th International Conference on Extending Database Technology (EDBT). OpenProceedings.org (2023)
16. Genossar, B., Shraga, R., Gal, A.: Flexer: flexible entity resolution for multiple intents. Proc. ACM Manag. Data **1**(1) (2023)
17. Isah, H., Abughofa, T., Mahfuz, S., Ajerla, D., Zulkernine, F., Khan, S.: A survey of distributed data stream processing frameworks. IEEE Access **7**, 154300–154316 (2019)
18. Kaji, N., Kobayashi, H.: Incremental skip-gram model with negative sampling. In: Proceedings of the 2017 Conference on Empirical Methods in Natural Language

Processing, Copenhagen, Denmark, pp. 363–371. Association for Computational Linguistics (2017)

19. Li, Y., Li, J., Suhara, Y., Doan, A., Tan, W.C.: Deep entity matching with pre-trained language models. Proc. VLDB Endow. **14**(1), 50–60 (2020)

20. Maharana, K., Mondal, S., Nemade, B.: A review: data pre-processing and data augmentation techniques. Glob. Transit. Proc. **3**(1), 91–99 (2022). https://www.sciencedirect.com/science/article/pii/S2666285X22000565. International Conference on Intelligent Engineering Approach (ICIEA-2022)

21. Mahdavi, S., Khoshraftar, S., An, A.: dynnode2vec: scalable dynamic network embedding arXiv:1812.02356 (2019)

22. Marz, N., Warren, J.: Big Data: Principles and Best Practices of Scalable Realtime Data Systems, 1st edn. Manning Publications Co., USA (2015)

23. Mikolov, T., Sutskever, I., Chen, K., Corrado, G., Dean, J.: Distributed representations of words and phrases and their compositionality. In: NIPS 2013, pp. 3111–3119. Curran Associates Inc., Red Hook (2013)

24. Misra, J., Gries, D.: Finding repeated elements. Sci. Comput. Program. **2**(2), 143–152 (1982)

25. Mudgal, S., et al.: Deep learning for entity matching: a design space exploration. In: Proceedings of the 2018 International Conference on Management of Data, Houston, TX, USA, pp. 19–34. ACM (2018)

26. Papadakis, G., Efthymiou, V., Thanos, E., Hassanzadeh, O., Christen, P.: An analysis of one-to-one matching algorithms for entity resolution. VLDB J. **32**(6), 1369–1400 (2023)

27. Peng, H., et al.: Incremental term representation learning for social network analysis. Future Gener. Comput. Syst. **86** (2017)

28. Pennington, J., Socher, R., Manning, C.D.: Glove: global vectors for word representation. In: Empirical Methods in Natural Language Processing (EMNLP), pp. 1532–1543 (2014)

29. Řehůřek, R., Sojka, P.: Software framework for topic modelling with large corpora. In: Proceedings of the LREC 2010 Workshop on New Challenges for NLP Frameworks, Valletta, Malta, pp. 45–50. ELRA (2010)

30. Ren, W., Lian, X., Ghazinour, K.: Online topic-aware entity resolution over incomplete data streams. In: Proceedings of the 2021 International Conference on Management of Data, SIGMOD 2021, pp. 1478–1490. Association for Computing Machinery, New York (2021)

31. Saeedi, A., Peukert, E., Rahm, E.: Incremental multi-source entity resolution for knowledge graph completion. In: Harth, A., et al. (eds.) ESWC 2020. LNCS, vol. 12123, pp. 393–408. Springer, Cham (2020). https://doi.org/10.1007/978-3-030-49461-2_23

32. Simonini, G., Zecchini, L., Bergamaschi, S., Naumann, F.: Entity resolution on-demand. Proc. VLDB Endow. **15**(7), 1506–1518 (2022)

33. Tu, J., et al.: Unicorn: a unified multi-tasking model for supporting matching tasks in data integration. Proc. ACM Manag. Data **1**(1), 1–26 (2023)

34. Wang, S., Zhou, W., Jiang, C.: A survey of word embeddings based on deep learning. Computing **102**(3), 717–740 (2020)

35. Zeng, X., Wang, P., Mao, Y., Chen, L., Liu, X., Gao, Y.: Multiem: efficient and effective unsupervised multi-table entity matching. In: 2024 IEEE 40th International Conference on Data Engineering (ICDE), pp. 3421–3434 (2024)

Cross-Modal Sequential Point-of-Interest Recommendation with Lightweight Hybrid Fusion Strategy

Tianxing Wang[1][(✉)], Can Wang[1][(✉)], Hui Tian[1], and Hong Shen[2]

[1] Griffth University, Gold Coast 4215, Australia
tianxing.wang@griffithuni.edu.au, {can.wang,hui.tian}@griffith.edu.au
[2] Central Queensland University, Brisbane 4000, Australia
h.shen@cqu.edu.au

Abstract. The sequential Point-of-Interest (POI) recommendation plays a vital role in today's location-based social network platforms. While numerous studies have proposed various promising solutions for improving POI recommendations, the field still faces key challenges: (1) Data sparsity is very high, and the number of user visit records is limited; (2) Most existing models ignore the power of different modalities to capture deep user-level preferences. In response to these issues, we propose a novel **C**ross-**M**odal Sequential **P**OI Recommender with **L**ightweight **H**ybrid Fusion Strategy (CMPLH), which is designed to enhance the performance and efficiency of sequential POI recommendations. By incorporating the Geo-Local Sensitive Hashing (Geo-LSH) attention mechanism, our model effectively fuses multiple modalities to capture dynamic user preference features from multi-modal historical check-ins, while reducing computational overhead. Furthermore, by leveraging a hybrid fusion strategy, CMPLH effectively integrates users' historical POI check-ins, categorical sequences, geographical information, and user-generated reviews. Extensive experimental results confirm that CMPLH surpasses existing state-of-the-art approaches, demonstrating the advantages of combining multiple modalities and hybrid fusion strategies in enhancing sequential POI recommendation systems.

Keywords: POI Recommendation · Sequential Recommendation · Cross-Modal Learning · Fusion Strategy

1 Introduction

Recently, location-based social networks (LBSNs) have become an integral part of today's digital community, bringing innovative and useful functions to their customers. A core feature of LBSNs is the sequential Point-of-Interest (POI) recommendation, which enhances user experience and fosters engagement by providing high-quality, tailored suggestions [16,21]. Consequently, sequential POI recommendation has become a prominent topic in both academia and industry. The domain of sequential POI recommendation has witnessed a variety

© The Author(s), under exclusive license to Springer Nature Switzerland AG 2026
C. K. Leung et al. (Eds.): DaWaK 2025, LNCS 16048, pp. 126–141, 2026.
https://doi.org/10.1007/978-3-032-02215-8_9

of innovative approaches. ST-RNN incorporates spatio-temporal influence into recurrent neural networks (RNNs) to infer the user dynamic interests [12]. Further advancements utilize the Long-Short-Term-Memory (LSTM) networks to enhance the recommendation quality by incorporating the gated units that explicitly model the time and distance information [19,25]. Following this, the self-attention mechanism leads to a new trend [7]. For example, STAN considers the relationships between non-adjacent POI check-ins to more precisely capture user preferences among POIs [14].

Despite the advancements in the sequential POI recommendation, this field still faces significant challenges: 1) High data sparsity is a major issue in the POI recommendation scenario. With a limited user visit records, it is challenging for the model to accurately capture user preferences within a sparse dataset; 2) The POI recommendations are often influenced by diverse factors such as contextual data, user reviews, and categorical information of POIs. However, many existing models mainly rely on the sequential-based data, and overlook the potential of incorporating information from various modalities. Addressing these issues is crucial for improving the accuracy and robustness of the sequential POI recommendation systems.

Recently, multi-modal (also known as cross-modal) learning has emerged as a compelling strategy in the item recommendation domain. By integrating auxiliary information like categorical data and user-generated content, cross-modal learning can discover potential relations between different modalities and infer users' preferences more effectively [6,26]. However, when employing cross-modal techniques in sequential POI recommendations, there are still some challenges. Firstly, the inherent nature of the modalities may affect the integrating strategy on different modalities. Some modalities offer sequential information (e.g., historical check-ins), while others provide item characteristics (e.g., user reviews). Simply adopting a single fusion strategy for these modalities may not be effective. For instance, early fusion overlooks the nuanced interactions within individual modalities, whereas late fusion inadequately captures the inter-modal interactions [4]. Secondly, introducing additional modalities increases computational demand. Each modality not only contributes more information but also add more computational operations within the model. Therefore, finding the right balance in merging these modalities is essential to fully leverage the advantages of cross-modal learning without compromising information coverage or unnecessarily increasing the computational demand.

Motivated by the preceding analysis, we propose the **C**ross-**M**odal Sequential **P**OI Recommender with **L**ightweight **H**ybrid Fusion Strategy (CMPLH). CMPLH integrates four key modalities: users' historical POI interactions to capture temporal dynamics, dense POI category representations to address sparsity issue, spatial information representing geographical distances between POIs, and user-generated POI reviews to incorporate additional user preference knowledge. Recognizing each modality's intrinsic nature, we employ a hybrid fusion strategy that utilizes intermediate fusion for processing user check-in information (historical check-ins, category sequences, and geographical distances) and modeling

POI reviews solely before feeding it into the late fusion layer. The reason of dealing with the review information individually is: in sequential POI recommendation, POI reviews act as static indicators of global preferences, remaining constant regardless of user visits. Users read reviews, and if they align with their preferences, they are more likely to visit the location. In contrast, check-in information reflects the spatio-temporal dynamics of users' historical interactions, providing engagement timelines and travel costs between POIs. By modelling these distinct characteristics, our hybrid fusion strategy in CMPLH can effectively balance modeling user-POI dynamics and global preferences.

Subsequently, the computational cost is an issue when introducing more modalities. Each modality not only contributes more information but also add more computational operations within the model. Therefore, we adopt a lightweight approach - the Local-Sensitive-Hashing (LSH) attention mechanism in the intermediate fusion layer. LSH attention is a lightweight version of the traditional self-attention [8]. Unlike cross-modal attention mechanisms [13], it significantly reduces computational overhead by focusing on similar information across modalities, minimizing unnecessary calculations To adopt the sequential POI recommendation scenario, we design the Geo-LSH attention layer that can dynamically cluster closely related temporal information from the historical user-POI interaction modality, the corresponding category modality, and consider the geographical information at the same time. Thus, the Geo-LSH attention layer jointly capture the user's preferences while achieving lower computational costs. The main contributions of this paper are:

- We introduce a novel cross-modal sequential POI recommendation model (CMPLH) that harnesses users' historical interactions, corresponding categorical sequences, spatial distance between check-in POIs, and POI reviews.
- We employ a hybrid fusion strategy, combining intermediate and late fusion approaches, to dynamically learn the information from different modalities.
- We design a Geo-LSH attention layer to dynamically clusters similar information from different modalities with the consideration of the geographical influence, which enables an efficient cross-modal interactions.
- Extensive experiments demonstrate that CMPLH surpasses state-of-the-art models in sequential POI recommendation, verifying the importance of leveraging multi-modal information in this field.

2   Related Works

2.1   Cross-Modal Recommendation

Multi-modal recommendation, also known as cross-modal recommendation, has emerged as a promising direction in the category of recommendation systems, such as multimedia and item recommendation [26]. This approach capitalizes on features from various modalities [2,5], enriching the recommendation process. Among these models, the strategy for modality fusion is the core part in the cross-modal recommendation models. The fusion operation could occur before, during or after the modalities being fed into the feature interaction layers, named as early fusion, intermediate fusion and late fusion, respectively [4], which will highly affect the recommendation performance. Liu et al. propose an early fusion model that fuses user ID, text, and image information before feeding them into the attention layers [11]. MUIR is a micro-video recommendation model that considers both the early fusion and late fusion strategies to optimize its performance [1]. Similarly, Ji et al. develop a transformer-based multi-modal recommender that employs a late fusion structure to generate a comprehensive prediction result from different data sources [6]. MMGCN and DualGNN both consider the late fusion strategy by utilizing a Graph Convolutional Network to model the relationship between different modalities [20]. MMSR discusses the fusion strategy debate and proposes a graph-based architecture to achieve an adaptive intermediate fusion process [4]. Despite these advances in item and multimedia recommendation fields, the application of multi-modality in sequential POI recommendations remains largely underexplored, highlighting a significant gap and an opportunity for innovative research in this area.

2.2 Sequential POI Recommendation

The conventional approach in the sequential POI recommendation adopts the Markov Chain technique. These models leverage the Markov process to capture the transition probability between sequential check-in activities, providing the foundational insights into user movement patterns among POIs [3]. As the deep learning technique evolved, the DNN-based models, such as RNNs, have attracted many attention due to their robust capability to capture complex user-POI interaction patterns from historical data [12,23]. After that, STGN proposes a LSTM model with the gated units to dynamically capture the time and distance information of the check-ins [25]. Sun et al. employ a twin-tower LSTM architecture to balance the long and short-term preferences [19]. In recent year, the attention-based model becomes a new trend in this field. Luo et al. proposes a self-attention POI recommender that consider the spatio-temporal relationships between the non-adjacent check-in POIs [14]. AutoMTN is a multi-modal POI recommender that combines the POI data and corresponding category information with the auto-correlation layer, showcasing the power of integrating diverse data streams [16]. GeoMixer introduces MLP layers to adaptively combine user check-in sequences with travel distance information [21]. MMPOI is a recent multi-modal content-aware framework designed for POI recommendation, which effectively integrates multiple modalities, including textual reviews, categorical data, and geographical information [22]. Compared with MMPOI, which relies on early fusion techniques for modality integration, our proposed CMPLH introduces the hybrid fusion strategy that combines both intermediate and late fusion techniques. This design ensures that modality-specific characteristics are preserved while enabling effective cross-modal interactions. Moreover, CMPLH employs a lightweight design, which not only captures the dynamic user preferences across temporal sequences but also reduces computational complexity, achieving a better balance between performance and efficiency compared to MMPOI's resource-intensive multi-modal attention modules.

3   Problem Formulation

We define the set of users, POIs, and categories as $\mathcal{U} = [u_1, u_2, u_3, \ldots, u_{|U|}]$, $\mathcal{P} = [poi_1, poi_2, poi_3, \ldots, poi_{|P|}]$, and $\mathcal{C} = [c_1, c_2, c_3, \ldots, c_{|C|}]$, where $|U|$, $|P|$, and $|C|$ represent the number of users, POIs, and categories, respectively. For cross-modal POI recommendation, the historical check-in sequence includes multiple modalities, formulated as: $T_u = \{(u, poi_t, c_t, a_t)|t \in (1, N)\}$, where $poi_t \in \mathcal{P}$, $c_t \in \mathcal{C}$, and a_t represents the corresponding reviews of poi_t. To facilitate model training, we adopt a sliding window method, taking the prior $|L|$ check-in records as the input and the following $|T|$ records as the prediction target. The problem can be defined as: given a user's historical POI check-in sequence T_u, the cross-modal sequential POI recommender can determine which POIs from the candidate pool the user is likely to visit in the future.

4   Our Model

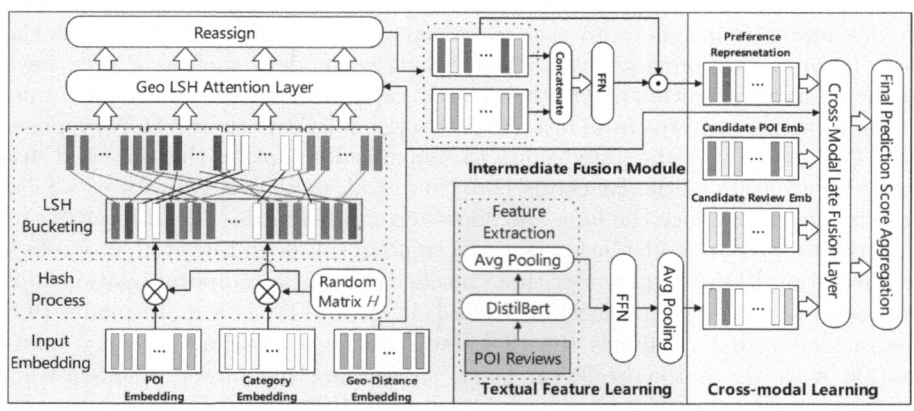

Fig. 1. The framework of CMPRH.

This section will discuss the intricacies of our proposed CMPLH. Figure 1 demonstrates the structure and main components of the CMPLH. As illustrated, CMPLH can be broken down into the following modules, which serve a pivotal role in the POI recommendation process: **1) Intermediate fusion module, 2) Textual Feature Learning Layer,** and **3) Cross-modal Learning Layer.** By utilizing the hybrid fusion, CMPLH adaptively models the multi-modal information and refines the recommendation quality.

4.1   Feature Extraction

Embedding Layer for Check-in Information

Sequential Embedding: Our model interprets user interactions with POIs and their associated categories as input sequences which reflect temporal user behavior patterns. In the embedding layer, we convert the fixed-length user historical POI interactions, represented as $S^P = \{p_1, p_2, \ldots, p_{|L|}\}$, and related category interactions $S^C = \{c_1, c_2, \ldots, c_{|L|}\}$, into the latent representation space. For each user, these embeddings are expressed as $E_p \in \mathbf{R}^{|L| \times d}$ for POIs and $E_c \in \mathbf{R}^{|L| \times d}$ for categories, where d is the dimensionality of the embedding space, and $|L|$ is the length of input sequence. Here, E_p captures the embedded representation of the user's interactions with POIs, and E_c encapsulates the category interactions.

Geo-distance Embedding: We employ a unit embedding layer for the dense representation of the geographical distance between POIs, using 100 m as the basic unit. The unit embedding layer multiplies the spatial intervals by a unit embedding vector, avoiding the problem of the sparse relation encoding [14]. Mathematically, the geographical distance embedding is denoted as $g_{i,j}^\Delta \in \mathbb{R}^d$, representing the geographical distance between two POIs or categories.

Textual Information Extraction. To derive meaningful insights from user-generated textual content, such as POI reviews, we employ DistilBERT, a streamlined variant of the BERT model, for extracting semantic embeddings from POI reviews [18]. DistilBERT offers several advantages for processing the POI reviews within our framework: 1) Efficiency: DistilBERT is engineered as a compact version of BERT. It retains most of the accuracy of BERT while being 40% smaller and 60% faster, which is able to save the computational resources; 2) Validated ability of sentiment analysis: compared with LTSM-based model, DistilBERT shows a superior capability on capturing sentiment nuances [15].

We first extract the textual embedding for each review per POI using DistilBERT. Then, these POI review embeddings of a POI are accumulated into a single semantic representation through average pooling:

$$r_{i,n} = \text{DistilBERT}(a^{p_{i,n}}), \quad e_n^t = \frac{1}{n} \sum_{i=1}^{n} r_{i,n} \tag{1}$$

where $a^{p_{i,n}}$ is the i-th review of a POI n, $r_{i,n} \in \mathbf{R}^{d_2}$ denotes the corresponding semantic embedding, and $e_n^t \in \mathbf{R}^{d_2}$ represents the aggregated single semantic embedding for a POI n, with d_2 being the embedding dimension of the POI review embedding. Finally, for each user, we generate $E_t = \{e_1^t, e_2^t, \ldots, e_{|L|}^t\}$, which aligns with the POI embedding E_p.

4.2   Intermediate Fusion for Check-in Information

The conventional multi-modal Transformer processes inputs from diverse modalities using multiple self-attention layers. Some state-of-the-art models adopt a multi-tower transformer structure with complexity $O(ML^2)$, where L is the input sequence length and M is the number of modalities [11,16]. Others concatenate input modalities along the sequence length, leading to a complexity of $O(M^2L^2)$ [6]. These approaches face scalability challenges as the number of modalities and input lengths increase, causing significant computational resource demands.

To address this inefficiency, we employ the Geo-LSH attention layer in the intermediate fusion layer. This lightweight design is based on LSH attention mechanism which reduces computational complexity from $O(L^2)$ to $O(L \log L)$, offering better scalability and lower computational costs in cross-modal POI recommendation [8]. Afterwards, a feed-forward network integrates the user's preferences over the modalities.

Geo-LSH Attention Module. The left part of the Fig. 1 illustrated the structure of Geo-LSH attention. Firstly, a random matrix $H \in \mathbf{R}^{d \times b/2}$ is generated, where b is the number of hash buckets. Following this, the vectors, e_p and $e_c \in \mathbf{R}^d$ from two input embeddings, E_p and E_c, respectively, are projected into the hash space by $h(x) = argmax([xH; -xH])$, where x can be e_p and e_c, and $[a; b]$ represents the concatenation operation. This step ensures that both e_p^h and e_c^h, the hashed vectors, are allocated into b fixed-length hash buckets, which groups similar embedding vectors into the same buckets to facilitate the in-bucket attention computations. For each bucket, the attention representation for each position i can be denoted as:

$$o_i = \sum_j Softmax(q_i \cdot k_j + g_{i,j}^\Delta)v_j, \qquad (2)$$

where q_i, k_j and v_j are the query, key, and value embedding vectors belonging to the same bucket, and $g_{i,j}^\Delta$ is the geographical distance embedding between positions i and j. As illustrated in the top-left of Fig. 1, the embedding vectors that interact within the Geo-LSH attention layer are reassigned and recombined back into their respective modalities after attention. To enhance the model's robustness and accuracy, a multi-round Geo-LSH attention process is employed, leveraging distinct random hash functions in each round to ensure diversified and comprehensive model learning. Through the Geo-LSH attention network, our model adeptly learns cross-modal information, enhancing its capability to capture users' behaviors via three modalities.

Integration by FFN. After the modality interaction within Geo-LSH attention module, we integrate the weighted POI and category representations to better capture the users' preference among POIs. To achieve this, we employ a double-layer feed-forward network (FFN), which introduces the non-linear transformations to the concatenated inputs, effectively capturing complex relationships from the prefrence representations. The formula is defined as:

$$Z_m = W_2 \cdot \sigma(W_1 \cdot [L_p; L_c]^T + b_1) + b_2, \qquad (3)$$

where $L_p \in \mathbf{R}^{|L| \times d}$ and $L_c \in \mathbf{R}^{|L| \times d}$ denote the weighted POI and category representations extracted from Geo-LSH attention module, respectively, W_1 and $W_2 \in \mathbf{R}^{d \times 2d}$, b_1 and $b_2 \in \mathbf{R}^{|L| \times d}$ are the learnable parameters of the MLP layer, $[;]$ is the concatenation operation, and $\sigma(\cdot)$ is the ReLU activation function. After the intermediate fusion process by Geo-LSH attention layer and the FFN layer, the model has adeptly learned the users' preference from check-in information, represented as $Z_m \in \mathbf{R}^{|L| \times d}$.

4.3   Textual Feature Learning

As we have discussed in the Introduction part, the POI review embedding and sequential POI (category) embedding have different intrinsic meanings. Modeling the textual modality individually before allocating it to the late fusion branch is a proper strategy to leverage the rich semantic potential of the textual information related to POIs. Giving a user's POI review embedding E_t aligning with the users historical POI embedding E_p, we introduce a two-layer FFN to delve into the deeper semantic information of the POI reviews. The process can be formulated as follows:

$$s_t = W_4 \cdot Dropout(\sigma(W_3 E_t + b_3)) + b_4, \tag{4}$$

where $W_3, W_4 \in \mathbf{R}^{d_2 \times d_2}$ and $b_3, b_4 \in \mathbf{R}^{|L| \times d_2}$ are the trainable parameters, $Dropout(\cdot)$ is the dropout operation to prevent overfitting by randomly omitting a subset of features, d_2 is the embedding dimension of the textual embedding. Through the FFN layer, we derive the $S_t \in \mathbf{R}^{|L| \times d_2}$ that represents the refined semantic understanding of the POI reviews.

4.4   Cross-Modal Learning and Model Optimization

Cross-Modal Late Fusion Layer. We design the cross-modal late fusion layer before merging the preference score, ensuring a comphrehensive information interaction between different modalities. Here we have users' preference from check-in information, represented as Z_m, and refined semantic representation of the POI reviews S_t, the late fusion strategy can be formulated as:

$$Z_m^{out} = Z_m \odot sigmod(W_5[Z_m; S_t] + b_5), \tag{5}$$

$$S_t^{out} = S_t \odot sigmod(W_6[Z_m; S_t] + b_6), \tag{6}$$

where $W_5 \in \mathbf{R}^{d \times (d+d_2)}, W_6 \in \mathbf{R}^{d_2 \times (d+d_2)}$, $b_5 \in \mathbf{R}^{|L| \times d}$, $b_6 \in \mathbf{R}^{|L| \times d_2}$, $\odot$ is the element-wise multiplication. Above, the late fusion process leverages a gated unit strategy to adaptively regulate the contribution of each modality. The gating mechanism allows the model to dynamically adjust the influence of different features, ensuring that important aspects from both check-in behaviors and textual representations are effectively retained while filtering less relevant information.

Preference Score Aggregation. Gathering the preference representations learned in both intermediate fusion module and late fusion module, we consider the following relations: 1) The correlation between user preference representation Z_m and the POI candidate, $T_{p,k} \in \mathbf{R}^{|L| \times d}$, with the consideration of the geographical distance embedding $g_{l,k}^{\Delta}$ between user's last check-in location l and the POI candidate k; 2) The correlation between learned POI review representation S_t and the candidate POI reviews, $C_{t,k} \in \mathbf{R}^{|L| \times d_2}$. the the Euclidean norm is used to measure the correlations. The final preference score can be calculated as follows:

$$\hat{y}_{u,k} = ||\frac{1}{|L|} \sum_{|L|} Z_m^{out} - T_{p,k} + g_{l,k}^{\Delta}||_2 + \eta||\frac{1}{|L|} \sum_{|L|} S_t^{out} - C_{t,k}||_2, \qquad (7)$$

where η represents a learnable parameter, $\frac{1}{|L|} \sum_{|L|} (\cdot)$ is the mean aggregation of the corresponding representations, and $|| \cdot ||_2$ is the Euclidean norm.

Model Optimization. To optimize our proposed CMPLH, we employ the Bayesian Personalized Ranking (BPR) [17] loss function which is normally used for handling the implicit feedback scenarios. The BPR function is defined as:

$$\mathcal{L}_{\text{BPR}} = - \sum_{(u,k) \in \mathcal{D}} \sum_{(u,j) \notin \mathcal{D}} \log \sigma(\hat{y}_{u,k} - \hat{y}_{u,j}) + \lambda ||\Theta||_2, \qquad (8)$$

where $\hat{y}_{u,k}$ and $\hat{y}_{u,j}$ denote the model prediction score of the observed (positive) POIs k and the unobserved (negative) samples j, respectively, $\sigma(\cdot)$ is the sigmoid activation function, λ is the regularization coefficient, and $||\Theta||_2$ is the L2 norm of the model parameters.

5   Experiment

5.1   Datasets

Two real-world LBSNs datasets are selected to evaluate our proposed CMPLH, which are widely used in the domain of the sequential POI recommendation [24]. These datasets comprise user check-in data from two well-known cities: New York and Tokyo, covering a period from 2012 to 2013. The detailed overview about the datasets is presented in Table 1. Considering that the original dataset does not contain any POI review data that is necessary in our model, we crawled the POI reviews by the Foursquare API[1]. Specifically, we collected the ten most popular reviews of each POI, which are the most representative insights of these locations and are also most visible for users. The popularity ranking of reviews relies on foursquare's internal algorithm that comprehensively considers the quality of the reviews, including: content richness, author trust, and user sentiment.

Within each dataset, we excluded users and POIs that have fewer than 5 interactions, focusing on only active participants. For training and evaluating

---

[1] https://developer.foursquare.com/.

our model, we keep the first 70% of users' check-in activities as the training set, the subsequent 10% of data as the validation set, and the final 20% of the records as the test set.

Table 1. The statistical information of the two datasets.

Dataset Name	Users	POIs	Sparsity	Avg. Interactions
NYC	1083	38333	99.45%	210.0
TKY	2293	61858	99.60%	250.2

5.2   Baseline Models

We select the following baseline models, categorizing to three segmentations: 1) classic POI recommendation models, which are based on matrix factorization technology: **GeoMF** [10] and **Rank-GeoFM** [9]; 2) sequential POI recommendations, that usually considers the spatio-temporal influence of POIs: **STGN** [25], **Flashback** [23], **STAN** [14], and **GeoMixer** [21]; 3) multi-modal sequential POI recommendation, which includes more input modalities such as category information: **AutoMTN** [16].

5.3   Metrics and Configurations

Performance Metrics.   To comprehensively evaluate the performance of our model, we adopt the following metrics: 1)**Hit Rate (HR@K)**: Measures whether the ground truth POI appears in the top-K recommended list; 2)**MR@K** offers a nuanced measure of the recommendation list quality by accounting for the rank of the first relevant item, reflecting the ranking quality of the result.

Model Configurations.   We conducted experiments using PyTorch on an NVIDIA GeForce GTX 4070 GPU. For all models, we applied the grid search strategy to determine the optimal hyperparameters for all models. The search space includes: the embedding dimension, the training batch size, the learning rate, the λ and the special hyperparameters according to their model configurations. The best configurations were selected based on performance on the validation set using HR@10 and nDCG@10 as metrics. For the CMPLH model, we set the input sequence length $|L|$ to 24 for NYC and TKY datasets, with number of candidates $|T| = 3$ for model learning. The dimensions d for the POI, category and geo-distance embedding were 150 for NYC and 160 for TKY. The textual embedding dimension d_2 was set to 768, following the DistilBert configuration. Training batch sizes were 1,000 for NYC and 800 for TKY. In the Geo-LSH attention layer, we employed a fixed-length bucket strategy [8], grouping 8 similar vectors from different modalities for interaction. We set the number of hashes to 2, applying two distinct hash functions and averaging the outputs to enhance modality interaction learning.

Table 2. The performance comparison.

	NYC				TKY			
	HR@5	nD@5	HR@10	MR@10	HR@5	MR@5	HR@10	MR@10
GeoMF	0.178	0.091	0.211	0.099	0.338	0.195	0.394	0.213
Rank-GeoFM	0.173	0.085	0.232	0.094	0.347	0.221	0.411	0.223
STGN	0.193	0.113	0.246	0.121	0.439	0.255	0.457	0.279
Flashback	0.211	0.115	0.273	0.119	0.449	0.257	0.471	0.282
STAN	0.229	0.128	0.301	0.138	0.442	0.301	0.473	0.311
GeoMixer	0.233	0.136	0.310	0.141	0.451	0.314	0.480	0.326
AutoMTN	0.242	0.145	0.322	0.149	0.465	0.307	0.532	0.322
CMPLH	**0.261**	**0.157**	**0.354**	**0.169**	**0.513**	**0.336**	**0.601**	**0.352**
Improv.	7.9 %	8.3%	9.9%	13.4%	10.3%	9.4%	13.0%	9.3%

5.4 Performance Comparison

Table 2 provides a detailed performance comparison of our CMPLH model against various baseline models. In our analysis, STGN and Flashback, both RNN (LSTM)-based POI recommenders, outperform two classic POI recommendation models, GeoMF and Rank-GeoFM. This enhancement is because STGN incorporates spatio-temporal context in the model, which dynamically adapt to users' spatio-temporal preferences within the POI recommendation scenario. Similarly, Flashback leverages the spatio-temporal information from all historical POIs to more accurately infer user interests of POIs. On the other hand, GeoMF and Rank-GeoFM performs worse compared with other models. Their limited performance is because these two models do not consider the sequential information, which omits the importance of the temporal dynamics of the user preference. Meanwhile, GeoSAN, STAN, and AutoMTN exhibit superior performance compared to the two RNN-based models, which use the spatio-temporal or category modalities with the transformer model. Such phenomenon indicates the effectiveness of self-attention mechanisms and validates the critical role of multi-modal information in enhancing POI recommendations.

CMPLH achieves an average 10% improvement over all baselines in terms of HR@K and MR@K. This result of performance can be attributes to several key points: 1) Unlike models that rely solely on historical POI sequence data, CMPLH's cross-modal architecture leverages additional information such as temporal dynamics, user reviews, and categorical data, offering a more complete view of user preferences; 2) CMPLH employs effective fusion strategies (intermediate and late fusion), to integrate cross-modal information when learning user preferences over POIs; 3) The model employs the Geo-LSH attention to model the historical user-POI interactions, corresponding category sequences, and the spatial distance information. This approach prioritizes salient features across modalities, enabling the model to emphasize the most influential factors in user decision-making processes. The detailed analysis and discussion of our model will be in the following sections.

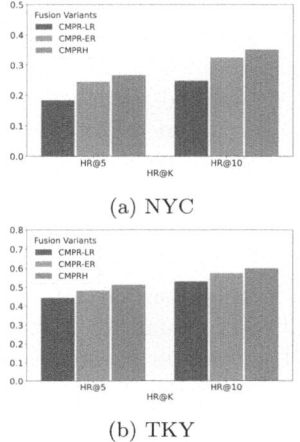

(a) NYC

(b) TKY

Fig. 2. The performance comparison of the fusion strategies.

Table 3. The performance impacted by different combinations of modalities.

	NYC		TKY	
	HR@5	HR@10	HR@5	HR@10
CMPLH$_{p+geo}$	0.226	0.311	0.475	0.562
CMPLH$_{p+geo+txt}$	0.239	0.323	0.478	0.583
CMPLH$_{p+c}$	0.243	0.321	0.489	0.578
CMPLH$_{p+c+geo}$	0.252	0.335	0.499	0.589
CMPLH	**0.261**	**0.354**	**0.513**	**0.601**

Table 4. The performance influenced by the input length.

	$|L|$	8	16	24	32	48	64
NYC	HR@5	0.246	0.248	*0.261*	0.238	0.218	0.221
	HR@10	0.332	0.341	*0.354*	0.320	0.311	0.301
TKY	HR@5	0.488	0.478	*0.513*	0.504	0.481	0.485
	HR@10	0.592	0.591	*0.601*	0.593	0.583	0.588

5.5   Fusion Strategy Exploration

We explore the fusion strategies based on our CMPLH framework, examining both on early and late fusion methods on the model performance. Below are the configurations for the CMPLH variants utilizing different fusion strategies:

- **CMPR-ER (Early Fusion Variant)**: This variant adopts the SASRec architecture but replaces the self-attention layer with the LSH attention layer. In this variant model, the embeddings of all modalities are merged at the beginning by concatenating the three input modalities before processing into the LSH attention layer. At the end, we employ the Euclidean distance to calculate the user preference score. However, this design may dilute the unique characteristics of each modality, as the interaction patterns among modalities are not explicitly modeled.
- **CMPR-LR (Late Fusion Variant)**: Also based on the SASRec architecture, this variant processes each modality independently through separate LSH attention layers followed by individual FFN networks. Subsequently, preference scores are calculated using Euclidean distance for each modality in the prediction layer. This design preserves the distinct contribution of each modality but may fail to capture inter-modal interactions that are crucial for comprehensive user preference modeling.

The results, shown in the Fig. 2, reveal that the original CMPLH model outperforms these variants across both datasets, as measured by HR@K. This superior performance indicates that neither early fusion nor late fusion can properly exploit and integrate the information across the four modalities. Early fusion blends the modalities at the initial stage, potentially ignoring their individual contributions on representing users potential interests of POIs, whereas late fusion may possibly miss the important inter-modal interactions essential for understanding user preference. Our findings underscore the importance of a balanced approach to fusion strategies which is able to deal with different types of input information and enhance cross-modal POI recommendation performance.

5.6   Ablation Study and Parameter Tuning

Ablation Study on Modalities. In this section, we examine four variants of our proposed CMPLH model to evaluate the impact of different modalities. In the Table 3, the notations p, geo, c, and $text$ represent the POI, geographical, category, and textual modalities, respectively. Our findings reveal that $CMPLH_{p+geo}$ and $CMPLH_{p+geo+txt}$ perform slightly worse than $CMPLH_{p+c}$ and $CMPLH_{p+c+geo}$, emphasizing the significance of the dense representation (the category modality). Additionally, the results show that the models including textual information ($CMPLH_{p+geo+txt}$, and CMPLH) enhance performance over the model without the textual modality ($CMPLH_{p+geo}$, and $CMPLH_{p+c}$). These results highlight the importance of each modality and the value of integrating them to capture a comprehensive view of user preferences.

Input Length. We examined the impact of input length $|L|$ on model performance by HR@K. Table 4 illustrates how the performance changes by the input length across both NYC and TKY datasets. For the NYC dataset, the model shows an optimal performance at an input length of $|L| = 24$. Both longer and shorter lengths reduce performance, suggesting that too-long sequences introduce irrelevant noise, while too-short sequences lack sufficient information. In the TKY dataset, HR@10 exhibits relatively stable across different input lengths, but we still observe a notably peak at $|L| = 24$ for HR@5, reflecting a similar pattern on the NYC dataset.

Embedding Dimension. Figure 3 showcases the influence of varying embedding dimensions on the performance of our proposed CMPLH. A dual y-axis presentation is employed to offer a clearer depiction of how performance metrics HR@K evolve in response to the changes in embedding dimensions. According to the figure, the model reaches its peak performance at an embedding dimension d of 150 for the NYC dataset and 160 for the TKY dataset. The result indicates the importance of optimizing the embedding dimension to better model various user contents and enhancing the performance.

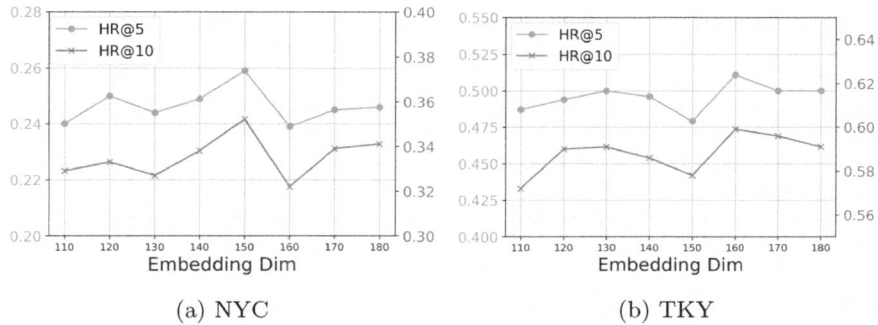

(a) NYC   (b) TKY

Fig. 3. The impact of the embedding dimension.

6   Conclusion and Future Work

This paper proposes the **C**ross-**M**odal Sequential **P**OI Recommender with **L**ightweight **H**ybrid Fusion Strategy (CMPLH), a novel approach that leverages cross-modal learning to enhance the effectiveness of sequential POI recommendation. CMPLH utilizes a hybrid fusion strategy, adeptly integrating multiple modalities, while considering the intrinsic nature of each input. Moreover, to reduce computational overhead from multiple modalities, we design the Geo-LSH attention mechanism in the intermediate fusion layer, efficiently capturing users' preferences by clustering similar temporal information, and reduce the computational cost at the same time. Experimental results validate CMPLH's superiority over existing approaches, highlighting its ability to balance performance with computational efficiency. This advancement demonstrates the potential of the cross-modal learning in enhancing POI recommendation systems. Future research could focus on exploring multi-modal learning on the few-shot POI recommendation, dealing with users with limited historical data in the LBSNs.

References

1. Chen, X., Liu, D., Xiong, Z., Zha, Z.J.: Learning and fusing multiple user interest representations for micro-video and movie recommendations. IEEE Trans. Multimedia **23**, 484–496 (2020)
2. Deldjoo, Y., Di Noia, T., Malitesta, D., Merra, F.A.: A study on the relative importance of convolutional neural networks in visually-aware recommender systems. In: Proceedings of the IEEE/CVF Conference on Computer Vision and Pattern Recognition, pp. 3961–3967 (2021)
3. He, J., Li, X., Liao, L., Song, D., Cheung, W.: Inferring a personalized next point-of-interest recommendation model with latent behavior patterns. In: Proceedings of the AAAI Conference on Artificial Intelligence, vol. 30 (2016)
4. Hu, H., Guo, W., Liu, Y., Kan, M.Y.: Adaptive multi-modalities fusion in sequential recommendation systems. In: Proceedings of the 32nd ACM International Conference on Information and Knowledge Management, pp. 843–853 (2023)

5. Hu, H., Pan, L., Ran, Y., Kan, M.Y.: Modeling and leveraging prerequisite context in recommendation. arXiv preprint arXiv:2209.11471 (2022)
6. Ji, W., Liu, X., Zhang, A., Wei, Y., Ni, Y., Wang, X.: Online distillation-enhanced multi-modal transformer for sequential recommendation. In: Proceedings of the 31st ACM International Conference on Multimedia, pp. 955–965 (2023)
7. Kang, W.C., McAuley, J.: Self-attentive sequential recommendation. In: 2018 IEEE International Conference on Data Mining (ICDM), pp. 197–206. IEEE (2018)
8. Kitaev, N., Kaiser, Ł., Levskaya, A.: Reformer: the efficient transformer. arXiv preprint arXiv:2001.04451 (2020)
9. Li, X., Cong, G., Li, X.L., Pham, T.A.N., Krishnaswamy, S.: Rank-geofm: a ranking based geographical factorization method for point of interest recommendation. In: Proceedings of the 38th International ACM SIGIR Conference on Research and Development in Information Retrieval, pp. 433–442 (2015)
10. Lian, D., Zhao, C., Xie, X., Sun, G., Chen, E., Rui, Y.: Geomf: joint geographical modeling and matrix factorization for point-of-interest recommendation. In: Proceedings of the 20th ACM SIGKDD International Conference on Knowledge Discovery and Data Mining, pp. 831–840 (2014)
11. Liu, F., Cheng, Z., Sun, C., Wang, Y., Nie, L., Kankanhalli, M.: User diverse preference modeling by multimodal attentive metric learning. In: Proceedings of the 27th ACM International Conference on Multimedia, pp. 1526–1534 (2019)
12. Liu, Q., Wu, S., Wang, L., Tan, T.: Predicting the next location: a recurrent model with spatial and temporal contexts. In: Proceedings of the AAAI Conference on Artificial Intelligence, vol. 30 (2016)
13. Lu, J., Batra, D., Parikh, D., Lee, S.: Vilbert: pretraining task-agnostic visiolinguistic representations for vision-and-language tasks. In: Advances in Neural Information Processing Systems, vol. 32 (2019)
14. Luo, Y., Liu, Q., Liu, Z.: Stan: spatio-temporal attention network for next location recommendation. In: Proceedings of the Web Conference 2021, pp. 2177–2185 (2021)
15. Pramanik, V., Maliha, M.: Analyzing sentiment towards a product using DistilBERT and LSTM. In: 2022 International Conference on Computing, Communication, and Intelligent Systems (ICCCIS), pp. 811–816. IEEE (2022)
16. Qin, Y., Fang, Y., Luo, H., Zhao, F., Wang, C.: Next point-of-interest recommendation with auto-correlation enhanced multi-modal transformer network. In: Proceedings of the 45th International ACM SIGIR Conference on Research and Development in Information Retrieval, pp. 2612–2616 (2022)
17. Rendle, S., Freudenthaler, C., Gantner, Z., Schmidt-Thieme, L.: BPR: Bayesian personalized ranking from implicit feedback. arXiv preprint arXiv:1205.2618 (2012)
18. Sanh, V., Debut, L., Chaumond, J., Wolf, T.: Distilbert, a distilled version of bert: smaller, faster, cheaper and lighter. arXiv preprint arXiv:1910.01108 (2019)
19. Sun, K., Qian, T., Chen, T., Liang, Y., Nguyen, Q.V.H., Yin, H.: Where to go next: modeling long-and short-term user preferences for point-of-interest recommendation. In: Proceedings of the AAAI Conference on Artificial Intelligence, vol. 34, pp. 214–221 (2020)
20. Wang, Q., Wei, Y., Yin, J., Wu, J., Song, X., Nie, L.: Dualgnn: dual graph neural network for multimedia recommendation. IEEE Trans. Multimedia **25**, 1074–1084 (2021)
21. Wang, T., Wang, C., Tian, H., Shen, H.: Geomixer: the MLP-based sequential poi recommender with travel routing modelling. In: 2023 IEEE International Conference on Data Mining (ICDM), pp. 1373–1378. IEEE (2023)

22. Xu, Y., Cong, G., Zhu, L., Cui, L.: Mmpoi: a multi-modal content-aware framework for poi recommendations. In: Proceedings of the ACM on Web Conference 2024, pp. 3454–3463 (2024)

23. Yang, D., Fankhauser, B., Rosso, P., Cudre-Mauroux, P.: Location prediction over sparse user mobility traces using RNNs. In: Proceedings of the Twenty-Ninth International Joint Conference on Artificial Intelligence, pp. 2184–2190 (2020)

24. Yang, D., Zhang, D., Zheng, V.W., Yu, Z.: Modeling user activity preference by leveraging user spatial temporal characteristics in LBSNs. IEEE Trans. Syst. Man Cybern. Syst. **45**(1), 129–142 (2014)

25. Zhao, P., et al.: Where to go next: a spatio-temporal gated network for next poi recommendation. IEEE Trans. Knowl. Data Eng. **34**(5), 2512–2524 (2020)

26. Zhou, H., Zhou, X., Zeng, Z., Zhang, L., Shen, Z.: A comprehensive survey on multimodal recommender systems: taxonomy, evaluation, and future directions. arXiv preprint arXiv:2302.04473 (2023)

Alternatives to Shallow Autoencoders for Collaborative Filtering

Mario Mallea$^{(\boxtimes)}$ [iD], Àngela Nebot[iD], and Francisco Mugica[iD]

Universitat Politècnica de Catalunya, Soft Computing Research Group at the
Intelligent Data Science and Artificial Intelligence Research Center, Barcelona, Spain
`mario.carlos.mallea@upc.edu`

Abstract. Collaborative filtering (CF) is a cornerstone of recommender
systems and plays a relevant role in many modern applications. CF uses
user-item interaction data to discover future preferences. Although deep
learning models have shown promise in CF, Embarrassingly Shallow
Autoencoders for Sparse Data (EASE) has gained attention for its outstanding ranking accuracy provided by its closed-form solution. EASE
relies primarily on relationships among items to fit a full-rank high-
dimensional linear mapping. We hypothesize that this design limits its
capacity to capture similar but not equivalent fine-grained user relationships, consequently limiting its recommendation accuracy. This paper
introduces an alternative formulation based on EASE that takes advantage of user-user information. Furthermore, we propose a hybrid model
that combines both user-user and item-item adjacency distributions.

Our experiments reveal that the proposed models outperform EASE
on well-known recommendation benchmarks, highlighting the significance of including the user alternative in shallow autoencoder studies
for CF.

Keywords: Autoencoder · recommender systems · collaborative
filtering

1 Introduction

In our digital age, filtering valuable information is a crucial challenge, as it helps
users navigate the overwhelming amount of available content. Recommender
systems play an important role in addressing this issue by identifying and prioritizing items that align with user preferences. Collaborative filtering (CF) [21]
methods are particularly noteworthy, as they focus on learning patterns and relationships from sparse historical user-item interactions itself. These interactions
are usually ratings, clicks, check-ins, or purchases. The main objective of these
types of systems is to provide personalized experiences to users [9,15,19].

In recent years, efficient machine learning methods have gained significant
attention due to their outstanding results through innovative approaches such
as infinitely wide random CNN [1], broad learning systems [2,10], and high-
dimensional or full-rank shallow linear approximations [27]. These methods often

© The Author(s), under exclusive license to Springer Nature Switzerland AG 2026
C. K. Leung et al. (Eds.): DaWaK 2025, LNCS 16048, pp. 142–156, 2026.
https://doi.org/10.1007/978-3-032-02215-8_10

employ a ridge regression formulation, which provides a closed-form solution, typically using the mathematical concept of pseudo-inverse [2,8]. The closed form simplifies model deployment and eliminates the need for extensive and iterative optimization, making such methods especially well suited for fast and effective CF tasks. One of the most representative methods in this domain is EASE [26]. EASE demonstrates exceptional effectiveness through a closed-form solution for a high-dimensional linear mapping. The performance of EASE relies on the Gram matrix or item-item adjacency distribution derived from the number of co-users consuming items. However, when datasets contain more users than items, more granular insights can be extracted by shifting to similar but not equivalent user-user relationships based on the number of co-liked items among users. Consequently, the resolution or dimensionality of the full-rank approximation becomes a critical factor in determining the most suitable approach, particularly because the learned linear mapping has the dimensionality of the total number of items or users. Indeed, we hypothesize that depending on the distribution of users and items, the EASE design could limit its ability to capture user relationships. For instance, we could have users with completely different behaviors, but their consumed items could still be linked through other users. Thus, the user Gram matrix reflecting the user similarities will mark zero interaction between them, whereas the item Gram matrix reflecting the item similarities might identify niches in terms of concepts, but this item co-occurrence will mask the lack of user interaction. As a result, in certain scenarios, a user-user approach will distinguish user communities that an item-item approach may fail to differentiate.

This research addresses the limitations of EASE by introducing alternative solutions that leverage both user-user and item-item information. Specifically, we propose a user-centric variant of EASE (u-EASE) and explore a novel hybrid approach that combines both perspectives (u-EASE-i). Despite the straightforward nature and fast training of the closed-form solutions of u-EASE and u-EASE-i, they can improve EASE and even achieve state-of-the-art recommendation results.

Our evidence is provided in terms of computational efficiency and performance in short and large rankings. We compared our contributions with well-known competitive models, including the most important deep, graph, variational, and nonlinear neural networks, as well as other recent shallow autoencoder approaches. The findings from our experiments indicate that the proposed models surpass EASE in performance on established recommendation benchmarks, for instance on Yelp2018 [28], and exhibit state-of-the-art outcomes on ML-1M [5]. This underscores the significance of integrating the user alternative within the framework of shallow autoencoders employed in collaborative filtering literature.

The remainder of this paper is organized as follows. In Sect. 2, we review related work on shallow autoencoders in the context of collaborative filtering. Section 3 provides the mathematical preliminaries necessary to understand EASE and its closed-form solution. In Sect. 4, we introduce a generalization of

EASE and our proposed models, u-EASE and u-EASE-i, and we describe their connections. Section 5 describes the experimental setup, including the datasets, evaluation metrics, and baseline comparisons. In Sect. 6, we present and analyze our results, highlighting the advantages of our approaches over EASE and their comparison with other recent recommenders. Finally, in Sect. 7, we conclude the study and present potential directions for future research.

2   Related Work

Collaborative filtering has been studied through techniques such as matrix factorization [19], graphs [6,22,28], deep neural networks [7,14] and autoencoders [8,20,26]. Recently, lightweight and efficient methods with closed-form solutions have garnered attention due to their potential to reduce computational overhead while achieving competitive or even state-of-the-art ranking accuracy through simple linear but high-dimensional approximations of the implicit feedback matrix [25].

EASE [26] introduced a full-rank autoencoder with a closed-form solution based on ridge regression, which requires tuning only one hyper parameter. EASE learns to reconstruct the interactions among users and items by learning a linear mapping among items, in particular, these trainable weights are defined by the Gram matrix or adjacency matrix among items.

Recent works study EASE properties in different contexts. [13] analyses theoretical insights of the relationships between linear autoencoders and matrix factorization. Also, [12] includes additional side information or metadata about items. The interpretability of EASE from the point of view of graphs was explored in [23], and the personality bias concerns behind their recommendations [16]. In addition, [24] studies the scalability of EASE and develops methods to parallelize it. Finally, its application to the discovery of sub communities is discussed in [4].

In terms of the recommender model, ∞-AE [20] employs the neural tangent kernel in a kernelized ridge regression to form a closed-form solution. Theoretically, this is equivalent to training an infinitely wide autoencoder for an infinite number of stochastic gradient descent steps. Unlike EASE, this model relies on the construction from the user-user perspective. SVD-AE [8] is an extension of EASE, which adopts a fast randomized truncated singular value decomposition as a low-rank inductive bias. Following this approach, they enhance the noise robustness, computation time, and accuracy, particularly in short rankings, as a consequence of preserving the top singular vectors and values.

On the other hand, graph convolutional networks have gained popularity for their ability to learn representations that capture complex relationships within the graph structure. These methods represent users and items as nodes in a graph, with edges representing interactions. Importantly, Graph filter based collaborative filtering (GF-CF) [22] developed a general graph convolution-based framework for CF, particularly they combined a linear filter and ideal low-pass filters, resulting a closed-form solution, that utilizes the item-item Gram matrix or adjacency of the graph and its top singular vectors version.

3   Preliminaries

We assume implicit feedback data represented by a sparse matrix $R \in \mathbb{R}^{|U| \times |I|}$, where $|U|$ and $|I|$ are the number of users and items, respectively. $R_{u,i} = 1$ indicates a user-item interaction for all $i \in I_u^+$ (the set of consumed items by a user).

The EASE model uses an item-item weight matrix $W \in \mathbb{R}^{|I| \times |I|}$ to predict scores for user u and item i as: $R_{u,\cdot}.W_{\cdot,i}$. To avoid a trivial solution, the diagonal of W is constrained to zero. The weights W are learned by minimizing the regularized Frobenius norm ($\|Z\|_F^2 = \mathrm{tr}(Z^t Z)$) of the approximation:

$$\min_{W} \quad \|R - RW\|_F^2 + \lambda_i \|W\|_F^2 \tag{1}$$

$$\text{s.t.} \quad Diag(W) = 0 \tag{2}$$

We introduce the suffix i or u to denote an element which corresponds with learnable weights in the space of items or users. λ_i is the regularization parameter.

The minimum of (1) is given by:

$$W_i = (G_i + \lambda_i I)^{-1} G_i, \tag{3}$$

where the item-item Gram matrix is $G_i = R^t R$. However, considering the constraint (2), to find the minimum, we define γ as the vector of Lagrangian multipliers representing the zero diagonal constrain. The new minimum is given by:

$$W = (G_i + \lambda_i I)^{-1}(G_i - DiagMat(\gamma)). \tag{4}$$

We still have to find γ. Let $P_i = (G_i + \lambda_i I)^{-1}$, and we can rewrite the solution as:

$$W = P_i(P_i^{-1} - \lambda_l I - DiagMat(\gamma)) \tag{5}$$

$$W = I - P_i \, DiagMat(\hat{\gamma}) \tag{6}$$

where $\hat{\gamma} = \lambda_i \mathbf{1} + \gamma$, with $\mathbf{1}$ denoting a vector of ones. Then $\hat{\gamma}$ is determined using the following:

$$0 = Diag(W) = \mathbf{1} - Diag(P_i) \odot \hat{\gamma}, \tag{7}$$

where $\odot$ denotes the product element wise.

Finally, the optimal closed form solution for EASE is given by:

$$W = I - P_i \, DiagMat\left(\mathbf{1} \oslash Diag(P_i)\right), \tag{8}$$

where $\oslash$ denotes the element wise division.

4 Proposal

We formulated an optimization problem to obtain an approximation based on both item-item and user-user perspectives. This approach preserves the one-shoot learning property of the closed-form solutions, while it makes both alternatives compatible. Figure 1 illustrates the idea of the approximation from the item's side (EASE), user's side (u-EASE), and both (u-EASE-i). Although the approach appears simple, some mathematical details are needed to preserve the good features of EASE and well-define all terms in the alternatives.

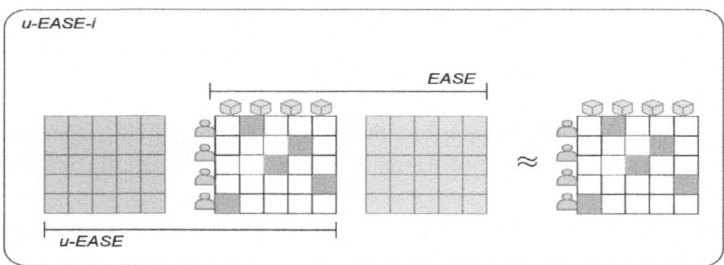

Fig. 1. Illustration of the model idea. EASE reconstructs the feedback matrix from the side of the items. We proposed an approximation from the user side with u-EASE. u-EASE-i combines both.

We are going to introduce a generalization of (3), then our approaches and EASE can be seen as special cases. Given matrices $Y \in \mathbb{R}^{n \times m}$, $A \in \mathbb{R}^{n \times d}$ and $B \in \mathbb{R}^{d \times m}$. We defined the following optimization problem:

$$\min_{W} \|Y - AWB\|_F^2.$$

To obtain the minimum W, we differentiate the norm (denoted f):

$$\frac{\partial}{\partial W} f(W) = -2A^t Y B^t + 2A^t A W B B^t.$$

Setting the derivative to zero, we find the closed form for W:

$$W = (A^t A)^{-1} A^t Y B^t (BB^t)^{-1}. \tag{9}$$

In fact, mathematically, this is a special case of a more general theorem of generalized inverses, known as the pseudo inverse of Moore-Penrose [18].

4.1 u-EASE

Note that we recover EASE by doing $Y = A = R$, and $B = I$, and considering the extra regularization of $Diag(W) = 0$. Equivalently, we can perform the symmetric version from the left instead of the right, that is, $Y = B = R$, $A = I$, and the regularization $Diag(W) = 0$. In other words, we introduced a novel user-centric variant of the EASE model, termed u-EASE, which symmetrically mirrors the item-centric approach. This allows us to exploit user-user relationships similar to how EASE leverages item-item interactions. Therefore, the complete formulation is:

$$\min_{W} \quad \|R - WR\|_F^2 + \lambda_u \|W\|_F^2 \tag{10}$$

$$\text{s.t.} \quad \text{diag}(W) = 0, \tag{11}$$

where λ_u denotes the regularization parameter. Equivalent to EASE, the closed-form solution without considering the constraint, for $W \in \mathbb{R}^{|U| \times |U|}$ is derived using the user-user Gram matrix $G_u = RR^t$:

$$W_u = (G_u + \lambda_u I)^{-1} G_u. \tag{12}$$

To obtain the complete solution, we must incorporate the Lagrangian multipliers γ. Let $P_u = (G_u + \lambda_r I)^{-1}$. Following the same steps as EASE, we can find the final closed form:

$$W = (G_u - DiagMat(\gamma)) P_u \tag{13}$$

$$W = (P_u^{-1} - \lambda_u I - DiagMat(\gamma)) P_u \tag{14}$$

$$W = I - DiagMat(\hat{\gamma}) \, P_u \tag{15}$$

$$W = I - DiagMat\left(\mathbf{1} \oslash Diag(P_u)\right) \, P_u \tag{16}$$

The last step is due to the commutative property of the element-wise product, so the computation of $\hat{\gamma}$ is equivalent to step (7). Figure 1 shows the difference between EASE and u-EASE, where we can visualize the effect of the resolution. Indeed, based on the amount of users and items of the problem, the decision of the alternative defines the number of parameters to learn, because we are going to fit the best linear mapping in the space of such dimensionality. Therefore, we could expect an impact on the model's accuracy, because the corresponding Gram matrix represents more grained relationships. Preliminary, we should take EASE if we take more items and u-EASE otherwise. However, experimentally, we found that u-EASE can enhance EASE even with fewer users than items.

4.2   u-EASE-i

We propose a combined optimization framework that jointly minimizes reconstruction error for both user-user and item-item interaction matrices. Considering the generalization, if we take $Y = R^t R R^t$, and $A = B = R^t$, then we are directly learning to reconstruct the matrix R. The optimization problem is as follows:

$$\min_{W} \|R^t R R^t - R^t W R^t\|_F^2 + Reg(W), \tag{17}$$

$$Reg(W) = \lambda_i \|R^t W\|_F^2 + \lambda_u \|W R^t\|_F^2 + \lambda_u \lambda_i \|W\|_F^2. \tag{18}$$

where the regularization terms are mathematically necessary to achieve the factorization. See the details in Appendix A. The closed form for $W \in \mathbb{R}^{|U| \times |I|}$ is the following simple expression:

$$W_{u,i} = W_u R W_i = P_u G_u R G_i P_i \tag{19}$$

This allows our model to capture a richer representation of user preferences and item relationships. Figure 1 shows the combination of the learned weights W_i, W_u, and $W_{u,i}$. We observed similar results with other possible alternatives such as $R \approx RWR$ or $RR^t R \approx RWR$. To end, note that Eqs. 3, 12, and 19 provide deterministic closed-form solutions to the recommendation problem.

4.3   Computational Cost

Training EASE, u-EASE, or u-EASE-i requires the expensive computation of the Gram matrix, however, as the closed-form solution depends on the Gram matrices, this computation can be performed on a big-data pre-processing system prior to the training phase [26]. The computational complexity of the training is determined by the inversion of the matrix, which is $O(X^3)$ using a basic approach, where X will be $|U|$ or $|I|$. Thus, depending on the number of users $|U|$ or items $|I|$, EASE or u-EASE can be prioritized. On the other hand, u-EASE-i needs both Gram matrices, thus its computational complexity will be defined by the possible dominant term.

In terms of scalability or memory concerns, future work could extend the efficient techniques already applied to scale EASE in our proposals, such as using lightweight methods to estimate a truncated singular value decomposition [8] or using a sparse approximate inversion framework [24].

5   Experimental Setting

5.1   Data

We evaluated our approaches in three datasets with more than 1 M interactions, their statistics are presented in Table 1. Gowalla [3] is a location-based social networking, where users share their location by checking-in. Yelp-2018 connects users with local businesses [28]. ML-1M [5] comes from a website that helps users find movies.

5.2   Models Used for Evaluation

We compared the performance of our proposed methods against some of the most relevant and recent recommendation models. The most related, EASE [26], and some other recent closed-form approaches, such as GF-CF [22], $\infty -$AE [20] and SVD-AE [8]. Additionally, we consider four important models: MF-BPR

Table 1. Statistics of datasets. We call the resolution of the problem to $\rho = |U|/|I|$.

Dataset	#User	#Item	#Interaction	Density	ρ
Gowalla	29,858	40,981	1,027,370	0.0008	0.73
Yelp2018	31,668	38,048	1,561,406	0.0013	0.83
ML-1M	6,040	3,706	1,000,209	0.0447	1.63

[19] is a matrix factorization model that leverages the Bayesian Personalized Ranking loss. NeuMF [7] combines MF (Matrix Factorization) embeddings of users and items with a nonlinear deep neural network. LightGCN [6] is a simplified graph convolutional network, which is designed to retain only the essential neighborhood aggregation component while removing feature transformation and non-linear activation. This linear propagation framework computes the final embeddings as a weighted sum across layers, significantly improving efficiency and recommendation performance. MultVAE [14] is a variational autoencoder model, which uses a multinomial likelihood to enhance the modeling capacity beyond linear factor models.

For a fair comparison, we used the same train, validation, and test splits [6,20,22] (80%/10%/10%). Consequently, we adopted a standardized evaluation procedure used for all those methods. Specifically, we followed the implementation of SVD-AE [8]. Our code is publicly available: https://github.com/MariodotR/Alternatives_EASE.git.

5.3   Metrics

We utilized the following metrics to evaluate the model's ability to recommend relevant items and position them highly in the ranked list. Following [8], we evaluated the performance in short and large rankings, taking the top $K = 10$ and $K = 100$.

- The hit rate or recall [17] measures the proportion of relevant items that are successfully recommended within the top ranked list.

$$HR@K = \frac{1}{|U|} \sum_{u \in U} \frac{|I_u^+ \cap \hat{R}_u|}{min(|I_u^+|, K)},$$

where $\hat{R}_u$ denotes the set of items recommended to the user.
- The normalized discounted cumulative gain [17] evaluates the ranking quality by considering the positions of relevant items in the list.

$$nDCG@K = \frac{1}{|U|} \sum_{u \in U} \frac{DCG_u}{IDCG_u}$$

$$DCG_u = \sum_{i=1}^{K} \frac{\mathbb{1}(\hat{R}_{u,i} \in I_u^+)}{log_2(i + 1)},$$

where $IDCG$ represents the best possible DCG at K per user, and the indicator function $\mathbb{1}$ is one if and only if the recommended item was consumed by the user.

– The propensity-scored precision [11] measures the precision of the recommendations considering a propensity (or popularity) correction term ϕ (see the details in [20]).

$$PSP@K = \frac{1}{|U|} \sum_{u \in U} \frac{uPSP_u}{mPSP_u},$$

$$uPSP_u = \frac{1}{K} \sum_{i=1}^{K} \frac{\mathbb{1}(\hat{R}_{u,i} \in I_u^+)}{\phi(\hat{R}_{u,i})},$$

where $mPSP$ represents the best possible $uPSP$ at K per user.

6   Results and Discussion

In u-EASE, the only hyper parameter to tune is the regularization parameter λ_u, similarly to EASE, we searched for a large regularization to avoid the trivial solution, we tuned it using a large range grid search $[10^1, 10^2, 10^3, 10^4, 10^5, 10^6]$ in the validation phase. Figure 2 shows the results, where we found 10^3 as the optimal regularization.

In u-EASE-i, we defined two regularization hyper parameters to tune λ_i and λ_u. We searched both in the grid search $[10^1, 10^2, 10^3, 10^4]$, exploring sixteen options. Figure 3 shows the results, the best (λ_i, λ_u) were $(10^2, 10^2), (10^3, 10^1)$, and $(10^2, 10^3)$ for Gowalla, Yelp2018, and ML-1M, respectively. We observed robust results, where $(10^1, 10^3)$ is usually competitive,in addition to an automatic correction behavior, for instance, when λ_i is larger, λ_u compensates with lower values, and vice versa.

Tables 2, 3, and 4 present the results for the Gowalla, Yelp2018, and ML-1M datasets, respectively. Our alternative formulations, u-EASE and u-EASE-i, demonstrated improvements over EASE on Yelp2018 and ML-1M. Specifically, on Yelp2018, u-EASE and u-EASE-i achieved a percentage improvement of 2.5% and 3.7% in terms of nDCG@100, respectively. On the ML-1M dataset, the enhancements were more pronounced, with u-EASE and u-EASE-i showing improvements of 6.0% and 6.3% in terms of nDCG@100, respectively. On ML-1M, where we have a positive resolution ρ, we achieved state-of-the-art results in short and large rankings, achieving even better results than other recent improvements over EASE, such as ∞-EASE and SVD-AE. Our alternatives also improve EASE's PSP@10, demonstrating that these enhancements do not overfit popularity bias. These results highlight the effectiveness of our proposed models in improving the ranking accuracy of the classic approach by simply selecting simple alternatives.

On Yelp2018 and Gowalla the best model is different based on the size of the ranking, for large rankings GF-CF is the best, while for short rankings is SVD-AE. Despite this, on Yelp2018 our proposals perform better than EASE for long rankings, and particularly with u-EASE-i in terms of HR@10. Besides, we achieved better metrics than MF-BPR, NeuMF, MultVAE, LightGCN (except by HR@100), and ∞-AE (except by PSP@10). On Gowalla, u-EASE still performs better than NeuMF, MultVAE, and ∞-AE (except by PSP@10).

In general, the relative performance gains seem to correlate with the resolution ρ, suggesting that user-user information becomes more impactful when the number of users is larger than the number of items. Although the results on Yelp2018 show that $\rho > 1$ is not exclusively necessary. Furthermore, in that scenario, by incorporating both user-user and item-item information, u-EASE-i achieves a richer representation of relationships within the data, leading to more accurate recommendations.

Finally, computational efficiency is an important concern in real-world applications. Therefore, we provide evidence to support this concern. We empirically measured the total wall clock time, including pre-processing (Gram matrix), its inversion, and evaluation. The Table 5 shows the comparison among the alternative approaches. The machine was equipped with 16 VCPUs and 8 GB of system memory per vCPU. All alternatives took a few minutes to complete the entire experiment, a great feature considering the well-known timescale required to train deep learning methods [8]. We observed that the u-EASE alternative is faster than EASE when $\rho < 1$, since we are considering a lightweight Gram matrix. Therefore, the performance drop in a data condition, such as Gowalla could be accepted, considering limited resources or the better computational efficiency of u-EASE. On the other hand, u-EASE-i consumed more computational resources, since it needs both Gram matrices. In particular, our state-of-the-art results on ML-1M were achieved in seconds.

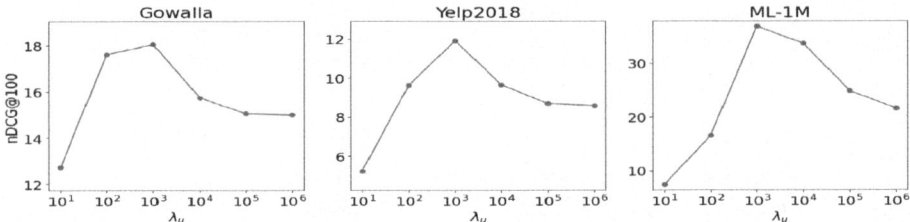

Fig. 2. Regularization's results for the u-EASE model in the validation set.

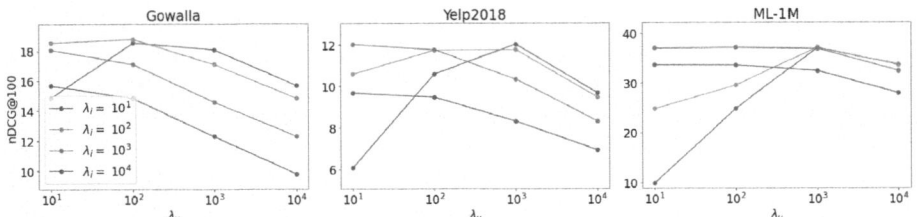

Fig. 3. Regularization's results for the u-EASE-i model in the validation set.

Table 2. Gowalla Results

Method	HR@10	HR@100	NDCG@10	NDCG@100	PSP@10
MF-BPR	12.08	32.84	12.09	18.44	1.92
NeuMF	8.87	27.25	8.27	13.96	1.37
LightGCN	14.00	37.40	13.50	21.04	2.26
GF-CF	14.08	**38.84**	13.71	**21.25**	2.47
MultVAE	11.88	33.56	11.30	18.11	2.09
∞-AE	11.77	34.20	10.84	17.97	2.02
SVD-AE	**14.40**	37.34	**13.94**	21.15	**2.48**
EASE	13.67	35.74	13.15	20.08	2.31
u-EASE	12.10	34.22	11.39	18.33	1.95
u-EASE-i	11.47	34.15	10.56	17.80	2.04

Table 3. Yelp2018 Results

Method	HR@10	HR@100	NDCG@10	NDCG@100	PSP@10
MF-BPR	3.86	16.97	3.70	8.50	0.34
NeuMF	3.57	14.39	3.08	7.16	0.27
LightGCN	4.32	19.01	4.19	9.33	0.39
GF-CF	4.87	**20.86**	4.66	**10.53**	0.44
MultVAE	4.31	18.75	4.10	9.37	0.43
∞-AE	4.62	18.33	4.48	9.02	0.43
SVD-AE	**4.90**	19.79	**4.74**	10.22	**0.45**
EASE	4.65	17.74	4.55	9.37	0.42
u-EASE	4.63	18.55	4.50	9.60	0.40
u-EASE-i	4.68	18.78	4.56	9.72	0.40

Table 4. ML-1M Results

Method	HR@10	HR@100	NDCG@10	NDCG@100	PSP@10
MF-BPR	28.49	57.21	29.84	39.47	2.89
NeuMF	27.95	54.24	29.36	37.98	2.75
LightGCN	29.07	57.62	30.33	39.95	3.17
GF-CF	28.60	59.10	30.32	42.00	3.13
MultVAE	27.86	57.67	28.44	39.34	3.16
∞-AE	31.15	60.75	32.27	42.54	3.22
SVD-AE	31.79	59.33	33.55	42.57	3.22
EASE	30.43	57.74	31.90	40.95	3.16
u-EASE	**32.34**	60.79	33.68	43.41	**3.37**
u-EASE-i	32.33	**60.97**	**33.71**	**43.51**	3.35

Table 5. Computation time in minutes.

Dataset	EASE	u-EASE	u-EASE-i
Gowalla	11.3	6.8	24.5
Yelp2018	9.9	8.0	23.7
ML-1M	0.1	0.3	0.4

7   Conclusion and Future Work

EASE has gained significant attention in the recommender systems community due to its simplicity, efficiency, and strong ranking accuracy. Those properties have been transferred to other important methods based on graph neural networks, singular value decomposition, or neural tangent kernel. However, other alternative high dimensional shallow linear autoencoders have not been explored.

In this research, we introduced two alternative approaches to EASE by leveraging user-user information, instead of the original item-item based method. Through comprehensive experiments on three benchmark datasets, we showed that our proposed methods can outperform EASE, particularly but not limited to datasets with a higher user-to-item ratio, notably achieving state-of-the-art results in ML-1M. Our proposals offer rapid deployment and streamlined, one shoot learning. Our research indicates that incorporating user-user relationships may improve the accuracy of collaborative filtering recommendations, particularly when dealing with a larger number of users than items.

Future work could explore the formulation of the dual-side factorization idea behind our proposal in a randomly generated low-dimensional embedding for users and items. In addition, we note that graph regularization is possible in all alternatives. We planned to investigate the impact of data characteristics such as sparsity or resolution over the decision of the alternatives, and develop robust strategies for selecting the most appropriate model. Finally, transferring methods of interpretability, local communities, or scalability created for EASE are relevant next steps.

Acknowledgments. Supported by the 'Siemens Energy AI Chair: Energy Sustainability for a Decarbonized Society 5.0' (TSI-100930-2023-5), funded by the Secretary of State for Digitalization and Artificial Intelligence through the ENIA 2022 Chairs call, and co-funded by the European Union-Next Generation EU.

A   Appendix: The Closed-Form Solution for u-EASE-i

In this section, we present the complete derivation of the closed-form solution (19) for the optimization problem (17) related to the u-EASE-i model. Firstly, we differentiate each term:

– First term: Using the generalized solution (9):

$$\frac{\partial}{\partial W}\|R^t R R^t - R^t W R^t\|_F^2 = -2R(R^t R R^t)R + 2(R R^t)W(R^t R)$$

– Second term:

$$\lambda_i \frac{\partial}{\partial W}\|R^t W\|_F^2 = 2\lambda_i R R^t W.$$

– Third term:

$$\lambda_u \frac{\partial}{\partial W}\|W R^t\|_F^2 = 2\lambda_u W R^t R.$$

– Fourth term:

$$\lambda_u \lambda_i \frac{\partial}{\partial W}\|W\|_F^2 = 2\lambda_u \lambda_i W.$$

The Gram matrices G_i and G_u are used to simplify the notation. Secondly, gathering all the derivatives and setting them equal to zero:

$$-2G_u R G_i + 2G_u W G_i + 2\lambda_i G_u W + 2\lambda_u W G_i + 2\lambda_u \lambda_i W = 0$$
$$G_u W G_i + \lambda_i G_u W + \lambda_u W G_i + \lambda_u \lambda_i W = G_u R G_i$$
$$(G_u + \lambda_u I)W(G_i + \lambda_i I) = G_u R G_i$$

After simplifying and factorizing the terms from left and right sides, the solution is:

$$W = (G_u + \lambda_u I)^{-1} G_u R G_i (G_i + \lambda_i I)^{-1}.$$

Finally, unconstrained solutions for EASE (3) and u-EASE (12) are used to recover the final closed form for u-EASE-i presented in (19).

References

1. Arora, S., Du, S.S., Li, Z., Salakhutdinov, R., Wang, R., Yu, D.: Harnessing the power of infinitely wide deep nets on small-data tasks (2019). https://arxiv.org/abs/1910.01663

2. Chen, C.L.P., Liu, Z.: Broad learning system: an effective and efficient incremental learning system without the need for deep architecture. IEEE Trans. Neural Netw. Learn. Syst. **29**(1), 10–24 (2018). https://doi.org/10.1109/TNNLS.2017.2716952

3. Cho, E., Myers, S.A., Leskovec, J.: Friendship and mobility: user movement in location-based social networks. In: Proceedings of the 17th ACM SIGKDD International Conference on Knowledge Discovery and Data Mining, pp. 1082–1090 (2011)

4. Choi, M., Jeong, Y., Lee, J., Lee, J.: Local collaborative autoencoders. In: Proceedings of the 14th ACM International Conference on Web Search and Data Mining, pp. 734–742 (2021)

5. Harper, F.M., Konstan, J.A.: The movielens datasets: history and context. ACM Trans. Interact. Intell. Syst. **5**(4) (2015). https://doi.org/10.1145/2827872

6. He, X., Deng, K., Wang, X., Li, Y., Zhang, Y., Wang, M.: Lightgcn: simplifying and powering graph convolution network for recommendation. In: Proceedings of the 43rd International ACM SIGIR Conference on Research and Development in Information Retrieval, SIGIR 2020, pp. 639–648. Association for Computing Machinery, New York (2020). https://doi.org/10.1145/3397271.3401063

7. He, X., Liao, L., Zhang, H., Nie, L., Hu, X., Chua, T.S.: Neural collaborative filtering. In: Proceedings of the 26th International Conference on World Wide Web, WWW 2017, pp. 173–182. International World Wide Web Conferences Steering Committee, Republic and Canton of Geneva, CHE (2017). https://doi.org/10.1145/3038912.3052569

8. Hong, S., Choi, J., Lee, Y.C., Kumar, S., Park, N.: SVD-AE: simple autoencoders for collaborative filtering. arXiv abs/2405.04746 (2024). https://api.semanticscholar.org/CorpusID:269626235

9. Hu, Y., Koren, Y., Volinsky, C.: Collaborative filtering for implicit feedback datasets. In: 2008 Eighth IEEE International Conference on Data Mining, pp. 263–272 (2008). https://api.semanticscholar.org/CorpusID:10537313

10. Huang, L., et al.: Broad recommender system: an efficient nonlinear collaborative filtering approach (2024). https://arxiv.org/abs/2204.11602

11. Jain, H., Prabhu, Y., Varma, M.: Extreme multi-label loss functions for recommendation, tagging, ranking & other missing label applications. In: Proceedings of the 22nd ACM SIGKDD International Conference on Knowledge Discovery and Data Mining, KDD 2016, pp. 935–944. Association for Computing Machinery, New York (2016). https://doi.org/10.1145/2939672.2939756

12. Jeunen, O., Van Balen, J., Goethals, B.: Closed-form models for collaborative filtering with side-information. In: Proceedings of the 14th ACM Conference on Recommender Systems, RecSys 2020, pp. 651–656. Association for Computing Machinery, New York (2020). https://doi.org/10.1145/3383313.3418480

13. Jin, R., Li, D., Gao, J., Liu, Z., Chen, L.C., Zhou, Y.: Towards a better understanding of linear models for recommendation. In: Proceedings of the 27th ACM SIGKDD Conference on Knowledge Discovery & Data Mining (2021). https://api.semanticscholar.org/CorpusID:235212063

14. Liang, D., Krishnan, R.G., Hoffman, M.D., Jebara, T.: Variational autoencoders for collaborative filtering. In: Proceedings of the 2018 World Wide Web Confer-

ence, WWW 2018, pp. 689–698. International World Wide Web Conferences Steering Committee, Republic and Canton of Geneva, CHE (2018). https://doi.org/10.1145/3178876.3186150

15. Mallea, M., Ñanculef, R., Parra, D.: Adversarial pairwise multimodal recommendation. In: 2024 International Joint Conference on Neural Networks (IJCNN), pp. 1–10 (2024). https://doi.org/10.1109/IJCNN60899.2024.10650977

16. Melchiorre, A.B., Zangerle, E., Schedl, M.: Personality bias of music recommendation algorithms. In: Proceedings of the 14th ACM Conference on Recommender Systems, RecSys 2020, pp. 533–538. Association for Computing Machinery, New York (2020). https://doi.org/10.1145/3383313.3412223

17. Parra, D., Sahebi, S.: Recommender systems: sources of knowledge and evaluation metrics. In: Advanced Techniques in Web Intelligence-2: Web User Browsing Behaviour and Preference Analysis, pp. 149–175. Springer (2013)

18. Penrose, R.: On best approximate solutions of linear matrix equations. Math. Proc. Camb. Philos. Soc. **52**, 17–19 (1956). https://api.semanticscholar.org/CorpusID:122260851

19. Rendle, S., Freudenthaler, C., Gantner, Z., Schmidt-Thieme, L.: BPR: Bayesian personalized ranking from implicit feedback. In: Proceedings of the Twenty-Fifth Conference on Uncertainty in Artificial Intelligence, UAI 2009, Arlington, Virginia, USA, pp. 452–461. AUAI Press (2009)

20. Sachdeva, N., Dhaliwal, M.P., Wu, C.J., McAuley, J.: Infinite recommendation networks: a data-centric approach. In: Proceedings of the 36th International Conference on Neural Information Processing Systems. NIPS 2022. Curran Associates Inc., Red Hook (2024)

21. Schafer, J.B., Frankowski, D., Herlocker, J.L., Sen, S.: Collaborative filtering recommender systems. In: The Adaptive Web (2007). https://api.semanticscholar.org/CorpusID:14231524

22. Shen, Y., et al.: How powerful is graph convolution for recommendation? In: Proceedings of the 30th ACM International Conference on Information & Knowledge Management, CIKM 2021, pp. 1619–1629. Association for Computing Machinery, New York (2021). https://doi.org/10.1145/3459637.3482264

23. Spišák, M., Bartyzal, R., Hoskovec, A., Peška, L.: On interpretability of linear autoencoders. In: Proceedings of the 18th ACM Conference on Recommender Systems, pp. 975–980 (2024)

24. Spišák, M., Bartyzal, R., Hoskovec, A., Peska, L., Tůma, M.: Scalable approximate nonsymmetric autoencoder for collaborative filtering. In: Proceedings of the 17th ACM Conference on Recommender Systems, pp. 763–770 (2023)

25. Steck, H.: Collaborative filtering via high-dimensional regression. arXiv preprint arXiv:1904.13033 (2019)

26. Steck, H.: Embarrassingly shallow autoencoders for sparse data. In: The World Wide Web Conference, WWW 2019, pp. 3251–3257. Association for Computing Machinery, New York (2019). https://doi.org/10.1145/3308558.3313710

27. Steck, H.: Autoencoders that don't overfit towards the identity. In: Advances in Neural Information Processing Systems, vol. 33, pp. 19598–19608 (2020)

28. Wang, X., He, X., Wang, M., Feng, F., Chua, T.S.: Neural graph collaborative filtering. In: Proceedings of the 42nd International ACM SIGIR Conference on Research and Development in Information Retrieval, SIGIR 2019, pp. 165–174. Association for Computing Machinery, New York (2019). https://doi.org/10.1145/3331184.3331267

Accurate Concept Drift Detection Without Updating Autoencoders

Taisei Takano$^{(\boxtimes)}$ and Hisashi Koga

University of Electro-Communications, Tokyo 182-8585, Japan
{takano,koga}@sd.is.uec.ac.jp

Abstract. As concept drifts change the feature of data streams, detecting concept drifts is significant for analyzing data streams. ABCD (Adaptive Bernstein Change Detector) is a drift detection method based on autoencoders and is one of the most accurate drift detectors. It learns the current property of a given stream by training an autoencoder (AE). Then, it alerts a concept drift when the reconstruction errors of AE grow large for recent data. To adapt to the new concept, ABCD updates the AE whenever a concept drift happens. This paper shows that, even without updating the AE, we can create a more precise drift detector than ABCD by coupling the non-updated autoencoder-based drift detector with another simple lightweight one.

1 Introduction

As IoT (Internet of Things) prevails, we need to handle many data streams to which new data are added continuously. Thus, pattern classification for data streams has been strongly demanded, e.g., anomaly detection from network traffic and the trend prediction for stock prices. As concept drifts change the underlying data distribution, they seriously damage the accuracy of classifiers that have learned the obsolete training data. Thus, detecting concept drifts timely for data streams is important.

Drift detection algorithms are called unsupervised, if they do not rely on the ground-truth class labels for the data in the stream. Unsupervised methods typically check if the current data distribution differs from the past one. ABCD (Adaptive Bernstein Change Detector) [4] is an unsupervised drift detector that exploits an autoencoder. ABCD uses the autoencoder to learn the current feature of the data stream. Then, for every arriving data, it measures the reconstruction error yielded by the autoencoder. Finally, ABCD alerts a concept drift, if the errors grow large for a while, because it indicates that new arriving data have changed from the past data used to train the autoencoder. Specifically, ABCD can detect concept drifts very correctly and achieves much higher F1 scores than other state-of-the-art methods [4].

This paper aims to enhance ABCD to realize more precise drift detection without sacrificing the processing speed. In general, to improve the detection accuracy, additional complex operations are required, which usually slows down

© The Author(s), under exclusive license to Springer Nature Switzerland AG 2026
C. K. Leung et al. (Eds.): DaWaK 2025, LNCS 16048, pp. 157–163, 2026.
https://doi.org/10.1007/978-3-032-02215-8_11

the detection speed. To retain the execution speed, we stop updating the autoencoder after detecting concept drifts and avoid the heavy cost to relearn the autoencoder. This strategy has an obvious drawback that the autoencoder cannot track the latest feature of the stream. We complement the drawback by assigning the yielded surplus time to a second lightweight drift detector.

By coupling the non-updated autoencoder-based drift detector with the second lightweight one, our method not only works more accurately than ABCD but also runs faster than ABCD for high-dimensional multivariate data streams.

2   Unsupervised Concept Drift Detection

The context of concept drift detection assumes some classification task with the data stream. Concept drift is a phenomenon that the property of the data stream changes as time elapses. For example, the deterioration of sensors may change the characteristics of data streams gradually. The concept drift does harm to the classifier, because the test data, i.e. the new data after a concept drift wear different features from the past data used to train the classifier. To maintain the classifier performance, one should detect concept drifts promptly.

Unsupervised concept drift detection methods operate without the correct class labels for any previously seen data. Let DS be a multivariate data stream. DS consists of a series of feature vectors $X_1, X_2, \ldots, X_t, \ldots$ that arrive ordered by time. X_t is a d-dimensional vector added to the stream at time t. The unsupervised drift detection usually considers virtual concept drifts that change the data distribution $P(X)$. Let $P_t(X)$ be the data distribution of X at t. Formally, if $P_t(X) \neq P_{t+\delta}(X)$, we say that a virtual concept drift occurs in the time interval $[t, t + \delta]$. Unsupervised drift detection methods are further divided into two types according to whether they depend on the classifier or not. ABCD [4], IKS [6], D3 [3], WATCH [2], and IBDD [7] are categorized as unsupervised drift detection methods that do not depend on the classifier.

Some unsupervised drift detection methods including ABCD rely on the autoencoder [4,5]. An autoencoder is a special neural network to learn a refined lower-dimensional representation for a group of higher-dimensional input vectors. It consists of an encoder ϕ and a decoder ψ that follows ϕ. Let X be an input vector whose dimension equals d. When X arrives, the encoder ϕ first transforms X to a d'-dimensional vector $\phi(X)$ where $d' < d$. At this moment, we may regard $\phi(X)$ as a low-dimensional essential feature in X. After this, the decoder ψ converts $\phi(X)$ back to a d-dimensional vector $X' = \psi(\phi(X))$. X' is called a reconstructed version of X.

The autoencoder configures the network parameters Θ by learning the training data $S = \{V_1, V_2, \cdots, V_n\}$, a set of n d-dimensional vectors. The learning process minimizes the loss function in Eq. (1) that sums up the reconstruction errors between V_i and V_i' for $1 \leq i \leq n$.

$$L(S) = \sum_{i=1}^{n} \text{MSE}(V_i, V_j') = \sum_{i=1}^{n} \sum_{j=1}^{d} (V_{i,j} - V_{i,j}')^2. \tag{1}$$

The autoencoder reconstructs vectors with little error if they are similar to the training data. The errors become larger, as they become more dissimilar to the training data. An autoencoder enables unsupervised drift detection in such a way that it has learned the current concept initially and alerts a concept drift when new arriving data increase the reconstruction errors.

3   ABCD (Adaptive Bernstein Change Detector) [4]

ABCD is an unsupervised drift detector based on the autoencoder. This paper treats ABCD as a baseline. ABCD can detect drifts very accurately. According to [4], ABCD outperforms its best competitor out of D3, WATCH, IKS, IBDD and AdwinK by up to 20% in terms of the F1 score for multiple datasets.

ABCD manages an adaptive window W whose length is variable. Suppose that W stores α recent vectors just before time t, that is, $W = \{V_1, V_2, ..., V_\alpha\}$, As V_α presents the most recent vector that arrived at time $t - 1$, $V_\alpha = X_{t-1}$. When $V_{\alpha+1} = X_t$ appears at t, ABCD appends $V_{\alpha+1}$ to W. Then, it divides W into the old front part $W_f = \{V_1, V_2, ..., V_m\}$ and the new rear part $W_b = \{V_{m+1}, V_{m+2}, ..., V_{\alpha+1}\}$, where m indicates a boundary index. ABCD expects that the reconstruction error will increase if a concept drift occurs. So it computes the sample mean errors $\hat{\mu}_f = \frac{1}{m} \sum_{i=1}^{m} \mathrm{MSE}(V_i, V_i')$ over W_f and $\hat{\mu}_b = \frac{1}{\alpha+1-m} \sum_{i=m+1}^{\alpha+1} \mathrm{MSE}(V_i, V_i')$ over W_b, and then conducts a statistical test to evaluate the null hypothesis H_0: $\mu_f = \mu_b$.

In the statistical test, Bernstein's inequality obtains the upper bound of the probability $P(|\hat{\mu}_f - \hat{\mu}_b| \geq \epsilon)$, where ϵ equals the value of $|\hat{\mu}_f - \hat{\mu}_b|$ actually observed. For the demanded significance level δ, H_0 is rejected if the upper bound falls short of δ. In this case, m becomes a candidate for the change point.

In practice, ABCD performs the statistical test for all the time indices in the range $1 \leq m \leq \alpha$. If H_0 is correct for any m, ABCD judges that a concept drift has not occurred. Otherwise, if H_0 is rejected for multiple values of m, ABCD chooses the time m^* that minimizes the upper bound of $P(|\hat{\mu}_f - \hat{\mu}_b| \geq \epsilon)$ as the unique change point, because the null hypothesis is rejected the most strongly there. Then, ABCD advances the left end of W to $m^* + 1$ and updates the autoencoder by learning the vectors that come from the new concept.

4   Our Method

By modifying ABCD, we aim to realize more precise drift detection than ABCD. We also expect that our method is comparable to ABCD in terms of the processing speed. In general, elaborated drift detection mechanisms require complicated operations and prolong the processing time. In order to improve the detection accuracy without damaging the execution speed, we propose to stop relearning the autoencoder even after a concept drift is detected. Since the autoencoder grows non-updated, our method is named as DDNAE (Drift Detection with Non-updated AutoEncoder).

By updating the autoencoder at every drift detection, ABCD must devote much time to train the autoencoder many times. By contrast, our DDNAE avoids the expensive relearning of the autoencoder. DDNAE exploits the surplus time without relearning autoencoders to run a new lightweight detector. Thus, DDNAE simultaneously runs two drift detectors (I) a detector based on the non-updated autoencoder and (II) a new lightweight detector. DDNAE alerts a concept drift, if either of these two models detects a concept drift.

Of course, the non-updated autoencoder has some risks: By updating the autoencoder, ABCD assures that the training data follows the same data distribution as the old concept. Thus, ABCD keeps the average reconstruction error small before the concept drift, and the error increases when a new concept emerges. In DDNAE, the quite past data used to train the non-updated autoencoder can correspond neither to the old concept nor to the new concept. As a result, the average reconstruction error may decrease, when the old concept is replaced with the new one. Fortunately, the previous concept drift detector IBDD [7] showed that the strategy without updating the reference model is promising. IBDD converts a window of the data stream to an image and detects drifts from the gap between the initial reference image and the latest image representing the current window. Though IBDD never updates the reference image, it succeeds in grasping the change of data distribution.

4.1   Details of DDNAE

Though ABCD rarely over-detects drifts, it often misses some concept drifts that do not affect the reconstruction errors of the autoencoder. Therefore, the new lightweight detector should operate independently of the primary autoencoder-based detector to find concept drifts overlooked by the primary detector.

Our lightweight detector computes the mean MV of the raw vectors that belong to the current concept and specifies it as the reference model. While the primary detector focuses on the reconstruction errors, the second lightweight detector is interested in how much the arriving vectors deviate from MV. Though the lightweight detector must update the reference model at every concept drift, the overhead to compute MV is negligibly smaller than the cost to train the autoencoder.

Let us denote the primary drift detector based on AE by DD_1 and the second one based on MV by DD_2. At the beginning, the initial training data trains the autoencoder AE and computes their mean vector MV. After that, DD_1 never updates AE.

Algorithm 1 explains how DDNAE behaves when a new vector X_t arrives at time t. X_t is reconstructed to X_t' by AE (the first line). At this point, we compute (1) $\mathrm{MSE}(X_t, X_t')$ associated with DD_1 and (2) $\mathrm{MSE}(X_t, MV)$ associated with DD_2 in the second line. The statistical test begins from the third line. Consider the window $W = \{V_1, V_2, ..., V_\alpha, V_{\alpha+1}\}$ as in Sect. 3. For the statistical test, DDNAE calculates the probability $P(|\hat{\mu}_f - \hat{\mu}_b| \geq \epsilon)$ at all the $\alpha+1$ time instances for each of DD_1 and DD_2: In the 3rd and 4th lines, DD_1 decides that $P(|\hat{\mu}_f -$

Algorithm 1. Algorithm DDNAE

Require: AE_1, MV, Significance level δ

1: $X_t^{'} \leftarrow AE(X_t)$

2: Compute $\mathrm{MSE}(X_t, X_t^{'})$ and $\mathrm{MSE}(X_t, MV)$

3: $p_1 = \min_{1 \leq m \leq \alpha+1} P(|\hat{\mu}_f^1 - \hat{\mu}_b^1| \geq \epsilon_1)$ with respect to DD_1

4: $m_1 = \mathrm{argmin}_{1 \leq m \leq \alpha+1} P(|\hat{\mu}_f^1 - \hat{\mu}_b^1| \geq \epsilon_1)$ with respect to DD_1

5: $p_2 = \min_{1 \leq m \leq \alpha+1} P(|\hat{\mu}_f^2 - \hat{\mu}_b^2| \geq \epsilon_2)$ with respect to DD_2

6: $m_2 = \mathrm{argmin}_{1 \leq m \leq \alpha+1} P(|\hat{\mu}_f^2 - \hat{\mu}_b^2| \geq \epsilon_1)$ with respect to DD_2

7: **if** $p_1 < \delta$ **then**

8:     Alerts a Drift at time m_1.

9:     Moves the left end of W to $m_1 + 1$.

10: **else if** $p_2 < \delta$ **then**

11:     Alerts a Drift at time m_2.

12:     Moves the left end of W to $m_2 + 1$.

13: **end if**

$\hat{\mu}_b| \geq \epsilon)$ takes the minimum value p_1 at time m_1. In the 5th and 6th lines, DD_2 identifies that $P(|\hat{\mu}_f - \hat{\mu}_b| \geq \epsilon)$ takes the minimum value p_2 at time m_2.

Specifically, DD_2 divides W into the old front part W_f and the new rear part W_b at t. Then, by gathering $\mathrm{MSE}(V_i, MV)$ for every arriving vector V_i for $1 \leq i \leq \alpha$, DD_2 derives the sample mean errors $\hat{\mu}_f^2 = \frac{1}{m} \sum_{i=1}^{m} \mathrm{MSE}(V_i, MV)$ over W_f and $\hat{\mu}_b^2 = \frac{1}{\alpha+1-m} \sum_{i=m+1}^{\alpha+1} \mathrm{MSE}(V_i, MV)$ over W_b. m presents the boundary index between W_f and W_b. The statistical test evaluates the null hypothesis $\mu_f^2 = \mu_b^2$.

DDNAE considers that a drift happens if either p_1 or p_2 lowers the significance level δ. Especially if $p_1 \leq \delta$, DDNAE reports the changing point m_1 found by DD_1 and advances the left end of W to $m_1 + 1$ in the 8th and 9th lines. Else if $p_2 \leq \delta$, the changing point m_2 discovered by DD_2 is adopted. Note that when $p_1 \leq \delta$ and $p_2 \leq \delta$, DDNAE trusts the primary detector DD_1 more than DD_2. DDNAE must update DD_2 at every detected drift. When a drift is detected at time m^*, the adaptive window shrinks to $W = \{V_{m^*+1}, V_{m^*+2}, ..., V_\alpha, V_{\alpha+1}\}$ and will store only the vectors generated by the new concept. Then, DD_2 updates the mean vector MV to the average of the $n_{\min}$ vectors belonging to the new concept. $n_{\min}$ specifies the minimum number of samples necessary to calculate the mean vector.

5   Experiments

We compare DDNAE with ABCD in terms of the drift detection accuracy and the total time to process the stream. With respect to ABCD, we utilized the python code developed by the authors of ABCD. We synthesize artificial dataset by using real-world classification datasets in the same way as the developer of ABCD [4]. Specifically, given a classification dataset, we create a multi-variate data stream by sorting the data by the class labels. In this data stream, we regard a change of class as a concept drift. For example, if a dataset consists of 10 classes,

Table 1. F1 (Precision, Recall)

Dataset	DDNAE	ABCD	NU-ABCD
HAR	0.949 (0.931, 0.969)	0.871 (0.906, 0.845)	0.941 (0.929, 0.954)
GAS	0.840 (0.854, 0.805)	0.697 (0.753, 0.659)	0.711 (0.764, 0.672)
MNIST	0.956 (0.956, 0.970)	0.869 (0.906, 0.839)	0.887 (0.905, 0.872)
FMNIST	0.970 (0.965, 0.977)	0.893 (0.919, 0.874)	0.905 (0.928, 0.886)
CIFAR10	0.655 (0.711, 0.616)	0.554 (0.669, 0.489)	0.562 (0.695, 0.485)
IMAGENET	0.799 (0.847, 0.763)	0.587 (0.750, 0.499)	0.568 (0.728, 0.477)

the data stream contains 9 concept drifts. We test the six datasets **HAR** [1], **GAS** [8], **MNIST, FMNIST, CIFAR10** and **IMAGENET** summarized in the left side of Table 2. From the IMAGENET, we select 10 classes and resize the images out of the 10 classes to $32 \times 32 \times 3$ color images. The 5 datasets except IMAGENET have also appeared in the original paper [4].

5.1   Drift Detection Accuracy

To evaluate the accuracy of detected concept drifts, we define TP (true positive), FN (false negative), and FP (false positive). The first drift detected within 500 samples after the genuine drift is counted as **TP**. Any detected drifts other than **TP** are categorized as **FP**. If no drift is detected over 500 samples after the genuine drift, **FN** increases by 1. For a drift detection algorithm, we access its precision $\frac{\#TP}{\#TP+\#FP}$, its recall $\frac{\#TP}{\#TP+\#FN}$ and its F1 score.

Table 1 compares DDNAE with ABCD in terms of the F1 score. The F1 scores in Table 1 are accompanied by the pair (Precision, Recall). To highlight the effect of DD_2 in DDNAE, we implement a method called NU-ABCD (Non-Updated ABCD) that detects concept drifts only by DD_1. Remarkably, DDNAE achieves higher recall and precision values than ABCD for every dataset and improves the F1 by 18.9% on average. Since ABCD has achieved the highest F1 scores of many state-of-the-art methods so far, this result shows that DDNAE may be the most accurate drift detector at the present time. Furthermore, DDNAE gets higher F1 scores than NU-ABCD. This result shows that DD_2 in DDNAE can detect concept drifts missed by DD_1.

5.2   Stream Processing Time

The right side of Table 2 shows the stream processing time and tells that DDNAE runs slightly slower than ABCD except for IMAGENET. This result is because DDNAE must execute the statistical tests twice as many times as ABCD. Since the time complexity of a statistical test is moderate, DDNAE spends a longer execution time than ABCD by at most 7 s. Since DDNAE achieves higher F1 scores than ABCD, it is tolerable that DDNAE increases the execution time by just a few seconds. Interestingly, for the IMAGENET dataset, DDNAE runs 2.4

Table 2. Dataset and Stream Processing Time(s)

Dataset	#classes	#features	DDNAE	ABCD	NU-ABCD
HAR	6	561	14.1	10.2	8.4
GAS	6	128	15.4	9.3	7.9
MNIST	10	784	15.9	12.3	10.8
FMNIST	10	784	15.5	12.4	10.4
CIFAR10	10	1024	16.5	15.9	12.4
IMAGENET	10	3072	27.2	66.4	29.7

times faster than ABCD, because the cost to retrain the autoencoder becomes enormous, if the data dimension is as high as 3072. Without DD_2 or the update of DD_1, NU-ABCD naturally runs the fastest.

Acknowledgments. This work was supported by JSPS KAKENHI Grant Number JP21K11901, 2025.

References

1. Anguita, D., Ghio, A., Oneto, L., Parra, X., Reyes-Ortiz, J.L., et al.: A public domain dataset for human activity recognition using smartphones. In: ESANN, vol. 3, p. 3 (2013)
2. Faber, K., Corizzo, R., Sniezynski, B., Baron, M., Japkowicz, N.: Watch: Wasserstein change point detection for high-dimensional time series data. In: 2021 IEEE International Conference on Big Data (Big Data), pp. 4450–4459. IEEE (2021)
3. Gözüaçık, O., Büyükçakır, A., Bonab, H., Can, F.: Unsupervised concept drift detection with a discriminative classifier. In: Proceedings of the 28th ACM CIKM, pp. 2365–2368 (2019)
4. Heyden, M., Fouché, E., Arzamasov, V., Fenn, T., Kalinke, F., Böhm, K.: Adaptive bernstein change detector for high-dimensional data streams. Data Mining Knowl. Discov. 1–30 (2024)
5. Menon, A.G., Gressel, G.: Concept drift detection in phishing using autoencoders. In: Thampi, S.M., Piramuthu, S., Li, K.-C., Berretti, S., Wozniak, M., Singh, D. (eds.) SoMMA 2020. CCIS, vol. 1366, pp. 208–220. Springer, Singapore (2021). https://doi.org/10.1007/978-981-16-0419-5_17
6. dos Reis, D.M., Flach, P., Matwin, S., Batista, G.: Fast unsupervised online drift detection using incremental kolmogorov-smirnov test. In: Proceedings of the 22nd ACM SIGKDD, pp. 1545–1554 (2016)
7. Souza, V.M.A., Parmezan, A.R.S., Chowdhury, F.A., Mueen, A.: Efficient unsupervised drift detector for fast and high-dimensional data streams. Knowl. Inf. Syst. **63**(6), 1497–1527 (2021). https://doi.org/10.1007/s10115-021-01564-6
8. Vergara, A., Huerta, R., Ayhan, T., Ryan, M., Vembu, S., Homer, M.: Gas sensor drift mitigation using classifier ensembles. In: Proceedings of the Fifth International Workshop on Knowledge Discovery from Sensor Data, pp. 16–24 (2011)

Graph Data Processing and Analytics

Graph Data Processing and Analytics

Parallel and Distributed SQL/PGQ Query Processing for Property Graphs

Kosuke Yamasaki[1] (ID), Tadashi Masuda[1,2] (ID), and Toshiyuki Amagasa[2(✉)] (ID)

[1] Graduate School of Science and Technology, University of Tsukuba,
Tsukuba, Japan
{kosuke.y,masuda}@kde.cs.tsukuba.ac.jp
[2] Center for Computational Sciences, University of Tsukuba,
Tsukuba, Japan
amagasa@cs.tsukuba.ac.jp

Abstract. This paper introduces an implementation for processing property graph path queries using a subset of SQL/PGQ on Apache Spark. Property graphs, which allow nodes and edges to have multiple labels and properties, are increasingly adopted across diverse domains. SQL/PGQ, an ISO-standard extension to SQL introduced in 2023, enables expressive graph pattern queries in combination with traditional relational operations. However, publicly documented approaches and performance baselines for its distributed execution on open-source platforms are still lacking. To address this, we implement a processor for this subset of SQL/PGQ on Apache Spark, with the goal of establishing a performance baseline for a direct implementation approach. We evaluate its performance by comparing query execution times against other systems. Our experimental results indicate that while the implementation shows notable overhead for queries with small result sets, presumably due to scan and communication costs, it can outperform existing systems for queries that produce large outputs.

Keywords: SQL/PGQ · Property Graph · Apache Spark · Parallel and Distributed Processing · Graph Query Processing

1 Introduction

In recent years, the increasing demand for analyzing big data and complex networks has brought heightened attention to graph data. Graph data enables the direct and flexible representation of relationships among entities through nodes and edges, making it applicable across diverse fields such as social networks, knowledge graphs, financial transaction networks, and pharmaceutical research [10, 12]. Effectively utilizing large-scale, heterogeneous datasets requires operations such as pattern matching, path queries [8], and community detection, which are challenging to perform with traditional relational databases.

To this problem, research on graph databases and graph analysis frameworks has advanced rapidly. Graph databases are designed to represent and store data

© The Author(s), under exclusive license to Springer Nature Switzerland AG 2026
C. K. Leung et al. (Eds.): DaWaK 2025, LNCS 16048, pp. 167–181, 2026.
https://doi.org/10.1007/978-3-032-02215-8_12

as nodes and edges in a natural and intuitive manner. By aligning data representation with the underlying structure, they enable complex operations—such as pattern matching and graph traversal—to be executed more intuitively and efficiently than in relational databases.

This study focuses on property graphs, one of the most widely used graph data models today, which support rich data representations. In property graphs, nodes and edges can carry labels and properties consisting of key-value pairs, facilitating the intuitive depiction of real-world entities and relationships.

So far, many systems and query languages have been used to query property graphs, such as Cypher in Neo4j [9] and GSQL in TigerGraph [5]. However, the lack of a standardized query language and the difficulty of integrating graph processing within SQL-based systems have posed challenges. To address these issues, ISO has recently standardized SQL/PGQ [7]. SQL/PGQ extends SQL with native support for property graph queries, allowing graph processing—such as pattern matching and path queries—to be performed directly within SQL.

Despite these advancements, a practical gap remains for processing SQL/PGQ within open-source, distributed data processing ecosystems. In the commercial space, major database systems like Oracle Database 23ai [4] and Google Cloud Spanner Graph [2] are integrating property graph queries into their SQL engines. While these are powerful systems, their nature as commercial products means their internal implementation details are generally not public, and their accessibility can be a consideration for academic research or open-source development. On the other front, DuckPGQ offers an open-source implementation for SQL/PGQ. However, it is built upon DuckDB, which is architected for high performance on a single node and is therefore not designed to operate on a multi-node, distributed cluster. As a result, there is a lack of publicly documented implementation approaches and corresponding performance baselines for SQL/PGQ on standard distributed frameworks.

To address this gap, this study presents an implementation of a path query processor for SQL/PGQ on Apache Spark, a de facto standard for large-scale data processing. The goal is to investigate the feasibility of this integration by directly translating PGQ constructs into Spark SQL operations, and to establish a performance baseline for this straightforward approach. This baseline provides a documented reference point for performance on a standard distributed platform, which can inform future research in this area.

To evaluate the performance of our implementation, we conducted experiments using two types of datasets. We measured query execution times to analyze the scalability of our approach and to discuss its potential performance bottlenecks. Furthermore, we compared its performance with Neo4j and a representative commercial system, hereafter referred to as "Commercial DB A", to position our results within the broader landscape of graph processing systems.

The remainder of this paper is organized as follows. Section 2 provides background information, Sect. 3 reviews related systems, Sect. 4 describes the proposed method, and Sect. 5 presents the results of our evaluation experiments. Finally, Sect. 6 concludes the paper and outlines future work.

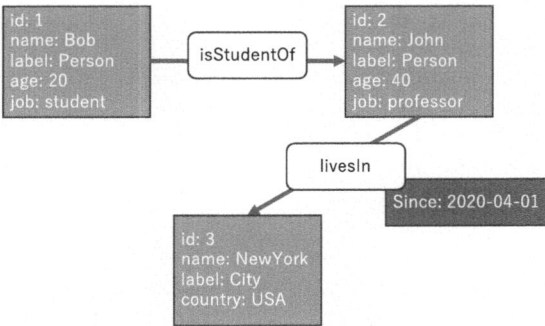

Fig. 1. An example of a property graph.

2   Preliminaries

In this section, we outline the fundamental concepts underlying this study. First, Sect. 2.1 describes the property graph, which serves as the target data model. Next, Sect. 2.2 introduces SQL/PGQ, an extension of SQL that enables operations on property graphs. Finally, Sect. 2.3 provides an overview of Apache Spark, the framework employed in this research.

2.1   Property Graph

A property graph is a data model that represents entities and their relationships using a graph structure. In a property graph, nodes represent entities or concepts of interest, while edges denote the relationships between them. Each node and edge can be assigned labels and properties. Labels specify the roles or categories of the node/edge within the graph, and properties express information associated with the node/edge in terms of a set of key-value pairs. Figure 1 illustrates an example of a property graph.

2.2   SQL/PGQ

Overview of SQL/PGQ. SQL/PGQ is an extension of SQL, the standard query language for relational databases (RDBs), that enables operations such as pattern matching and path exploration on property graphs. This extension allows a single query to seamlessly operate on both relational tables and property graphs, thereby integrating different data models in one operation. Consequently, SQL/PGQ has the potential to greatly enhance development efficiency and maintainability in large-scale data processing and complex data analysis scenarios.

Example Syntax of SQL/PGQ. Below is a basic example of an SQL/PGQ query:

```
SELECT person1, relation, person2
FROM GRAPH_TABLE(network
   MATCH (p1:Person WHERE p1.name = 'Bob')-[e]->(p2:Person)
   COLUMNS(p1.name AS person1,
           e.label AS relation,
           p2.name AS person2)
);
```

This query searches for two nodes labeled `Person`: one node (`p1`) with `name` equal to "Bob" and another node (`p2`) that is reachable from `p1` via a single hop, along with the connecting edge (`e`). It then retrieves the names of both nodes and the label of the edge.

Following conventional SQL syntax, the `GRAPH_TABLE` clause specifies the target graph, while the `MATCH` clause describes the graph pattern in an intuitive manner. Within the `GRAPH_TABLE` clause, additional keywords such as `WHERE` and `COLUMNS` facilitate data extraction and transformation from the matched graph pattern.

Moreover, SQL/PGQ allows the definition of path-length ranges in the graph pattern through the use of quantifiers. For example, the following `MATCH` clause includes the quantifier `{1,5}`:

```
MATCH (p1:Person)-[:Knows]->{1,5}(p2:Person)
```

In this example, a node `p1` labeled `Person` can reach another node `p2` (also labeled `Person`) by traversing edges labeled `Knows` for between 1 and 5 hops. There are four types of quantifiers in total: `*` (zero or more hops), `+` (one or more hops), and numeric specifications such as `{5}` (exactly 5 hops). Notably, `*` can be rewritten as `{0,}`, `+` as `{1,}`, and `{5}` as `{5,5}`. These expressions, which are components of Regular Path Queries (RPQs) [8,13], render the queries both concise and flexible. In this paper, we refer to queries with a fixed path length as "fixed-length queries" and those specifying a range of path lengths as "variable-length queries."

2.3   Distributed Processing Framework: Apache Spark

Overview of Spark. Apache Spark [1] is a distributed processing framework for large-scale data processing. Spark deploys one or more executors on each node in a cluster and parallelizes tasks by partitioning the dataset. Each executor can concurrently execute multiple tasks within its allocated memory and CPU resources. In this study, we leverage Spark SQL and its DataFrame API, which allow for flexible and high-performance parallel processing.

Spark SQL. Spark SQL is one of the interfaces provided by Spark and is utilized in this research for query processing. It supports various file formats (e.g., CSV, JSON, Parquet) and storage systems (e.g., HDFS, S3), enabling the integration and analysis of heterogeneous data sources via SQL. The Spark

SQL ecosystem also encompasses the DataFrame API, which allows developers to either write SQL queries directly or construct equivalent logic using method chains on DataFrames. Regardless of the approach, Spark internally constructs a logical execution plan, which is then optimized by an engine called the Catalyst optimizer before the tasks are executed.

DataFrame. A DataFrame is a high-level interface for working with structured data, offering a concise and flexible means of manipulating tabular data. Its structure consists of rows and columns, allowing operations similar to those on SQL tables while also enabling developers to programmatically perform operations such as joins, filtering, and aggregation in an intuitive manner. Furthermore, a single column in a DataFrame can store data in a struct format, allowing multiple pieces of related information to be grouped together while preserving nested or hierarchical data structures. In this study, we leverage this feature by storing each node or edge's information in a struct format within a single column.

3 Related Work

In this section, we present two systems as related work. Section 3.1 discusses GraphFrames, a Spark-based graph processing library, while Sect. 3.2 describes Neo4j, one of the most widely used graph database systems.

3.1 GraphFrames

GraphFrames [6] is a graph processing library built on top of Spark's DataFrame API. It is highly optimized to leverage the scalability and other advantages of DataFrames. GraphFrames supports a variety of graph algorithms and, in particular, facilitates motif discovery by allowing users to provide simple triple patterns—such as `(s)-[e]->(t)`—to the `find` function. However, this function focuses solely on enumerating fixed-length, simple triple patterns. It does not support advanced filtering conditions on labels or properties, nor does it accommodate variable-length path expressions, resulting in a notable lack of expressiveness and flexibility compared with the functionalities demanded by modern graph queries. Furthermore, the project's development focus appears to have shifted from new feature implementation to maintenance. An analysis of the official GitHub repository's commit history shows that activity in 2023 and 2024 was limited (10 and 6 commits, respectively), with changes primarily addressing compatibility with new Spark versions and minor bug fixes. While commit activity increased in the first half of 2025, it has continued to center on maintenance tasks rather than the introduction of new functionalities or significant enhancements to its query capabilities.

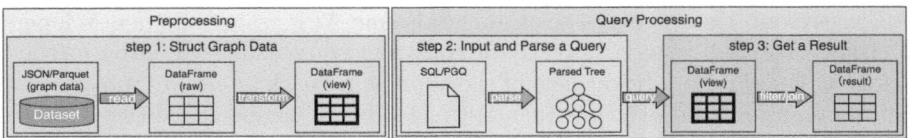

Fig. 2. Overall flow of the process.

3.2   Neo4j

Neo4j [3] is one of the most widely adopted systems in the commercial graph database market. It provides native graph storage and advanced graph manipulation capabilities through the Cypher query language [9]. On the other hand, Neo4j was originally designed to excel in single-cluster ACID-compliant environments and is therefore considered less effective in supporting smooth operations in large-scale distributed settings. Although Neo4j has recently introduced the Fabric feature to enable unified management of large-scale data, its deployment still requires careful operational and design considerations. Furthermore, features such as clustering, enhanced security, and high availability are available only in the commercial edition, posing cost-related challenges.

4   Proposed Method

In this section, we describe our proposed method. This study aims to execute queries on property graphs using SQL/PGQ on Apache Spark. Our approach consists of three main stages: a preprocessing stage in which the graph data is loaded and views are created; a parsing stage during which the received query is parsed; and a query processing stage in which the parsed query is executed. In this section, we first provide an overview of the entire process in Sect. 4.1 and then describe each processing stage in detail. Section 4.2 explains the schema of the input data and the table structure after loading, followed by Sect. 4.3, which details the targeted SQL/PGQ syntax and explains how the parsed query is used to manage the data. Finally, Sect. 4.4 describes the specifics of query execution.

4.1   Flow of the Proposed Method

Figure 2 illustrates the overall flow of our proposed method. The process begins with loading the dataset as a preprocessing step (step 1). After being read as a DataFrame, the dataset undergoes a column restructuring to match the expected format for query processing. The resulting DataFrame is then registered as a view. Once the view registration is complete, query processing can commence. When a query is submitted, it is first parsed, and the extracted data along with the necessary information about the graph pattern are stored as objects (step 2). After constructing the query object, join and filter operations are executed on the view prepared in step 1 (step 3). Through this sequence, the query result is returned as a DataFrame.

4.2   Graph Data Structure

Table 1. Schema of the available input datasets

Target	Attribute	Type
Node	id	Int | Long | String
	label	Array[String]
	property	Map[String, Any]
Edge	id	Int | Long | String
	label	Array[String]
	src	Any
	dst	Any
	property	Map[String, Any]

Table 1 shows the schema assumed for the input data. Both node and edge IDs can be of type Int, Long, or String; however, the type of a node's ID must match that of the src and dst fields in edges. Moreover, property values are restricted to scalar types.

After reading the node and edge data as DataFrames, all schemas are consolidated into a single structured column, converting them into two DataFrames named `nestVertices` and `nestEdges`. Next, using the src and dst fields from `nestEdges` and the id field from `nestVertices`, we construct three additional DataFrames: `nestSrcEdge` (joining source nodes and edges), `nestEdgeDst` (joining edges and destination nodes), and `nestTriplets` (joining source nodes, edges, and destination nodes). These five DataFrames—`nestVertices`, `nestEdges`, `nestSrcEdge`, `nestEdgeDst`, and `nestTriplets`—are all registered as views and cached. Furthermore, users can optionally specify the creation of 2-hop and 3-hop views. These views enumerate all paths of 2 or 3 hops starting from each node and are constructed by joining `nestTriplets` with `nestEdgeDst`. While this approach greatly reduces the number of joins required during query processing, enumerating all paths may result in massive data sizes, potentially increasing memory usage and degrading performance.

4.3   SQL/PGQ Parsing Process

To process basic path queries in SQL/PGQ, we implemented a parser that supports the targeted syntax. We utilized the recursive descent parser combinator library provided in Scala 2 [14].

Except for the `GRAPH_TABLE` clause, the syntax is identical to standard SQL. Therefore, the parser focuses on the SQL/PGQ-specific `GRAPH_TABLE` clause, first splitting the query into three parts: the `GRAPH_TABLE` clause itself and the segments before and after it. The parser then processes the `GRAPH_TABLE` clause and registers the resulting table as a view, enabling the remainder of the SQL query to be directly processed by Spark SQL.

Syntax of the MATCH Clause. The MATCH clause supports the specification of either individual nodes or path patterns that alternate between nodes and edges. Both nodes and edges are defined by three components: a "variable", a "label", and a "condition". In the example query provided in Sect. 2.2, identifiers such as p1 and e serve as the "variable", :Person denotes the label, and WHERE p1.name ='Bob' represents the "condition". Although conditions can generally include criteria equivalent to those specified in a standard SQL WHERE clause, in this study we support only a representative subset (including comparison, logical, and set operations). Furthermore, while all three components are optional in principle, a "variable" must be specified when a "condition" is used.

Edges have their direction indicated using an ASCII-art syntax. SQL/PGQ supports directed, bidirectional, and undirected edges, as well as specifications such as "undirected or leftward edges". A shorthand notation also exists for edges when none of the three components is specified. In this study, we consider only directed edges and the abbreviated notation.

Syntax of the WHERE Clause. Within the GRAPH_TABLE clause, the WHERE clause permits condition specifications identical to those of standard SQL. Since these conditions are shared with the "condition" components in the graph pattern, the parser employs a common processing routine for both, supporting the same set of operators. Note that the WHERE clause is optional.

Syntax of the COLUMNS Clause. The COLUMNS clause enables the selection of required data from the intermediate result obtained from the graph pattern, and allows the definition of alias names for formatting the data. In this study, we support only the format "variable.property AS alias", where "variable" refers to an identifier that appears in the MATCH clause and "property" must be a key contained in the properties of the corresponding node or edge. Multiple specifications can be provided, separated by commas.

Internal Data Structures of the Parsing Result. The data parsed from the GRAPH_TABLE clause is managed using custom classes corresponding to each keyword. The graph pattern specified in the MATCH clause is stored in a GraphPattern class as an array consisting of triples (which may include quantifiers) and standalone nodes. Each triple is managed by an Edge class, which comprises the three components as well as the source node, destination node, and any quantifier. Nodes are managed by a Vertex class, which holds the three components; this class is used both for standalone node specifications and for representing source and destination nodes within an Edge. In these classes, the "label" and "condition" fields are used as filter criteria during query processing. Since Spark's filter function accepts arguments of type Column, these "label" and "condition" components are stored in custom classes that provide methods to convert them into Column types.

The data for the `WHERE` clause is parsed using the same process as that for the graph pattern's "condition" (as described in Sect. 4.3) and is managed similarly using a custom class that converts the data into a `Column` type.

Since the `COLUMNS` clause consists of triples—"variable name", "property name", and "alias name"—its data is stored as an array of these triples. During query processing, the DataFrame's `select` function is used (which accepts arguments of type `Column`), so these triples are also managed using a custom class that provides a method for converting them into `Column` types.

4.4 SQL/PGQ Query Processing

Query processing is carried out by repeatedly applying filter and join operations on the five views prepared in Sect. 4.2. As described in the internal data structure section, the graph pattern is stored as an array of triples and standalone nodes. During processing, we sequentially extract each element from this array and apply the "label" and "condition" of the nodes and edges it contains as filter criteria to the corresponding view. The resulting DataFrame is then updated by renaming columns based on the "variable" and joining it with the previous result DataFrame (`resultDF`). Note that if multiple elements in the graph pattern share the same "variable", the same condition is applied; therefore, before processing each element, we check whether the node with that "variable" has already been processed and skip it if so.

The selection of the target view is performed as follows:

– For an unprocessed standalone node, `nestVertices` is selected. After renaming and filtering, the result is integrated with `resultDF` via a cross join.
– For an edge where both the source and destination have already been processed, `nestEdges` is selected. After renaming and filtering, the result is integrated with `resultDF` by performing an inner join on the edge's source and destination.
– For an edge where the source has already been processed, `nestEdgeDst` is selected. After renaming and filtering, the result is integrated with `resultDF` by performing an inner join on the edge's source.
– For an edge where the destination has already been processed, `nestSrcEdge` is selected. After renaming and filtering, the result is integrated with `resultDF` by performing an inner join on the edge's destination.
– For an edge where neither the source nor the destination has been processed, `nestTriplets` is selected. After renaming and filtering, the result is integrated with `resultDF` via a cross join.

If a triple contains a quantifier, join operations are repeated until the maximum path length specified by the quantifier is reached. During this process, the edges between the source and destination of each path are aggregated into an array and consolidated into a single column, which is then renamed to the "variable" of the edge, thereby enabling the triple (source, edge, destination) to be obtained as a DataFrame. The operations corresponding to the edge-based conditions described earlier are then applied to this DataFrame, and the result is integrated with `resultDF`. For triples with variable-length quantifiers, all intermediate results obtained during the repeated joins are recorded to enable processing.

When 2-hop and 3-hop views are available from preprocessing, the number of join operations can be significantly reduced. If the path length l is a multiple of 3, the use of the 3-hop view reduces the number of joins to $l/3 - 1$. Otherwise, after performing $\lfloor l/3 \rfloor - 1$ joins using the 3-hop view, the remaining path length is processed by joining with `nestEdgeDst` or a 2-hop view to obtain the final result. This method is applicable only to fixed-length queries and cannot be used when all 1-hop paths within a specified range are required, as in variable-length queries.

5   Experiments

In this section, we describe the evaluation experiments conducted to assess the performance of our proposed method. Our experiments compare two comparison systems using both a synthetic dataset and a real-world dataset to examine query execution speeds. In addition, we evaluate scale-out performance by varying the number of Spark worker nodes.

5.1   Experimental Environment

Experiments were conducted on AWS EMR (emr-7.5.0). Both the primary instance and the core instances were r5.4xlarge (16 vCPUs, 128 GiB memory), with an additional st1 500 GiB EBS volume attached.

Our proposed method was implemented in Scala (2.12) and executed using Apache Spark (3.5.2). The primary instance served as the master node and the core instances as worker nodes. To evaluate scale-out performance, the number of core instances was increased from one to four, one at a time. Additionally, each node was allocated three executors, each with 5 cores and 34 GiB of memory.

As comparison systems, we selected Neo4j (4.4.0 Community Edition) and Commercial DB A, a commercial system that supports the SQL/PGQ features under comparison. Neo4j was executed on the primary instance. For Commercial DB A, we utilized its publicly available version, which operated within a predefined execution environment subject to resource constraints that we could not alter. Consequently, the results from Commercial DB A are presented as a functional reference point rather than a direct performance benchmark under equivalent conditions to our Spark cluster.

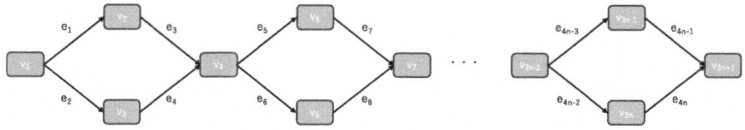

Fig. 3. Example of a Diamond graph.

5.2 Datasets

We employed two datasets that were used in the study by Benjamín et al. [8].

Diamond. Diamond is a synthetic dataset originally used in the study by Martens et al. [13]. As shown in Fig. 3, it has a diamond-like structure such that the number of paths doubles with every increase of 2 in the path length. In our experiments, we used a graph with 3,001 nodes and 4,000 edges. The queries applied include both fixed-length queries (using quantifiers) and variable-length queries, where the starting point is specified and the maximum path length is increased in steps of 2 (i.e., 2, 4, 6, ...). These queries are designed to evaluate processing performance under rapidly increasing result sizes.

Pokec. Pokec is a social network from Slovakia provided by SNAP [11]. It represents user follow relationships using directed edges, with the edge label being `"follows"` exclusively. The graph comprises approximately 1.6 million nodes and 30 million edges, and its diameter is 11. For this dataset, we used fixed-length queries and variable-length queries similar to those for Diamond, with the path length incremented by 1 (starting at 1). The starting node was chosen following the approach of Benjamín et al., selecting a node whose centrality is near the median. These queries evaluate the processing performance on real-world data.

5.3 Evaluation Metrics

Performance is evaluated by comparing the maximum processable path length and the corresponding query execution times for each dataset. The reported execution time is the average of four runs (excluding the first run) out of five executions for each query. A timeout of 180 s was set; if any execution exceeded 180 s, the query was deemed unprocessable, and no further queries with longer path lengths were executed.

5.4 Experimental Results

In addition to the fixed-length query processing experiments presented below, we also conducted variable-length query experiments for both datasets. However, since the trends observed in the variable-length queries were consistent with those obtained in the fixed-length queries, only the representative fixed-length query results are reported.

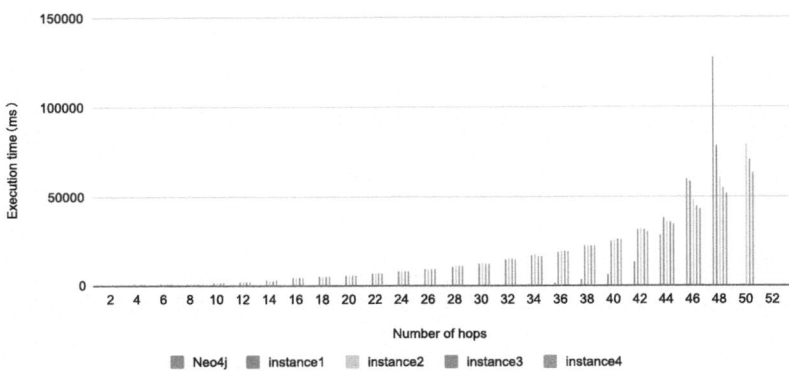

Fig. 4. Fixed-length query results using 1-hop joins on the Diamond.

Diamond. Due to its relatively small size, it was possible to construct 3-hop views for the Diamond dataset. Accordingly, for fixed-length queries, we evaluated both the performance of 1-hop join operations and that of using the 3-hop view. In contrast, the Free version of Commercial DB A could not handle queries with a path length greater than 10; hence, Diamond was compared only against Neo4j.

Fixed-Length Query Processing via 1-Hop Joins. Figure 4 shows the fixed-length query processing results using 1-hop join operations for the Diamond dataset alongside the execution times for Neo4j. For queries with a path length up to 20, Neo4j completed execution in under 10 ms. In contrast, our proposed method exceeded 1,000 ms once the path length reached 10, with little difference across various instance configurations. This is likely because the result set is small at shorter path lengths, making inter-instance communication a bottleneck. However, as the path length increases beyond 20, Neo4j's execution time increases exponentially, timing out at a path length of 50. In contrast, our method demonstrates stable scalability for these longer paths, becoming more efficient than Neo4j beyond a path length of 46. The benefits of scale-out are also clearly demonstrated, as performance differences across instance counts become more pronounced.

Fixed-Length Query Processing via 3-Hop View Joins. Figure 5 presents the results obtained when using the 3-hop view. With the 3-hop view, execution times remained below 1,000 ms for queries with a path length up to 42. While a clear performance difference exists between a single instance and configurations with multiple instances, differences among configurations with 2 to 4 instances are less pronounced. Notably, for a query with a path length of 68, the number of results exceeded 1.7 billion. This suggests that the resource constraints of a single instance cause a performance bottleneck, whereas in configurations with two or more instances, the explosive growth in data volume outweighs the bene-

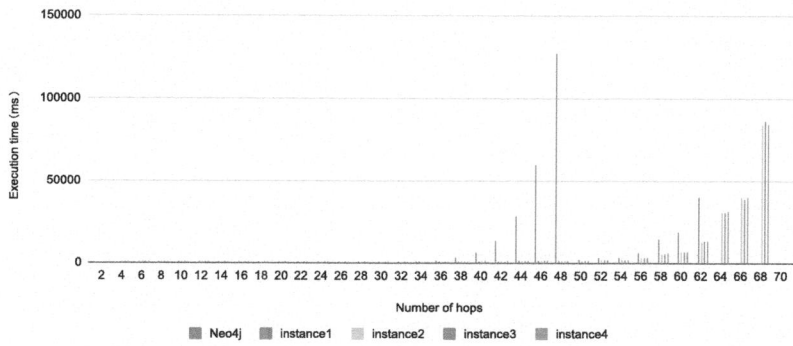

Fig. 5. Fixed-length query results using 3-hop view joins on the Diamond.

fits of increased memory and parallelism as network communication and shuffle I/O become significant overheads.

Pokec. For the Pokec dataset, the size of the 3-hop view exceeded 100 billion entries, rendering it impractical. Therefore, we compared fixed-length query processing using 1-hop join operations across our proposed method, Neo4j, and Commercial DB A. In addition, to assess the impact of differences in the underlying graph storage structures, we also compared performance for full enumeration of simple triplets.

Fixed-Length Query Processing. Figure 6 shows the fixed-length query processing results for each system on Pokec. For all systems, queries with a path length of 6 or greater did not complete within 180 s. Table 2 indicates that the number of results increases dramatically with path length; it is estimated that for a path length of 6 the result count exceeds several hundred million. Moreover, given that the original graph contains 30 million edges, the massive data volume likely contributed to the inability of some systems to process the queries. The scale-free nature of the Pokec network may also have caused skew during join operations. However, increasing the number of instances significantly improved performance. For example, when the path length was 5, the execution time nearly halved when moving from one to two instances, confirming the advantages of parallel and distributed processing.

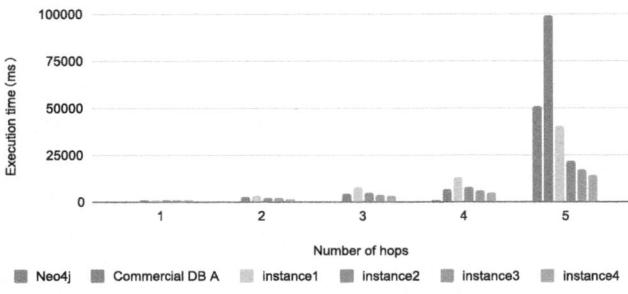

Table 2. Fixed-length Query Result Counts

Hops	Count
1	11
2	448
3	23,104
4	1,290,530
5	77,035,089

Fig. 6. Fixed-length query processing results on the Pokec.

6   Conclusions

In this study, we addressed the lack of a documented performance baseline for distributed SQL/PGQ processing on open-source platforms. By implementing and evaluating a direct, non-optimized approach on Apache Spark, this study establishes a crucial performance baseline for such processing. Our implementation processes both fixed- and variable-length queries by translating them into iterative join operations, and also explores performance gains from pre-calculated 3-hop views.

Our evaluation, comparing this baseline against established systems like Neo4j and Commercial DB A, characterized its performance profile. The results highlighted a key architectural trade-off: while our approach shows higher overhead for small queries, its design demonstrates effective scalability for large result sets. Furthermore, our experiments showed that pre-computation via 3-hop views can be a highly effective optimization strategy. We also validated the clear benefits of Spark's scale-out capability, confirming the advantage of using a distributed framework for large-scale tasks. This work provides a foundational reference point and critical insights for future research into optimized, large-scale SQL/PGQ query processing systems.

Future work will focus on three directions. First, we plan to further tune our method. This will involve first verifying suspected bottlenecks like scan and communication costs through detailed profiling, and then using those insights to guide the tuning of Spark's numerous parameters, such as executor resources and data partitions, based on dataset characteristics. Second, we will examine data layout optimizations. Our current approach stores edge, node, and triplet data in single, large DataFrames, and query processing relies on repeated join operations. This can lead to redundant scans and increased communication costs during shuffling, thereby creating bottlenecks. Therefore, we aim to explore techniques such as appropriate index construction and data partitioning to reduce the costs associated with data access, scanning, and shuffling. Third, we intend to expand the functionality of our system. The current system supports only the basic syntax for fixed-length and variable-length path queries. In the future, we plan to incorporate additional query constructs—such as path modes and shortest path processing—as well as to extend the range of conditions that can be specified in the WHERE clause and provide more options for specifying the output graph table format.

Acknowledgements. This paper is based on results obtained from the project, "Research and Development Project of the Enhanced infrastructures for Post-5G Information and Communication Systems" (JPNP20017), commissioned by the New Energy and Industrial Technology Development Organization (NEDO), JST CREST Grant Number JPMJCR22M2, and JSPS KAKENHI Grant Number JP23K24949.

References

1. Apache Spark. http://spark.apache.org. Accessed 25 Dec 2024
2. Google Cloud Spanner Graph. Accessed 7 June 2025
3. Neo4j. https://neo4j.com/. Accessed 25 Dec 2024
4. Oracle Database Free. https://www.oracle.com/jp/database/free/. Accessed 25 Dec 2024
5. TigerGraph. https://www.tigergraph.com/. Accessed 25 Dec 2024
6. Dave, A., Jindal, A., Li, L.E., Xin, R., Gonzalez, J., Zaharia, M.: GraphFrames: an integrated API for mixing graph and relational queries. In: Proceedings of the Fourth International Workshop on Graph Data Management Experiences and Systems, pp. 1–8 (2016)
7. Deutsch, A., et al.: Graph pattern matching in GQL and SQL/PGQ. In: Proceedings of the 2022 International Conference on Management of Data, pp. 2246–2258 (2022)
8. Farías, B., Martens, W., Rojas, C., Vrgoč, D.: PathFinder: a unified approach for handling paths in graph query languages (2024). https://arxiv.org/abs/2306.02194
9. Francis, N., et al.: Cypher: an evolving query language for property graphs. In: Proceedings of the 2018 International Conference on Management of Data, SIGMOD 2018. Association for Computing Machinery (2018). https://doi.org/10.1145/3183713.3190657
10. Hogan, A., et al.: Knowledge graphs. ACM Comput. Surv. (CSUR) **54**(4), 1–37 (2021)
11. Leskovec, J., Krevl, A.: SNAP Datasets: Stanford Large Network Dataset Collection (2014). http://snap.stanford.edu/data
12. Li, X., et al.: FlowScope: spotting money laundering based on graphs. In: Proceedings of the AAAI Conference on Artificial Intelligence, vol. 34, pp. 4731–4738 (2020)
13. Martens, W., Niewerth, M., Popp, T., Vansummeren, S., Vrgoc, D.: Representing paths in graph database pattern matching. arXiv preprint arXiv:2207.13541 (2022)
14. Moors, A., Piessens, F., Odersky, M.: Parser combinators in Scala. CW Rep. **54** (2008)

Graph Constraint Language for Industrial Knowledge Graphs and Machine Learning

Zhuoxun Zheng[1,2](✉)(iD), Ognjen Savković[3](iD), Baifan Zhou[2,4](iD),
Antonis Klironomos[1,6](iD), Evgeny Kharlamov[1](iD), and Ahmet Soylu[2,5](iD)

[1] Bosch Center for AI, Renningen, Germany
zhuoxun.zheng@de.bosch.com
[2] University of Oslo, Oslo, Norway
{baifanz,ahmets}@ifi.uio.no
[3] Free University of Bozen-Bolzano, Bozen-Bolzano, Italy
[4] OsloMet – Oslo Metropolitan University, Oslo, Norway
[5] Kristiania University of Applied Sciences, Oslo, Norway
[6] University of Mannheim, Mannheim, Germany

Abstract. Knowledge graphs (KGs) are seeing widespread application across diverse domains, including querying, search, machine learning (ML), and stream data processing. In all these areas, ensuring data consistency and quality is of critical importance. For instance, erroneous attributes in industrial KGs, such as those describing machinery, can lead to unexpected machine downtime or unreliable analytical predictions. Several frameworks for specifying integrity constraints have been proposed, such as SHACL and ShEX for RDF, and PG-schema for property graphs. However, the challenge of defining comprehensive constraints remains only partially addressed, particularly with respect to combining open- and closed-world assumptions and handling numerical constraints. To this end, we propose a graph constraint language, GraphCo, composed of three key components: first, it supports integrity constraints under both open and closed reasoning when the KG is defined over an underlying ontology; second, it offers a set of constraints that are both expressive and tractable, addressing the requirements of diverse industrial applications, for example, in verifying the structural properties of industrial machines or ML data pipelines; third, it supports data-intensive validation tasks involving numerical transformations such as aggregation and comparison. We implemented GraphCo using Answer Set Programming and Apache Spark to evaluate its scalability, and the results are promising.

Keywords: Constraint languages · Machine learning · Knowledge graphs · SHACL

1 Introduction

Knowledge graphs (KGs) represent information as graph structures, where nodes correspond to real-world entities, such as objects, events, situations, or concepts,

© The Author(s), under exclusive license to Springer Nature Switzerland AG 2026
C. K. Leung et al. (Eds.): DaWaK 2025, LNCS 16048, pp. 182–199, 2026.
https://doi.org/10.1007/978-3-032-02215-8_13

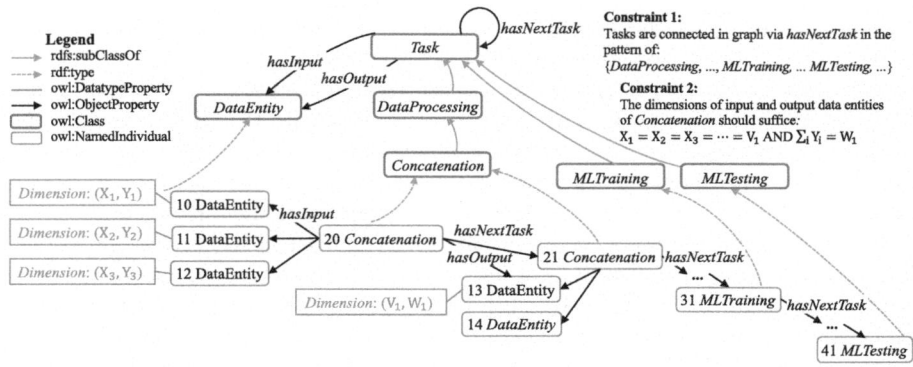

Fig. 1. KG and constraint examples of an ML pipeline. Each node has one identifier, e.g., 10, 11, and has one or more classes, e.g., *Concatenation*, *DataProcessing*. The nodes are affiliated with datatype properties, e.g., *Dimension: X_1, Y_1*.

and edges capture the relationships between them [13]. KGs gained an increasing attention and are reaching widespread applications in industry [11,20,21], from traditional problems such as querying and search [27] to new domains such as machine learning (ML) [19,29] and stream data processing [24], due to the ease of data integration and flexibility of data representation, as well as semantic transparency to human users. Ontologies, in this context, provide machine-processable descriptions of the domains represented by KGs, serving as both a structural framework and a schema. They guide the organization of graph topology as well as the interpretation of data values. Moreover, ontologies enable reasoning over KGs, allowing implicit information to be inferred. In practical applications, there is a high importance in ensuring consistency and quality of KGs. In the database community this is typically addressed with the help of integrity constraints (ICs) and following this trend a number of IC languages have been recently proposed for KGs. Indeed, some of the most prominent approaches include SHACL [8] and ShEX [25] for RDF, and PG-Schema for property graphs [3].

In the following, we discuss several challenges associated with integrity constraints for knowledge graphs, which we have encountered in industrial scenarios and which are common across other data-intensive, practical applications. These challenges serve as the primary motivation for the IC language we propose in this paper. Consider an example of ontology, KG and constraints in Fig. 1. The ontology (in blue) encodes (a fragment of) the ML domain knowledge applied in the industry, and the KG (in dark and light grey) is an instance of the ontology and encodes one concrete ML pipeline. Note that it is a recent trend to represent ML pipelines in KGs [5,17,19,29] so that both the data and ML processing of it can be represented in a uniform and consistent way, thus ensuring standardisation and accessibility of both data and analytical solutions. The KG encodes an ML pipeline, where the KG comes with an ontology, e.g., each *Concatenation* is a *DataProcessing*, which is again subclass of *Task*. On top of the ontology, we need to check some constraints. For instance, the *Constraint 1* states that the tasks are connected via *hasNextTask* in graphs in the pattern of {*DataProcessing*, ..., *MLTraining*, ..., *MLTesting*}, where "..." means repetitions of the previous task. The *Constraint 2* is about the dimensions of the input and output data entities of the *Concatenation*, which stipulates that the first elements of dimensions of input data entities should always be equal, and the second elements should sum to the second element of the output data entity.

In this example we see a common pattern of industrial KGs with ontologies. Note that by the example we do not lose generality, as similar structures exist in a wide range of industrial KGs [6,7,12]. KGs typically comprise two components: one is small in size but structurally complex, representing the topology of equipment or processes (as illustrated by the dark grey part in our example); the other is large in volume but structurally simpler, containing extensive numerical data such as sensor readings or the results of data processing (depicted in light grey). Consequently, constraint languages for such scenarios should satisfy three key requirements: (1) they must bridge both types of KGs-handling large volumes of numerical data on one hand, and accounting for complex topologies on the other; (2) they should support reasoning over both ontologies and data, thereby accommodating both open- and closed-world semantics (OWA and CWA); and (3) they should strike a balance between expressiveness and computational complexity, enabling both practical applicability and efficient validation.

Contributions. To address the above mentioned issues we propose a new language for KG constraints, named GRAPHCO. Note that our intention is not a new comprehensive IC language for graphs but an attempt for addressing several issues of constraints that frequently found industrial KG with a theoretically solid IC language. In particular, our contributions are as follows:

- We propose a novel IC language GRAPHCO for KGs and ontologies that supports open and closed reasoning when KGs are defined over underlying ontologies. Moreover, GRAPHCO provides 2 sets of constraints for 2 kinds of verification.
  1) Constraints with comparative expressivity to SHACL, Shex and PG-schema where we keep complexity polynomial in combined and sub-polynomial in data.
  2) Constraints that are intended for handling operations on numerals, such as aggregation and comparisons, for which execution has to allow a high-level of parallelisation and efficiency.
- We provide a negative result of intractable reasoning using standard OWL 2 semantics, and positive results with GRAPHCO. In particular, the validation in our language runs in polynomial time for combined complexity and for data in some cases in even sub PTIME.
- As implementation, we provide techniques for our language based on Answer Set Programming (ASP) and Apache Spark, where ASP is used to deal with ontological reasoning and part on the topological check, and SPARK is used to calculate the operations on numerals in a highly parallelisable fashion.

– As proof, we provide experiments on our implementation. We show that checking structural properties behaves linear. We also show significant run-time improvement using PySpark verse Python implementation on the numerical computation.

The remainder of the paper is structured as follows. Section 2 outlines the motivation and requirements driving our work. Section 3 introduces the GRAPHCO constraint language, beginning with relevant preliminaries and followed by a detailed account of its syntax and semantics. Section 4 presents our implementation approach and the results of the evaluation. Related work is discussed in Sect. 5, and Sect. 6 finally concludes the paper.

2   Motivation and Requirements

Industrial KG. Industrial data is normally organised in a graph-like repetitive structures. Examples of such data include sensor measurements, operation/transaction logs, tables in ML or data pipelines, that are commonly stored in relational databases; on top of which, these data are connected via a set of relations, such as the spatial/temporal/meronym relation between the sensors and logs, the ML or data pipelines, that form a graph-like structure. These industrial data can be represented as industrial KGs where some nodes in the KGs are linked to relational tables.

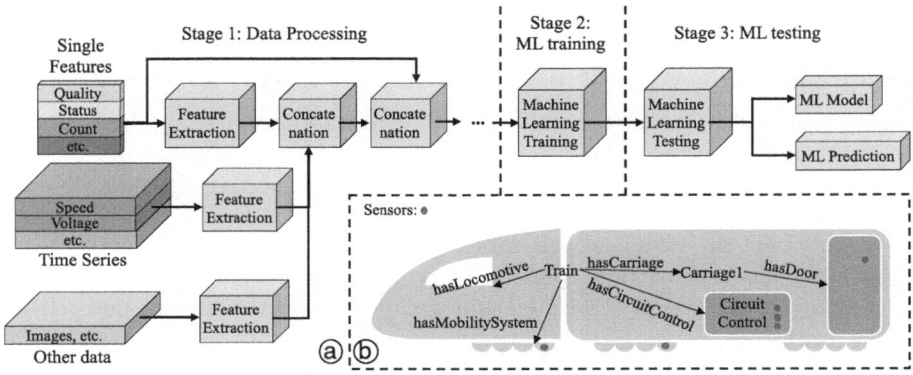

Fig. 2. Two common types of industrial data about (a) ML data pipeline with three stages; (b) sensors installed on a train. Both data can be represented with a common structure: a graph-like structure (ML pipeline in a, network of sensors in b) that organises data in repetitive structures (ML input data and predictions in a, sensor measurements in b).

We illustrate this commonality of industrial KG with two real-world examples. Figure 2a depicts a typical ML pipeline that processes single features, time series, and other data into ML models and predictions through three stages:

data processing, training, and testing. These stages consist of functions that transform inputs into outputs based on specified arguments. The ML pipeline forms a graph-like structure linking input, output, and potentially intermediate data (not shown), often represented as relational tables due to the repetitive nature of industrial data processing. As in [29], such pipelines can be modelled as industrial KGs, enhancing transparency, standardization, and accessibility in ML practices. Figure 2b shows another example of sensor data on a train (it can also be a machine, a factory, etc.), which has many sensor measurements connected by spatial or temporal relations. These measurements are also in repetitive structures as the sensors steadily collect measurements in pre-defined formats (e.g., time as the primary key and measurements as attributes), and the connecting relations are in a graph-like structure. These data can also be expressed as an industrial KG where some nodes denoting sensors are linked to relational tables.

Constraints on Industrial KG. In the both examples, some constraints are of interest to validate the data integrity. The constraints are limited on scope or subset of data which is relevant to the constraints, and the limit on syntax and semantics which stipulate certain things are legal and prohibit otherwise. For example, in the ML pipeline case, the input data need to satisfy some constraints on dimensions and data types, and the computing tasks need to satisfy some constraints on arguments, otherwise the ML pipeline will not execute, and instead return bugs or erroneous predictions that are not reliable at all. In the train sensor measurement example, the sensor data need to satisfy a set of constraints based on engineering knowledge expressed in mathematical formulas, otherwise the train can malfunction and consequences arise in the e.g., power system, security system, that result in shutdown of the train or even safety issues.

Requirements. Based on the motivating examples, the language is required as:

- *descriptive*: the language enables describing data in graph and relational formats, allowing users to limit the scope of constraint validation. This descriptive power is essential for users and applications to access and validate constraints.
- *prescriptive*: the language prescribes the limits to be enforced on the data, such as quantification of participation roles and cardinality constraints, thereby ensuring that structural and semantic expectations are met within the graph.
- *efficient in handling numerals*: the language should be able to express limits on numerical values both directly and following numerical operations, including aggregation and comparison. This, in turn, requires the language to be easily parallelizable and optimizable, as it must support KGs involving intensive numerical computations.

3   Graph Constraint Language

3.1   Preliminaries

We assume a signature Σ that consists of three infinite countable sets of *constants*, that correspond to entities or values; *classes* of unary predicates, that correspond to types; and *properties* or binary predicates that connects constants and a special predicate "a" that essentially labels entities with classes. We also consider an infinite countable *domain* Δ of entities and number values. For simplicity of the presentation we do not distinguish between object properties and data properties.

OWL 2: We briefly recall the syntax and semantics of OWL 2 QL relying on the Description Logics *DL-Lite$_R$* [4] that is behind this profile. (Complex) classes and properties in OWL 2 QL are recursively defined as follows:

$$B:: = A \mid \exists R \mid \exists R.A, \ C:: = B \mid \neg B, \ R:: = P \mid P^-, \text{ and } E:: = R \mid \neg R,$$

where $A \in \Sigma$ is called an *atomic* class or concept, $P \in \Sigma$ is a property, and P^- is the inverse of P. B is called a *basic* class or concept. A *DL-Lite$_R$* ontology is a finite set of axioms of the form $B \sqsubseteq C$ or $R \sqsubseteq E$.

Knowledge Graph (KG) is a pair $\mathcal{K} = \langle \mathcal{O}, \mathcal{G} \rangle$ of an ontology $\mathcal{O}$ (Tbox) and a graph $\mathcal{G}$ (Abox). The formal semantics of *DL-Lite$_R$* is given in terms of standard first-order logic interpretations $\mathcal{I} = (\Delta, \cdot^{\mathcal{I}})$ over Δ in the standard way. We use $adom(\mathcal{K})$ to denote the *active domain* of $\mathcal{K}$, i.e., all constants that appear in $\mathcal{K}$.

Example 1. Consider some axioms in the OWL 2 QL from our initial example

> Concatination $\sqsubseteq$ DataProcessing, DataProcessing $\sqsubseteq$ Task,
>
> $\exists$hasInput $\sqsubseteq$ Task, $\exists$hasOutput $\sqsubseteq$ Task, Task $\sqsubseteq$ $\exists$hasNextTask.Task

that describes Concatenation is subclass of DataProcessing and DataProcessing is subclass of Task. Further it says that properties hasInput and hasOutput have Task as the domain, and that for each task there is a next task. $\square$

A useful property of *DL-Lite$_R$* is the existence of a so-called *canonical model* that exist for any satisfiable KG $\langle \mathcal{O}, \mathcal{G} \rangle$ that can be obtained by applying a *chase* procedure. Canonical model has a property that it can be homomorphically mappe to any other model of $\mathcal{K}$, and because of that, all answers of a conjunctive query (CQ) over the canonical model also answers of an arbitrary model. We observe that this property holds for CQs, but does not for more general queries, e.g., queries with negation or aggregation. In this work, we consider *restricted chase* model, denoted as $\mathcal{I}^*$, that produces a minimal extension according to ontological axioms, and introduces so-called "labelled-nulls" only when it is necessary [9]. We also note that, the chase procedure can in general create infinite models. In this paper, we consider the fragment of *DL-Lite$_R$*, called *DL-Lite$_R^-$* that does not allow cycling dependencies between concepts that include $A \sqsubseteq \exists R$ in the chain. It is not hard to show that the restricted chase for *DL-Lite$_R^-$* is guaranteed to be finite. We also recall that OWL 2 EL, based on DL $\mathcal{EL}$, allows for the concepts in the form $C:: = A \mid \top \mid C_1 \sqcap C_2 \mid \exists R.C$ and axioms of the form $C_1 \sqsubseteq C_2$ [14]. The same properties for canonical models also hold for OWL 2 EL.

3.2   Syntax and Semantics of GRAPHCO

We then presents our Graph Constraint language, GRAPHCO. Each expression in the language consists of three parts:

- **Deduce** formed as Φ, deduces new facts in the KGs by applying so-called *open-world assumption* (OWA), like in OWL, via query answering over ontologies.
- **Mandate** formed as Ψ, imposes *close-world assumption* (CWA), like cardinality constraints or functional dependencies.
- **Compare** formed as Ω, formulates complex range constraints, which compare property values of possible different individuals by transforming and aggregating.

We first give two examples as below.

Example 2. The following constraint limits the *DataProcessing* tasks with *SingleFeature* inputs should have at least two inputs, and the sums of all the input dimensions equal to the output dimensions of these tasks.

```
DEDUCE  Concatenation.hasInput.SingleFeature
MANDATE [2,*] hasInput.SingleFeature
COMPARE hasInput.Dimension :: Agg ::
    Comp(=,hasOutput.Dimension) :: LogAnd
```

To express such a constraint GRAPHCO first applies the DEDUCE part to deduce all instances of class *Concatenation* from the KGs which have single features as inputs. Then in the MANDATE part the rule verifies the input number. At last in the COMPARE part the literals (dimensions) of the input and output data are verified.

Example 3. Another typical example limits the combination of task sequences. Specifically, a valid ML pipeline consists of a sequence of *DataProcessing* tasks, followed by a sequence of *MLTraing* tasks, and finally, a sequence of *MLTesting* tasks. The sequence of *MLTraing* tasks can be optional, while the other two sequences are mandatory.

```
DEDUCE  MLPipeline
MANDATE [1,*](hasNextTask.DataProcessing).
 [1,*](hasNextTask.MLTesting) OR
(hasNextTask.DataProcessing)[1,*].
  (hasNextTask.MLTraining)[1,*].(hasNextTask.MLTesting)[1,*].
```

This constraint for instance does not have the COMPARE part. At this point, we observe that [2,*] talks about number restriction on the outgoing edges, while [1,*] (written as suffix) limits how many times a pattern under the scope repeats. Here, * stand for that there is no upper-bound.

Next we provide the syntax and formal semantics of the language. We use $[\![\cdot]\!]_{\mathcal{K}}$ to denote the evaluation of the GRAPHCO expression given a KG $\mathcal{K}$. (for simplicity $[\![\cdot]\!]$ when $\mathcal{K}$ is clear from the context).

DEDUCE. In this part, we allow Φ to be a union of path queries. In our setting, a path query obtained by simply concatenating properties and classes names. As a result DEDUCE returns all tuples that satisfy at least one of the paths. The Φ is evaluated over the model $\mathcal{I}^*$ for the underlying KG $\mathcal{K} = (\mathcal{O}, \mathcal{G})$. In particular

$$\Phi = path_1, \ldots, path_n, \quad path{::} = A \mid R \mid A.path \mid R.path,$$

Here we interpret them as: $[\![\Phi]\!] = \cup_i [\![path_i]\!]$, and

- $[\![A]\!] = \{(a, \bot) \mid A(a) \in \mathcal{I}^*\}, \qquad [\![R]\!] = \{(a, v) \mid R(a, v) \in \mathcal{I}^*\}$
- $[\![A.path]\!] = \{(a, v) \mid (a, v) \in [\![path]\!], (a, \_) \in [\![A]\!]\}$
- $[\![R.path]\!] = \{(a, v') \mid (a, v) \in [\![path]\!], (v, v') \in [\![R]\!]\}$

we introduce $\bot$ to make the composition of formula correct, and symbol $\_$ denotes an arbitrary constant when it is not important for the context.

Example 4. The following clause collects all tasks that have an input whose data format is array: DEDUCE Task.hasInput.Array.
Assume KG with axioms $\mathcal{G} = \{\texttt{Concatenation}(c_1), \texttt{hasInput}(c_1, i_1), \texttt{Array}(i_1),$ $\texttt{MlTesting}(m_1), \texttt{hasInput}(m_1, i_2), \texttt{Array}(i_2)\}$. Then the result of DEDUCE is

$$[\![\texttt{Task.hasInput.Array}]\!] = \{(c1, i1), (c2, i2)\}$$

since both Concatenation and MlTesting are indirect and direct subclasses of Task.

MANDATE. We define MANDATE as an expression with the following grammar:

$$\Psi ::= A \mid \texttt{[m,n]}\, R \mid \texttt{[m,n]}\, R.\Psi \mid (path)\texttt{[m,n]} \mid$$
$$\text{EQ}(path_1, path_2), \mid \Psi_1 \,\text{And}\, \Psi_2 \mid \Psi_1 \,\text{Or}\, \Psi_2 \mid \text{Not}\, \Psi$$

where R is a property, $m, n \in \mathbb{N}$; And, Or, Not for conjunction, disjunction and negation respectively, "$\texttt{[m,n]}\, R.\Psi$" for "must have at least n and at most m-successors in the graph verifying Ψ" (where "n=*" means that there is no upper-bound), "EQ($path_1, path_2$)" means that "$path_1$ and $path_2$ successors of a node must coincide".
Besides the definitions in DEDUCE, we interpret

- $[\![\texttt{[m,n]}\, R.\Psi]\!] = \{(k, \bot) \mid k \in adom(\mathcal{K}), k$ has at least m and at most n R-successors v such that $(v, \_) \in [\![\Psi]\!]\}$ (omitted Ψ mean no constraints on v)
- $[\![(path)\texttt{[m,n]}]\!] = \{(k, v) \mid$ such that there is a path from k to v in $\mathcal{I}^*$ via path $path^i$ where i ranges from m to $n\}$
- $[\![\text{EQ}(path_1, path_2)]\!] = \{(k, \bot) \mid k \in adom(\mathcal{K})$ and all nodes reachable via $path_1$ are also reachable via $path_2$ starting from k in $\mathcal{I}^*\}$

- $[\![\Psi_1 \text{ And } \Psi_2]\!] = \{(k,\bot) \mid (k,\_) \in [\![\Psi_1]\!] \wedge (k,\_) \in [\![\Psi_2]\!]\}$, $[\![\Psi_1 \text{ Or } \Psi_2]\!] = \{(k,\bot) \mid (k,\_) \in [\![\Psi_1]\!] \vee (k,\_) \in [\![\Psi_2]\!]\}$, $[\![\text{Not } \Psi]\!] = \{(k,\bot) \mid k \in adom(\mathcal{K}), (k,\_) \notin [\![\Psi]\!]\}$

$path^i$ denotes an expression obtained by concatenating expression $path$ exactly i-times.

Example 5. The following expression ensures that tasks should have at least two inputs, i.e., connected to two or more <relation, object> pairs via the relation *hasInput*: MANDATE [2,*] hasInput.

over $\mathcal{I}^* = \{\text{hasInput}(c_1,i_1), \text{hasInput}(c_1,i_2), \text{hasInput}(c_2,i_3)\}$. We get:

$$[\![\,[2,*]\ \text{hasInput}]\!] = \{(c_1,\bot)\}$$

COMPARE. In Compare part we can formulate the expressions of the form

$$\Omega ::= path :: \Lambda \mid \Lambda,\ \Lambda ::= \text{FUNCT} \mid \Lambda_1 :: \Lambda_2.$$

Here, symbol "::" represents that data transformed from the left-hand side and provided as an input for the operation on the right-hand side, $path$ is a path expression as defined above, and FUNCT denotes a built-in functions for numerical calculation. Let $K_0 = [\![\Phi]\!]$ is going to a first set of tuples over which COMPARE applies the functions by transforming set of tuples into a new set of tuples starting from K_0. Alternatively, one uses only the first component, so-called *keys*, of tuples in K_0 and with $path$ access new set of tuples with the same key with expression $path :: \Lambda$. After it applies the list of transforming FUNCT for which we define the semantics as follows:

- $[\![path(K)]\!] = \{(k,v) \mid (k,\_) \in K \wedge v \text{ is reachable from } k \text{ via } path \text{ in } I^*\}$
- $[\![\text{Agg}(K)]\!] = \{(k,v_k) \mid k \in K, v_k = \sum_{(k,v)\in K} v\}$
- $[\![\text{Count}(K)]\!] = \{(k,v_k) \mid k \in K, v_k = \#\{v \mid (k,v) \in K\}\}$
- $[\![\text{Comp}(\odot, path)]\!] = \{(k, v_1 \odot v_2) \mid (k,v_1) \in K, (k,v_2) \in [\![path]\!]\}$, here $\odot \in \{=,\neq,>,<,\geq,\leq\}$ and $v_1 \odot v_2 = 1$ if $v_1 \odot v_2$; otherwise 0
- $[\![\text{Transform}(K)]\!] = \{(k, f(v)) \mid (k,v)\}$ where f is a arithmetic expression on v
- $[\![\text{Filter}(K)]\!] = \{(k,v) \mid (k,v) \in K, v \odot r\}$ for some number r
- $[\![\text{LogAnd}(K)]\!] = 1$ if for every $(k,v) \in K, v \neq 0$; otherwise 0.
- $[\![\text{LogOr}(K)]\!] = 1$ if for at least one $(k,v) \in K, v \neq 0$; otherwise 0.

and then the composition is defined as: $[\![\Omega_1 :: \Omega_2]\!] = [\![\Omega_2]\!]([\![\Omega_1]\!])$.

Example 6. The following clause ensures that the aggregation of the input dimensions for the tasks should be equal to the output dimension:

```
COMPARE hasInput.Dimension {:}{:} Agg {:}{:}
       Comp(=,hasOutput.Dimension) {:}{:} LogAnd
```

Assume KG $\mathcal{G}$ = {Concatenation(c_1), hasInput(c_1, i_{11}), Array(i_{11}), hasInput (c_1, i_{12}), Array(i_{12}), hasOutput(c_1, i_{13}), Array(i_{13}), Concatenation(c_2), hasInput (c_2, i_{21}), Array(i_{21}), hasInput(c_2, i_{22}), Array(i_{22}), hasOutput(c_2, i_{23}), Array(i_{23})}

with literals Dimension$(i_{11}, 5)$, Dimension$(i_{12}, 4)$, Dimension$(i_{13}, 9)$, Dimension$(i_{21}, 3)$, Dimension$(i_{22}, 10)$, Dimension$(i_{23}, 13)$. Then the following will be returned:

$$[\![\text{hasInput.Dimension}(\mathcal{G})]\!] = K_1 = \{(c_1, 5), (c_1, 4), (c_2, 3), (c_2, 10)\}$$

$$[\![\text{Agg}(K_1)]\!] = K_2 = \{(c_1, 9), (c_2, 13)\}$$

$$[\![\text{Comp}(\text{=,hasOutput.Dimension})(K_2)]\!] = K_3 = \{(c_1, 1), (c_2, 1)\}$$

$$[\![\text{LogAnd}(K_3)]\!] = K_4 = \{1\}$$

Combining the above three types of expressions, we can represent the data constraints in the executable KG. To **validate** the constraints, we say that a constraint DEDUCE Φ MANDATE Ψ COMPARE Ω is *valid* over a KG $\mathcal{K} = (\mathcal{O}, \mathcal{G})$ if for every $(k, v) \in [\![\Phi]\!]$, there exists some $(k, v') \in [\![\Psi]\!]$ and $[\![\Omega([\![\Phi]\!])]\!] = 1$.

4 Implementation and Evaluation

4.1 Implementation Technique

The implementation of prototypical validator of GRAPHCO combines two technologies and thus divided into two parts where : the Deduce, and Mandate are implemented as answer set programs with DLV system [1], and Compare as functional programs with Apache Spark [28]. In the following, we provide more details on the implementation.

Algorithm 1: CHECKING CONSTRAINT

Input: A KG $\mathcal{K}$ and a constraint $C =$ DEDUCE Φ MANDATE Ψ COMPARE Ω.

1: A set of facts $S \leftarrow Answer\text{-}Set(\Pi_K)$
2: **if** $fail \in Answer\text{-}Set(S \cup \Pi_{\Phi, \Psi})$ **then**
| **return** false
3: $K_0 = \{(k, v) \mid deduce(k, v) \in Answer\text{-}Set(S \cup \Pi_{\Phi, \Psi})\}$
4: **if** *verify_compare is not empty* in $\Pi_{\Omega}(K_0)$ **then**
| **return** false
5: **return** *true*

Answer-Set Programming. We consider rules of the form $A \leftarrow B_1, \ldots, B_n, not\ F_1, \ldots, not\ F_m$, where A is a head of rule (the consequence of the rule application) and $B_1, \ldots, B_n, F_1, \ldots, F_m$ are predicates that apply join or aggregate function that filters out the results. Our programs are such that they can have recursion but they do have a single *stable model*. For the theory of answer set programs we refer to [15]. With *Answer-Set*(Π) we refer as the set of answer-sets of program Π.

Deduce is part that requires reasoning on chase procedure. ASP tools have very direct way of expressing it with functional symbols with restricted chase. For instance, $Task \sqsubseteq \exists hasId$ can be represented as a rule

$$AuxhasId(X) \leftarrow Task(X), hasId(X, Y), adom(Y)$$
$$hasId(X, f(X)) \leftarrow Task(X), not\ AuxhasId(X)$$

where $AuxhasId$ is an auxiliary predicate that checks if $Task(X)$ has an id or not, and $adom(Y)$ is to filter that such condition is checked only over the active domain. Similarly, we introduce other rules for axioms in OWL 2 QL. Let us denote such program with $\Pi_{\mathcal{K}}$. Then, $\Phi = path_1, \ldots, path_n$ can be trivially encoded as set of rules, where we introduce rule for each $path_i$. To encode number restrictions we use aggregates in ASP. For instance, DEDUCE Concatenation MANDATE [2,*] hasInput. SingleFeature is encoded as

$$MandateAux(X, Y) \leftarrow hasInput(X, Y), SingleFeature(Y)$$
$$Mandate(X, bot) \leftarrow Concatenation(X), \#count\{Y \mid MandateAux(X, Y)\} >= 2.$$

To express * one can naturally rely on the recursion in ASP. Finally, we introduce a rule that checks if there is an instance for which **Mandate** fails with

$$fail \leftarrow deduce(X, \_), not\ mandate(X, Y), adom(Y).$$

With $\Pi_{\Phi, \Psi}$ we denote the program that contains all the rules described above.

Table 1. Statistics on the dataset. The KGs share the same schema with 95 classes, 7 object properties, and 60 data properties. ρ_p: ratio of valid ML pipelines, ρ_t: ratio of valid tasks.

	KG1	KG2	KG3	KG4	KG5	KG6	KG7
#Individual	7606	37264	49679	74079	123979	248713	495891
#Triple	29473	144024	192104	286451	479313	961951	1918112
#Literal	13999	68177	90986	135681	226998	455906	909241
#Pipeline	300	1500	2000	3000	5000	10000	200000
#Task	1880	9300	12339	18448	30791	61779	123367
ρ_p		84.69%	70.59%	74.51%	85.29%	84.69%	84.69%	84.69%
ρ_t		80.39%	69.31%	79.21%	85.71%	79.59%	86.16%	82.59%

We briefly discuss our implementation of Ω as an Apache Spark program. Spark is using distributed object types called RDD to store so-called *key-valued* pairs (k, v). For instance, the **Compare** part of $\mathbf{C_6}$ is encoded in PySPark as shown in Fig. 3, where first two lines read the input on given paths and store them as pairs. Then `rdd_verify` copies all pairs from `rdd_weightIn`; performs aggregate per key; does the join with `rdd_weightOut` by creating an array of length 2 on the second component. Finally, `filter` function compares those key-values pairs that satisfy the condition and leaves those that do not. Hence, `verify_compare` has at least one element, thus violate the condition `COMPARE`. Other build-in functions from Λ can be implemented similarly, and for their implementation we refer to our supplement. Let us denote such implementation with Π_{Ω}. In Algorithm 1, we summarize the implementation.

```
rdd_weightIn=spark.read.csv(weightedInput_path).rdd
rdd_weightOut=spark.read.csv(weightedOutput_path).rdd
rdd_verify = rdd_weightIn. reduceByKey(lambda a,b : a+b).
    join(rdd_weightOut).filter(lambda x: x[1][0]!=x[1][1])
verify_compare = rdd_verify.first()
```

Fig. 3. PySpark code used to validate the `Compare` part of $\mathbf{C_6}$.

4.2 Evaluation

To evaluate the implementation of GRAPHCO, we verify our problem by varying sizes of KGs and constraints, and record the runtime.

Data Description. The KGs used in our evaluation represent the industrial ML pipelines. The KGs are based on the same ML ontology [29], which is depicted in Fig. 1. It models the machine learning pipeline as a sequence of `Tasks`, each of which has certain `Method` and some *DataEntity* as input and outputs some *DataEntity*. Based on this ontology, the KGs are randomly created. In particular, each KG contains a fixed random number of ML pipelines, including various kinds of the tasks in each ML pipeline. Then we use another set of random values to decide whether these tasks are connected with proper data entities and literals, i.e., whether these tasks and the pipeline are valid. In this way, we created 7 KGs, the details of which are listed in Table 1. From KG1 to KG7, the number of triples increases exponentially from about 30K to 2M. These KGs include various types and amounts of errors in ML practice. The ratios of error are controlled in around 85%. Some errors are about the incomplete structure in the ML pipeline, such as the missing inputs in specific tasks. Some other errors are related to improper data used in the pipeline, such as the unmatched data dimension or number.

Evaluation Setting. In the evaluation, we verify the randomly generated KGs based on 6 constraints (see Table 2) expressed in GRAPHCO and to find the corresponding errors in KGs. These constraints are collected from the collaboration with our industrial partners and are typical in industrial ML practice.

We used datalog and PySpark for the implementation (datalog for DEDUCE and MANDATE, while PySpark for COMPARE), and recorded the runtime in KGs for different sizes.

Table 2. Example constraints used in the evaluation.

C1: Each *Task* in the ML pipeline should have at least one *DataEntity* as input.

```
DEDUCE Task
MANDATE [1,*] hasInput.DataEntity
```

C2: a valid ML pipeline should be the combination of a sequence of *DataProcessing* tasks, a sequence of *MLTraing* tasks and a sequence of *MLTesting* tasks. The *MLTraing* tasks are optional, while the other two sequences are mandatory.

```
DEDUCE MLPipeline
MANDATE [1,*](hasNextTask.DataProcessing).[1,*](hasNextTask.MLTesting)
  OR (hasNextTask.DataProcessing)[1,*].(hasNextTask.MLTraining)[1,*].(
  hasNextTask.MLTesting)[1,*]
```

C3: All the input *DataEntity* of *Concatenation* tasks should be the same subclass of DataStructure, which means that they should all be *Arrays* or *SingleValues*.

```
DEDUCE Concatenation
MANDATE (NOT [1,*](hasInput.SingleFeature) AND [1,*]hasInput.
SingleFeature) OR (NOT [1,*](hasInput.Array) AND [1,*]hasInput.Array)
```

C4: All the input dimensions of the *WeightedSum* tasks should be same.

```
DEDUCE WeightedSum
COMPARE hasInput.hasDimension :: Comp(=, hasInput.hasDimension) :: LogicAnd
```

C5: The number of inputs of the *WeightedSum* tasks should not be greater than the length of the weight vector for the tasks.

```
DEDUCE WeightedSum
COMPARE hasInput :: Count :: Comp( >=, hasWeightVector.Length)
```

C6: The *Concatenation* tasks should have at least two *SingleFeature* as inputs, and the aggregation of the input dimension should be equal to the output dimension.

```
DEDUCE Concatenation.hasInput.SingleFeature
MANDATE [2,*] hasInput.SingleFeature
COMPARE hasInput.Dimension :: Agg :: Comp(=,hasOutput.Dimension) :: LogAnd
```

In addition, we evaluate how GRAPHCO's support for parallelization offers significant advantages in terms of scalability, we compared the runtime of literal verification in COMPARE using PySpark and the traditional Python method. In particular, we first read tuples, which are generated in DEDUCE parts and denote the correspondence between instances and literals in the KGs (see Example 3 in Sect. 3.1). Then in COMPARE part we use PySpark to process these tuples and to verify whether the constraints on literals are satisfied. We also do the same verification using traditional Python method. The runtime of both kinds of methods are recorded and compared in varying data sizes in terms of tuple numbers. The experiments were conducted with Intel(R) Core(TM) i7-10710U Processor, 16 GB of RAM.

Results and Discussion. Figure 4 demonstrates the linear scalability of GRAPHCO, as the average runtime per triple remains consistent (in C1, C2, and C3) or decreases (in C4, C5, and C6) as the number of triplets in KGs increases exponentially from KG1 to KG7. Besides, it was observed that the average runtime of COMPARE is relatively large (about 0.3 ms) when the KG is small, while it decrease significantly to under 0.05 ms as the KG size increases over 50 times, from KG7 to KG1. We postulate the reason behind is that the startup time for Pyspark used in COMPARE cannot be neglected and remains relatively constant across the varying KG sizes, while the built-in parallelization in COMPARE leads to sub-linear increase in runtime as the KG size increases. This property of COMPARE makes GRAPHCO especially suitable for verifying big KGs. On the other hand, we observe the runtime of DEDUCE and MANDATE increases linearly with the size of the KG. It is worthy to note these parts are currently implemented using ASP, which sacrifices execution speed for ease of use. By utilizing graph databases or SPARQL engines instead in DEDUCE and MANDATE, the execution of GRAPHCO can be more efficiency and scalable in dealing with large datasize.

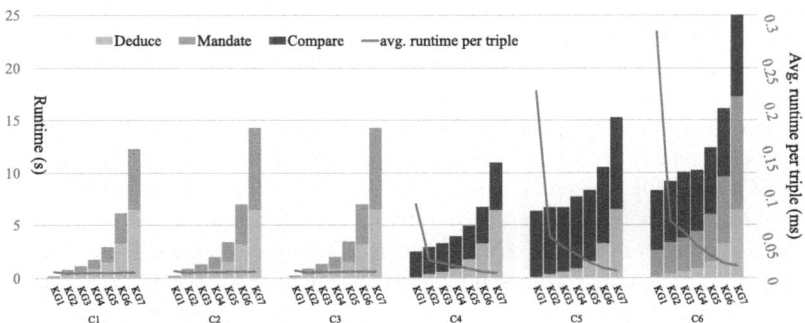

Fig. 4. Runtime of verifying constraints C1-6 in KG1-7. Each runtime is split into DEDUCE, MANDATE and COMPARE. From KG1 to KG7, the number of triples increases exponentially. Based on the triple numbers, we get the average runtime per triple for each KG, which remains consistent or decreases as the KG is enlarged, indicating the good scalability of GRAPHCO.

Figure 5 demonstrates good scalability of GRAPHCO in verifying KGs' literals in COMPARE, as the runtime remains polynomial as the number of tuples increases. In contrast, the traditional Python method experiences a significant increase in runtime when dealing with a large number of tuples. The tuples denote the correspondence between individuals and literals and are generated in DEDUCE. COMPARE is implemented using PySpark as GRAPHCO is designed in such a way to support highly parallelizable implementation, which makes it scalable for processing big data.

5   Related Work

In the context of constraint languages for graph, we first consider the classical semantic web stack with RDFS and OWL [16]. These formalisms model incomplete information under the OWA but are not ideal for constraint modelling. While expressive fragments of OWL 2 can represent many constraints, they are highly complex. To address this, some approaches combine OWL expressions, treating certain constraints under OWA and others under the CWA. However, validation in such settings becomes complex even for simple OWL fragments, requiring careful handling [18, 26].

Recently, several constraint languages for RDF have been proposed to mitigate this issue, with SHACL [8] and ShEx [25] being the most prominent. Both adopt the CWA, with SHACL becoming a W3C standard. SHACL can "deduce" nodes for constraint checking via KG queries but currently supports reasoning only over RDFS type hierarchies through *targets*. Compared to these, our MANDATE matches the expressivity of SHACL's non-recursive core fragment, though SHACL and ShEx allow constraint compositionality via *shape* names. Recursion in SHACL remains an open issue despite theoretical proposals [22], and while SHACL includes built-in functions for String operations, numeral support is limited.

Another prominent graph formalism are property graphs [2,3] which is gaining popularity especially with a strong support from existing tools such as Neo4j and TigerGraph [10]. Recently, there have been several proposals for constraint languages over property graphs. In particular, PG-Schema [3] is a comprehensive language for constraints over graphs, with a strong influence from conceptual modeling. PG-Schema introduces new features such as abstract types and inheritance, and a very expressive extension to model key constraints over graphs [3]. This proposal is relatively new and SHACL excels in terms of diversity of modelling features. Another proposal that attempts to introduce SHACL on property graph is ProGS [23].

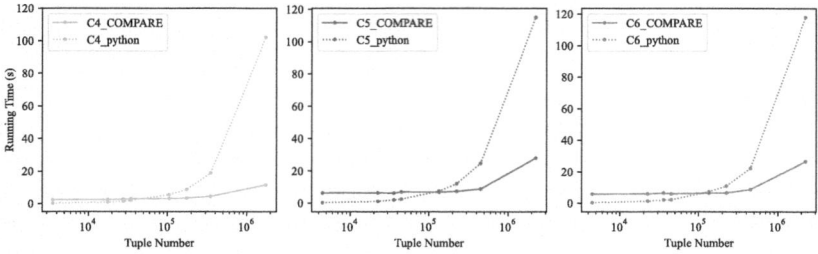

Fig. 5. Runtime of verifying literals in constraints C4-6 as the tuple increased. The tuples are generated in DEDUCE and denote the correspondence between individuals and literals (see Example 3 in Sect. 3.1). We compare the runtime of COMPARE in GRAPHCO using PySpark and traditional method using Python, indicating good scalability of COMPARE.

Table 3. Comparison of GraphCo to other graph constraint languages. '✓✓': supported, '✓': partially supported, '-': not supported.

	OWA	UIE	ABS	OCC	OPC	RCC	LRC	MCN	RCV	PVA
GraphCo	✓✓	✓✓	✓✓	✓	-	-	✓✓	✓✓	✓✓	✓✓
RDFS	✓✓	✓	-	-	-	✓✓	-	-	-	✓✓
OWL	✓✓	✓✓	✓	✓	✓	✓✓	✓✓	✓✓	✓✓	✓✓
SHACL	✓	✓✓	✓	✓✓	✓✓	✓	✓✓	✓✓	✓✓	✓✓
ShEx	-	✓✓	✓	✓	✓✓	✓	✓✓	✓✓	✓✓	✓✓
PG-Schema	-	✓✓	✓✓	✓✓	✓✓	-	-	✓✓	-	✓✓
Pro-GS	-	✓✓	✓	✓	✓	✓✓	✓✓	✓✓	✓✓	✓✓

We summarise the comparison between the languages and GraphCo in Table 3. We analyse if a language allows for the following: 1. OWA inference; 2. specifying union or intersection of constraint expressions (UIE); 3. abstraction of constraints (ABS); 4. constraints to be open and closed (OCC); 5. picking certain properties to be optional or not (OPC); 6. specifying constraints recursively (RCC); 7. limiting the allowed property values by providing range constraints (LRC); 8. specifying mandatory constraints for certain nodes (MCN); 9. specifying range for values (RCV, like [m,n] in GraphCo); and 10. partial validation (PVA) that is validation not all constraints are required to be validated.

6   Conclusion and Outlook

In this paper, we presented a new constraint language, GraphCo, for knowledge graphs, designed to bridge the gap in reasoning scenarios that require both open- and closed-world assumptions, as well as the ability to express constraints over numerical values. We demonstrated that GraphCo achieves a good balance between expressiveness and computational complexity, ensuring that reasoning remains tractable. Furthermore, we developed a prototypical implementation using Answer Set Programming and Apache Spark, which shows promising scalability in practice.

We believe that the ideas demonstrated by our language can be further incorporated into future versions of standard graph constraint languages such as SHACL, both in terms of expressiveness and implementation approach. As a next step, we plan to study the computational complexity of GraphCo when extended with more expressive features, such as named constraints and recursion. In addition, we aim to analyze support for string operators, which are also crucial in many practical applications.

Acknowledgements. This work was partly funded by the National Research Foundation of Korea (NRF) under grant number RS-2023-00268071.

References

1. Adrian, W.T., et al.: The ASP system DLV: advancements and applications. KI Künstliche Intelligenz **32**, 177–179 (2018). https://doi.org/10.1007/s13218-018-0533-0
2. Angles, R., et al.: PG-keys: keys for property graphs. In: International Conference on Management of Data (SIGMOD 2021), pp. 2423–2436. ACM (2021)
3. Bonifati, A., et al.: PG-Schema: Schemas for Property Graphs. arXiv preprint arXiv:2211.10962 (2022)
4. Calvanese, D., De Giacomo, G., Lembo, D., Lenzerini, M., Rosati, R.: Tractable reasoning and efficient query answering in description logics: the DL-lite family. J. Autom. Reason. **39**(3), 385–429 (2007)
5. Cao, X., et al.: DEKR: description enhanced knowledge graph for machine learning method recommendation. In: 44th International ACM SIGIR Conference on Research and Development in Information Retrieval (SIGIR 2021), pp. 203–212. ACM (2021)
6. Charron, B., Hirate, Yu., Purcell, D., Rezk, M.: Extracting semantic information for e-commerce. In: Groth, P., et al. (eds.) ISWC 2016. LNCS, vol. 9982, pp. 273–290. Springer, Cham (2016). https://doi.org/10.1007/978-3-319-46547-0_27
7. Civili, C., et al.: MASTRO STUDIO: managing ontology-based data access applications. Proc. VLDB Endow. **6**(12), 1314–1317 (2013)
8. Corman, J., Reutter, J.L., Savković, O.: Semantics and validation of recursive SHACL. In: Vrandečić, D., et al. (eds.) ISWC 2018. LNCS, vol. 11136, pp. 318–336. Springer, Cham (2018). https://doi.org/10.1007/978-3-030-00671-6_19
9. Deutsch, A., Nash, A., Remmel, J.: The chase revisited. In: 27th ACM SIGMOD-SIGACT-SIGART Symposium on Principles of Database Systems, pp. 149–158. ACM (2008)
10. Deutsch, A., Xu, Y., Wu, M., Lee, V.: TigerGraph: A Native MPP Graph Database. arXiv preprint arXiv:1901.08248 (2019)
11. Hogan, A., et al.: Knowledge graphs. ACM Comput. Surv. **54**(4), 1–37 (2021)
12. Hubauer, T., Lamparter, S., Haase, P., Herzig, D.M.: Use cases of the industrial knowledge graph at siemens. In: ISWC 2018 Posters & Demonstrations, Industry and Blue Sky Ideas Tracks. CEUR Workshop Proceedings, vol. 2180. CEUR-WS.org (2018)
13. Ji, S., Pan, S., Cambria, E., Marttinen, P., Philip, S.Y.: A survey on knowledge graphs: representation, acquisition, and applications. IEEE Trans. Neural Netw. Learn. Syst. **33**(2), 494–514 (2021)
14. Konev, B., Ludwig, M., Walther, D., Wolter, F.: The logical difference for the lightweight description logic EL. J. Artif. Intell. Res. **44**, 633–708 (2012)
15. Leone, N., Ricca, F.: Answer set programming: a tour from the basics to advanced development tools and industrial applications. In: Faber, W., Paschke, A. (eds.) Reasoning Web 2015. LNCS, vol. 9203, pp. 308–326. Springer, Cham (2015). https://doi.org/10.1007/978-3-319-21768-0_10
16. McBride, B.: The resource description framework (RDF) and its vocabulary description language RDFS. In: Handbook on Ontologies, pp. 51–65 (2004)
17. Mohamed, A., Abuoda, G., Ghanem, A., Kaoudi, Z., Aboulnaga, A.: RDF-Frames: Knowledge Graph Access for Machine Learning Tools. arXiv preprint arXiv:2002.03614 (2020)
18. Motik, B., Horrocks, I., Sattler, U.: Bridging the gap between OWL and relational databases. In: Web Semantics: Science, Services and Agents on the World Wide Web, vol. 7, no. 2, pp. 74–89 (2009)

19. Nicholson, D.N., Greene, C.S.: Constructing knowledge graphs and their biomedical applications. Comput. Struct. Biotechnol. J. **18**, 1414–1428 (2020)
20. Noy, N., Gao, Y., Jain, A., Narayanan, A., Patterson, A., Taylor, J.: Industry-scale knowledge graphs: lessons and challenges: five diverse technology companies show how it's done. Queue **17**(2), 48–75 (2019)
21. Pan, J.Z., Vetere, G., Gomez-Perez, J.M., Wu, H.: Exploiting Linked Data and Knowledge Graphs in Large Organisations. Springer, Cham (2017). https://doi.org/10.1007/978-3-319-45654-6
22. Reutter, J., Soto, A., Vrgoc, D.: Recursion in SPARQL. Semantic Web **12**(5), 711–740 (2021)
23. Seifer, P., Lämmel, R., Staab, S.: ProGS: property graph shapes language. In: Hotho, A., et al. (eds.) ISWC 2021. LNCS, vol. 12922, pp. 392–409. Springer, Cham (2021). https://doi.org/10.1007/978-3-030-88361-4_23
24. Shiralkar, P., Flammini, A., Menczer, F., Ciampaglia, G.L.: Finding streams in knowledge graphs to support fact checking. In: IEEE International Conference on Data Mining (ICDM 2017), pp. 859–864. IEEE (2017)
25. Staworko, S., Boneva, I., Gayo, J.E.L., Hym, S., Prud'Hommeaux, E.G., Solbrig, H.: Complexity and expressiveness of ShEx for RDF. In: 18th International Conference on Database Theory (ICDT 2015). Schloss Dagstuhl (2015)
26. Tao, J., Sirin, E., Bao, J., McGuinness, D.L.: Integrity constraints in OWL. In: AAAI Conference on Artificial Intelligence (AAAI 2010), pp. 1443–1448. AAAI (2010)
27. Xiao, G., Ding, L., Cogrel, B., Calvanese, D.: Virtual knowledge graphs: an overview of systems and use cases. Data Intell. **1**(3), 201–223 (2019)
28. Zaharia, M., et al.: Apache Spark: a unified engine for big data processing. Commun. ACM **59**(11), 56–65 (2016)
29. Zheng, Z., Zhou, B., Zhou, D., Soylu, A., Kharlamov, E.: ExeKG: executable knowledge graph system for user-friendly data analytics. In: 31st ACM International Conference on Information & Knowledge Management (CIKM 2022), pp. 5064–5068. ACM (2022)

SemViSG: Semantic Enrichment and Visualization of Software Graphs

Sami M'hamdi[1,2](✉) ⓘ, Hamamache Kheddouci[1] ⓘ, Damien Charlemagne[2], and Olivier Bonsignour[2]

[1] University Claude Bernard Lyon 1, Lyon, France
sami.mhamdi@etu.univ-lyon1.fr
[2] CAST Software Intelligence, Meudon, France

Abstract. Visualizing software graphs efficiently is crucial for understanding complex software architectures. Traditional techniques struggle with large, sparse, heterogeneous graphs containing semantically diverse relationships. This paper presents SemViSG, an approach that combines semantic-enriched hierarchical community detection, and custom layout algorithms to enhance software graph comprehension. We introduce multi-layered software graph representation and semantic enrichment techniques that leverage source code semantics to create functionally coherent communities. The custom layout algorithm preserves directional flows while minimizing edge crossings. Validation on large-scale systems demonstrates substantial improvements in clarity and usability, providing software engineers with enhanced architectural insights.

Keywords: Knowledge Graphs · Visualization · Community · Semantic

1 Introduction and Related Work

Understanding large-scale software systems is challenging due to their inherent complexity. Software applications produce vast graphs with heterogeneous nodes (classes, methods, database tables, ...) connected by semantically diverse relationships (CALL, SELECT, BELONG_TO, ...). General-purpose visualization systems such as Gephi [4], Pajek [5], and Tulip [3] demonstrate efficiency in analyzing standard networks, with Tulip offering support for certain semantic formats. However, they are not designed for software graphs' heterogeneous node types, multiplex relationships, sparse connectivity, and strong directional flows. Software-specialized commercial solutions like Structure101 [2], CodeSee [1], and CodeSonar [7] focus on specific languages and compositional dependencies, often reducing complexity through node or edge removal, leading to information loss. CodeSonar scale to large codebases, but as others remain language-dependent and lack graph-theoretic foundations, limiting architectural insights across complex systems.

ⓒ The Author(s), under exclusive license to Springer Nature Switzerland AG 2026
C. K. Leung et al. (Eds.): DaWaK 2025, LNCS 16048, pp. 200–206, 2026.
https://doi.org/10.1007/978-3-032-02215-8_14

Traditional graph visualization optimization focuses on two key aspects: reducing the amount of information displayed and improving how it is presented. Clustering approaches [8] rely solely on structural connectivity to group nodes, missing all the intrinsic semantic. Layout algorithms present significant limitations too when applied to software graphs. Force-directed methods [6] fail to reflect functional directional flows, while hierarchical approaches [9] often misplace functionally related nodes, disrupting execution flow representation. The core limitation across existing approaches lies in ignoring the semantic richness embedded within software architectures, applying generic techniques rather than leveraging domain-specific knowledge about functional relationships, execution flows, and architectural patterns.

To address these limitations, we propose SemViSG, a semantic-enriched visualization approach that enhances software graph comprehension through three key innovations: (1) multi-layered software graph representation, (2) semantic enrichment techniques that leverage source code semantics to create functionally coherent meta-node communities, and (3) custom layout algorithms that preserve functional flow while optimizing readability. The approach includes interactive capabilities enabling dynamic exploration, hierarchical drill-down, and user-driven community refinements for comprehensive software analysis.

SemViSG has been empirically evaluated on large-scale applications, demonstrating substantial improvements in visual clarity and structural interpretability, providing software engineers with enhanced tools for system understanding and analysis. In the following sections, we present our multi-layered graph representation (Sect. 2), semantic-enriched community detection (Sect. 3), custom layout algorithm (Sect. 4), and comprehensive evaluation (Sect. 5) demonstrating effectiveness on real-world software systems.

2   Multi-layered Software Graph Representation

Our approach models software systems as multi-layered graphs representing heterogeneous and multiplex networks. These graphs integrate various relationships within the same node set, conceptually organized into three layers: (1) *Call graph* – captures functional flow where functions invoke others or query database tables, (2) *Ownership graph* – represents hierarchical structure showing how functions are encapsulated within classes, and (3) *RELY ON graph* – models type definitions and dependencies between methods.

As shown in Fig. 1, these layers create a multiplex structure where nodes are interconnected by different edge types. For community detection, we primarily focus on the call graph layer for operational insights while incorporating ownership and type dependencies as contextual factors to refine groupings.

To manage complexity, we define *Transactions* as subgraphs capturing complete functional flows from entry points (UI elements) to exit points (database tables), and *DataGraphs* as backward traces from exit points to identify all connected entry points. This partitioning enables focused exploration of specific workflows or data dependencies without overwhelming users with full graph complexity.

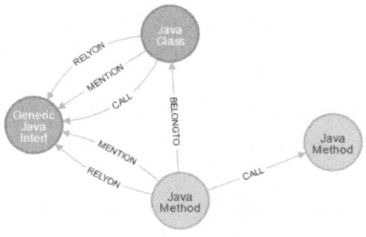

Fig. 1. Multi-layered structure showing how layers coexist within a single graph.

3   Semantic-Enriched Hierarchical Community Detection

Our community detection methodology enhances software graph comprehensibility by organizing nodes into functionally coherent groups, reducing cognitive overload for users exploring complex architectures. Unlike traditional methods that rely solely on structural connectivity, we integrate semantic elements by leveraging domain-specific information embedded in software graphs.

Hierarchical Community Detection and Meta-node. We identify communities exhibiting strong functional relationships through the call graph layer, representing functionally related components that collaborate to implement features, work within modules, or support workflows. Each community is abstracted into a *meta-node* serving as higher-level functionality representation.

The hierarchical approach produces multi-scale representations where higher-level communities contain sub-communities, mirroring software architectures from broad components to granular submodules. This enables exploration at different abstraction levels, from system-wide overview to detailed implementation interactions.

Meta-nodes require meaningful names for interpretability. We employ two independent strategies adaptable to privacy requirements: (1) TF-IDF-based keyword extraction identifying representative terms, $\text{TF-IDF}(t,d) = \text{TF}(t,d) \times \log\left(\frac{N}{\text{DF}(t)}\right)$ where $\text{TF}(t,d)$ is term frequency, $\text{DF}(t)$ is document frequency, and N is total documents, and (2) LLM-based naming providing contextually rich names by analyzing community composition. Naming is applied hierarchically, using sub-community names as input for parent meta-node naming.

Meta-node connectivity preserves essential relationships from the original structure. An edge exists between meta-nodes M_a and M_b representing communities C_a and C_b if and only if nodes $i \in C_a$ and $j \in C_b$ exist such that $\exists (i,j) \in G, A_{ij} \neq 0$ where A_{ij} denotes the adjacency matrix. This maintains semantic flow and dependencies in higher-level abstractions.

Semantic Enrichment Process. Our key innovation semantically enriches graph structure before community detection. Traditional algorithms fail on software graphs by treating all edges equally, ignoring semantic meaning in different relationship types as well as hidden semantics not represented by links.

Stage 1: Source Code Semantic Extraction. We extract semantic information from source code through preprocessing, cleaning, and normalization. Each node's code transforms into numerical representation using TF-IDF vectorization, capturing functional intent. For node n with keywords $\{t_1, t_2, ..., t_m\}$, the semantic vector is $v_n = \frac{1}{m} \sum_{i=1}^{m} \text{TF-IDF}(t_i)$.

Stage 2: Multi-layer Semantic Context Integration. While call graphs provide structural foundation, ownership and type dependency layers carry crucial semantic information influencing functional groupings. We incorporate contextual information from these layers to refine node representation. For instance, if function f belongs to class C, the semantic vector of C influences f's representation. The enhanced vector integrates multi-layered information:

$$v_n^{\text{final}} = (1 - \alpha)v_n + \frac{\alpha}{|\mathcal{N}(n)|} \sum_{j \in \mathcal{N}(n)} v_j \tag{1}$$

where α (set as 0.3) balances intrinsic semantics with contextual information from ownership and type dependency layers, and $\mathcal{N}(n)$ represents semantically related neighbors.

Stage 3: Semantic Similarity Link. We create semantic similarity links by computing pairwise cosine similarity between enriched vectors, $\text{sim}(i, j) = \frac{v_i \cdot v_j}{\|v_i\| \|v_j\|}$. When $\text{sim}(i, j)$ exceeds threshold τ (set as 0.8), we add weighted semantic edges, revealing latent functional relationships invisible from structure alone.

The enriched graph contains original structural relationships and discovered semantic associations. Applying hierarchical community detection to this enhanced structure produces communities capturing structural, functional, and latent semantic relationships, resulting in coherent partitioning aligned with software's true functional organization.

4   Custom Layout Algorithm

Even with meta-node abstraction, node arrangement remains crucial for clarity and usability.

Coordinate Assignment Algorithm. Our layout methodology aligns with the intrinsic directional structure of software graphs. Nodes are positioned vertically according to their functional role: in Transaction graphs, hierarchy is built starting from entry nodes, with subsequent levels positioned based on shortest distance from entry points. Conversely, in DataGraphs, hierarchy is established by anchoring exit nodes, with nodes positioned progressively based on distance from endpoints. This ensures that meta-nodes representing functional modules equidistant from entry or exit points are placed at the same hierarchical level, reflecting their similar roles in execution flow.

The y-coordinates encode hierarchical flow levels, while x-coordinates organize nodes within each level for optimal readability. For each hierarchical level, nodes are distributed evenly across the available horizontal space. Each node's x-coordinate is then determined by computing a barycenter based on connected nodes in previous levels:

$$x_n = \frac{1}{|\mathcal{P}(n)|} \sum_{p \in \mathcal{P}(n)} x_p \tag{2}$$

where $\mathcal{P}(n)$ represents the set of direct parent nodes and x_p denotes parent coordinates. Nodes are sorted by barycenter values and assigned the closest available position, ensuring balanced distribution while maintaining logical connection flow and avoiding edge crossings.

Edge Overlap Resolution. Edge overlaps are resolved through localized adjustments: horizontal ones shift nodes along y-axis, vertical ones along x-axis. High-degree nodes are prioritized to minimize movements and additional crossings. Multiple edges between nodes use bent arrows to preserve coherence.

5   Evaluation and Conclusion

We evaluate SemViSG on two real-world applications: Shopizer (e-commerce platform, 55K nodes, 570K edges) and Adempiere (ERP system, 138K nodes, 815K edges), assessing both semantic enrichment quality and visual clarity.

Semantic Enrichment Results. We evaluate community quality using Semantic Cohesion (average pairwise similarity within communities), Inter-Cluster Separation (distinctness between communities), and Silhouette Score (overall clustering quality). Software graphs' inherent sparsity challenges traditional community detection, yielding weakly cohesive communities with poorly defined boundaries (pre-enrichment Silhouette Score). Our semantic enrichment reveals latent functional groupings, achieving more than 50% increase in Semantic Cohesion with notable improvements across all metrics (Table 1).

Table 1. Community Quality: Semantic Enrichment Impact (Avg. DG and Trans.)

Metric	Shopizer		Adempiere	
	Before	After	Before	After
Semantic Cohesion ↑	0.42	0.64	0.40	0.60
Silhouette Score ↑	0.02	0.41	−0.04	0.25
Inter-Cluster Separation ↑	0.32	0.40	0.32	0.36

Visual Clarity Assessment. Figure 2 illustrates our three-stage visualization improvement: raw graph, semantic communities (force layout), and complete SemViSG with custom hierarchical layout. Quantitative visual improvements are measured using node and edge reduction across subgraphs from our initial partitioning step (DataGraphs and Transactions). Results (Table 2) demonstrate

88% node and 77% edge reduction on Shopizer, with even greater reductions (99%) on larger applications like Adempiere. Meta-node abstraction reduces displayed complexity while interactive drill-down preserves complete information accessibility without typical information loss.

Fig. 2. Visualization progression showing: (Left) Raw graph, (Middle) Semantic community meta-nodes (force layout), (Right) Complete SemViSG approach with custom hierarchical layout.

Table 2. Visual Clarity Metrics (Avg. DG and Trans.)

Metric	Shopizer		Adempiere	
	Raw Graph	SemViSG	Raw Graph	SemViSG
Nb of Nodes ↓	628	77	13,483	103
Nb of Edges ↓	1,738	395	36,612	520

SemViSG integrates multi-layered graph representation, semantic-enriched hierarchical community detection, and custom layout algorithms to address sparse software graph challenges. The interactive environment supports hierarchical exploration through drill-down capabilities and dynamic link type selection for real-time community experimentation, providing software engineers with effective tools for architectural analysis.

References

1. Codesee. https://www.codesee.io
2. Structure101. https://structure101.com
3. Auber, D., et al.: Tulip III. In: Encyclopedia of Social Network Analysis and Mining. Springer, Cham (2014)
4. Bastian, M., et al.: Gephi: an open source software for exploring networks. In: International AAAI Conference on Weblogs and Social Media (2009)

5. De Nooy, W., et al.: Exploratory Social Network Analysis with Pajek. Cambridge University Press (2011)
6. Fruchterman, T., et al.: Graph drawing by force-directed placement. Pract. Experience Softw. (1991)
7. GrammaTech: Codesonar.   https://www.grammatech.com/our-products/codesonar/
8. Newman, M., et al.: Finding and evaluating community structure in networks. Phys. Rev. E (2004)
9. Sugiyama, K., et al.: Methods for visual understanding of hierarchical systems. IEEE Trans. Syst. Cybern. (1981)

Data management and Indices

Certainty Attacks Using Explainability Preprocessing

Carina Newen[1,2]([✉]) [iD], Sofia Vergara Puccini[1], and Emmanuel Müller[1,2] [iD]

[1] Tu Dortmund University, Dortmund, Germany
[2] Research Center Trustworthy Data Science and Security, Dortmund, Germany
carina.newen@cs.tu-dortmund.de

Abstract. With the importance of machine learning rising over the years, learners have been perfected according to their performance on unseen data without considering that the data to be classified could be actively manipulated by an adversary to cause misinterpretations. In this work, we propose to attack the certainty of models. We consider this an important attack angle, as a lot of countermeasures for detection rely on certainty metrics. The new aspect of this paper is that we optimized our attack on four key aspects: The success rate, confidence in the misclassification, transferability of attacks to other models, and the image quality of the generated adversarial. We are introducing this as a means to improve attacks in general regarding key aspects such as certainty of the attack and other desirable attack metrics that do not limit themselves to accuracy. The code can be found at https://github.com/KDD-OpenSource/Certainty-Attacks.git.

1 Introduction

In this paper, we studied the effect of preprocessing adversarials using explanation methods on the certainty of networks. The goal of our method is to determine whether preprocessing of adversarials using explainability methods such as LIME [27] increases metrics such as the certainty of our successful adversarial attacks. This grants us more inconspicuous attacks due to less change of the input and higher model confidence in the misclassification itself. This setup is particularly relevant because standard defence techniques rely on metrics such as certainty to determine adversarial attacks [33]. While we will be using LIME [27] to evaluate our approach, other XAI methods might be feasible. The reasons for choosing LIME over methods such as Shapley values [16] or GRAD-CAM [30] are elaborated in detail in Sect. 1.3. We propose a framework for increasing the certainty of a network in misclassifications while preserving image quality and transferability and maintaining good success rates for the attack. In the past, similar approaches have focused on the transferability of feature-aware attacks [37] or developed algorithms with which the attack direction of the nearest counterfactual explanation could be found. However, no one has determined the influence of preprocessing adversarials using explainability methods

© The Author(s), under exclusive license to Springer Nature Switzerland AG 2026
C. K. Leung et al. (Eds.): DaWaK 2025, LNCS 16048, pp. 209–224, 2026.
https://doi.org/10.1007/978-3-032-02215-8_15

and then measuring the change of certainty in those adversarial examples. We show in the following that our method balances certainty, transferability, attack success rate, and image quality of produced adversarials, giving good results for all of the above. In contrast, competing methods perform poorly in at least one or two aspects.

1.1 Adversarial Machine Learning

Adversarial examples were first introduced by Szegedy et al. in 2013 [35], where it was discovered that targeted attacks were both feasible and easy to construct in a targeted manner using slight perturbations on the input data for any deep learners. While there is still no clear answer to when and why this phenomenon occurs, there have been countless attempts in the past to understand adversarial attacks and their causes. For example, Goodfellow et al. [10] claimed them to be caused by the linear nature of those networks. Since then, there have been discussions about the transferability of those attacks and their effectiveness in black-box settings [15], as well as different attack strategy suggestions [3, 24]. To go on, there have been several attempts at finding out the causes for transferability [22, 40–42]. There have also been several attempts to find countermeasures for those attacks, such as defensive distillation [25] or adversarial training [10]. However, these counterattacks prove ineffectual on larger scales [13]. This is why studying the properties of such attacks is important for future work. In this paper, we aim to build an attack that utilizes the benefits of explanations to generate adversarial examples that optimize the certainty in the misclassifications and the perturbation size to remain as inconspicuous as possible.

In this paper, we will show that our attack is feasible for emotion detection disruption using the datasets JAFFE [17] and FER13 [9], as well as standard datasets like [12] and high dimensional datasets such as Imagenet [29], by using 12 different models and regarding the attack strategies FGSM [10], BIM [14], Deepfool [19] and our own approach. A derived strategy from FGSM [10], capable of having a higher success rate, is named the Basic Iterative method (BIM) [14], which is why we will be testing both. Another technique in the literature known for its simplicity and high success rate is Deepfool [19]. This strategy is based on the iterative linearization of the target model, projecting a clean image towards the next closest hyperplane that produces a different label. Moosavi-Dezfooli et al. [19] compared the manipulations generated by Deepfool and FGSM and observed that Deepfool produces perturbations that are hardly perceptible compared to FGSM while having the same misclassification rate. We will be comparing our approach to these attack methods.

Additionally, to reduce the perturbation needed for fooling a model and predicting a specific (target) label, the Jacobian Based Saliency Map attack [39] focuses on measuring the impact the pixels have on the final classification using forward derivatives and a saliency map. However, [3] mentions that due to the approach of searching only a pair of pixels that increases the likeliness of the target class and decreases it for the rest, the strategy is computationally expensive and cannot be directly employed on data sets such as ImageNet. We, therefore, do not consider this strategy, as it can hardly be performed on high-resolution datasets without considerable computation effort. Furthermore, the white box gradient-based attack proposed by [3] is one of the strongest attacks found in the literature. The C.W. strategy [3] formulates the search for the optimal distortion as an optimization problem that seeks to minimize the distance between the original image and the perturbed one subject to the constraint that the model misclassifies the manipulated instance. This method has shown to be effective against defence mechanisms, but it is susceptible to hyperparameter tuning [43]. Its computation is, however, much less efficient than that of the proposed competitors. We compared attacks with low computation time and also evaluated high dimensional data such as Imagenet [29]. We do not consider Universal Adversarial Examples [18], as they are based on Deepfool but optimized for their transferability, which is only one of the criteria specified. We will later see Deepfool's poor performance in terms of attack confidence.

1.2   Explainable A.I. and Combinations with Adversarials and Uncertainties

The amount of research conducted on the explainability of artificial intelligence models has gained recognition during the last years, given the development of modern machine learning techniques capable of solving real-life problems with high precision but at the same time having a complex logic unintelligible for humans [36]. Several distinguished visualization techniques exist. While it started with LIME [27], it was soon extended into Anchors [28] and followed by entirely new approaches such as SHAPex [16]. Extensions of explainable A.I. to uncertainty measures include Unsupervised DeepView [21], Counterfactual Latent Uncertainty Explanations [1], or other efforts to generate uncertainty estimates for explainable A.I. methods [2,20]. FIMAP provides a reconciliation between adversarial learning and counterfactual explanations [6]. This method computes minimal adversarial perturbations as useful explanations. Other approaches include adversarial detection using Shapley values [8] or attacks on explainable A.I. [32]. None of these studies focus on the effect of uncertainties when enhancing adversarials with explainability preprocessing techniques.

1.3   Motivation for the Choice of LIME

For this evaluation, we were aiming for a broader idea: Does preprocessing with explainable A.I. impact desirable attack metrics, such as certainty in the misclassification, that are not commonly considered when evaluating attack types? This means that our main contribution does not focus on using a single XAI algorithm. However, we decided to focus on a reasonably efficient process in the sense that it could still be feasibly computed when calculating several iterations of it. We need those iterations if no new label can be obtained, for instance, after calculating an adversarial pattern for the preprocessed instance, which means that choices such as Shapely values [16] would have been too inefficient as the main framework due to its quadratic runtime. It has also been proven to be an NP-hard problem [11]. On the other hand, we aimed for a model-agnostic approach that excludes approaches like CAM [23] or GRAD-CAM [30] because we wanted it to be as applicable to different attack approaches and models as possible in real-life scenarios and not just limit ourselves to specific network types such as convolutional neural networks. On the other hand, we also wanted to test this with one of the more well-known algorithms, as the exact choice of algorithm to use is not really important for the aimed contribution. While it would be interesting to see how other XAI approaches influence metrics such as the certainty in the misclassification, this is therefore out of the scope of this paper.

2 Suggested Approach

Algorithm 1. Adversarial on a Subset of the Feature Space

Require: A classifier f, the pair $(\mathbf{x}, y)$, a subset of the original image $\mathbf{x}^p$, the number of iterations T, the vector of perturbations $\boldsymbol{\alpha}$, and the decay factor μ

Ensure: The intermediate adversarial image $\mathbf{x}^*$ and its label $\hat{y}^*$
   $g_0 \leftarrow 0$
   $\mathbf{x}_0^{p*} \leftarrow \mathbf{x}^p$
   $y^p \leftarrow \operatorname{argmax}(f(\mathbf{x}^p))$ $\triangleright$ y^p is assumed to be the true label of the subset image
   $\hat{y}_0^* \leftarrow y$
   $\mathbf{z} \leftarrow \mathbf{x} - \mathbf{x}^p$ $\triangleright$ The complement of the subset $\mathbf{x}^p$ in the original image
   α^t is the t-th element of vector $\boldsymbol{\alpha}$
   $t \leftarrow 0$
   while $\hat{y}_t^* == y$ **and** $t < T$ **do**
      $\boldsymbol{\eta} \leftarrow \nabla_{\mathbf{x}^p} J(\boldsymbol{\theta}, \mathbf{x}_t^{p*}, y^p)$
      $g_{t+1} \leftarrow \mu \cdot g_t + \frac{\boldsymbol{\eta}}{\|\boldsymbol{\eta}\|_1}$
      $\mathbf{x}_{t+1}^{p*} \leftarrow \mathbf{x}_t^{p*} + \alpha^t \cdot \operatorname{sign}(g_{t+1})$
      $\mathbf{x}_{t+1}^* \leftarrow \mathbf{z} + \mathbf{x}_{t+1}^{p*}$
      $\hat{y}_{t+1}^* \leftarrow \operatorname{argmax}(f(\mathbf{x}_{t+1}^*))$
      $t \leftarrow t + 1$
   end while
   return $\mathbf{x}^*, \hat{y}^*$

Our approach aims to uncover the question of whether the combination of explainable A.I. with adversarial attacks influences certainty. We create a successful adversarial example based on the features that have the highest importance in the original classification of a sample. Thus, a perturbation is generated only in a subset region of a clean instance. For an image classification task with k classes using a model f whose output is the k-dimensional vector containing the probability of each class, the generation of an adversarial exploiting only a subset of the sample image is accomplished by Algorithm 1. Let $\mathbf{x}$ be a clean sample image with ground truth label y and predicted label $\hat{y}$, obtained by getting the class with the highest probability produced by a classifier f with a loss function J and parameters $\boldsymbol{\theta}$. An adversarial image is only generated if the model can identify the correct label of a clean instance. $\mathbf{x}^p$ is a binary mask of pixels deemed as important by LIME. For this subset image $\mathbf{x}^p$ of $\mathbf{x}$, including only the p features that contribute the most for obtaining $\hat{y}$, the predicted class of $\mathbf{x}^p$ is obtained by $\operatorname{argmax}(f(\mathbf{x}^p)) = \hat{y}^p$. This is assumed to be the original class of $\mathbf{x}^p$, following $\hat{y}^p = y^p$ only for non-perturbed subsets. In the first step of the algorithm, the direction of the most significant change is calculated using only the subset image $\mathbf{x}^p$ and its original label y^p. This means that the gradient is only computed for the most important pixels in $\mathbf{x}^p$. Similar to BIM [14], we calculate the gradient for the subset image. The subset adversarial image $\mathbf{x}^{p*}$ is calculated by multiplying the sign of the velocity vector by the previously defined size of the perturbation α, such that:

$$\mathbf{x}^{p*}_{t+1} = \mathbf{x}^{p*}_t + \alpha^t \cdot \operatorname{sign}(g_{t+1}) \tag{1}$$

The entire image is then computed with the adversarial subset image $\mathbf{x}^{p*}_t$ and the rest of less important features $\mathbf{z} = \mathbf{x} - \mathbf{x}^p$, creating the full adversarial image as $\mathbf{x}^*_t = \mathbf{z} + \mathbf{x}^{p*}_t$. At each iteration, the label $\hat{y}^*_t$ of the adversarial is calculated using f, thus stopping the algorithm if a new label different from y has been attained. If the strategy has not found a different label for the subset $\mathbf{x}^p$, and the T iterations are completed, the number of most relevant features p used in the image increases (Algorithm 2), and the process is repeated.

The generation of the subset image $\mathbf{x}^p$ passed as input to Algorithm 1 uses an explainability method as depicted in Algorithm 2.

The Euclidean distance d between the subset $\mathbf{x}^p$ and $\mathbf{x}$ is calculated, and if this is greater than a threshold r, the algorithm keeps looking for a subset image, and consequently an adversarial. The maximal possible distance is computed using a picture of full black pixels as the baseline against the U pixels of image x. The initial choice of p is founded on experimental comparison to BIM, but the number of features increases using Algorithm 2 if no adversarial is found. This way we ensure comparable amount of iterations and a general setup that works in practice. The perturbation sizes were set to 0.001 for FGSM and BIM as these are typical α choices in practice, and we compared for the same and differing αs. The subset image $\mathbf{x}^{p_i}$ is passed to Algorithm 1 to compute the adversarial image $\mathbf{x}^*_i$ and its predicted label $\hat{y}^*_i$.

Algorithm 2. Generation of Adversarial Image with LIME

Require: Parameters as in Algorithm 1, N (neighborhood size of $\mathbf{x}$), the k labels, the threshold r, the constant ϵ, and optional parameters w for LIME. U represents the number of pixels in the image.

Ensure: The final adversarial image $\mathbf{x}^*$ and its label $\hat{y}^*$

$\mathrm{d}_{\max} \leftarrow \sqrt{\sum_{u=0}^{U}(x_u - 0)^2}$

$d_0 \leftarrow \mathrm{d}_{\max}$

$p_0 \leftarrow 10$

$\mathbf{x}_0^* \leftarrow \mathbf{x}$

$\hat{y}_0^* \leftarrow y$

$\mathbf{temp}_{\text{before}} \leftarrow \mathbf{0}$ ▷ Array with same dimensions as $\mathbf{x}$

Calculate full explanation:

    $\mathbf{temp} \leftarrow \mathrm{LIME}_{\text{explain}}(\mathbf{x}, f, N, k, w)$ ▷ LIME explanation

while $d_i > \mathrm{d}_{\max} \cdot r$ **and** $\hat{y}_i^* == y$ **do**

    $\mathbf{x}^{p_i} \leftarrow \mathbf{temp}(p_i)$ ▷ Subset image from the most important p_i features

    if $\mathbf{x}^{p_i} \neq \mathbf{temp}_{\text{before}}$ **then**

        $\mathbf{x}_{i+1}^*, \hat{y}_{i+1}^* \leftarrow$ Algorithm 1 $(f, J, \mathbf{x}, y, \mathbf{x}^{p_i}, \boldsymbol{\alpha}, \mu, T)$

        $\mathbf{temp}_{\text{before}} \leftarrow \mathbf{x}^{p_i}$

        $d_{i+1} \leftarrow \sqrt{\sum_{u=0}^{U}(x_u - x_u^{p_i})^2}$

        $p_{i+1} \leftarrow p_i + \epsilon \cdot z$

        $i \leftarrow i + 1$

    else

        $\mathbf{x}^l \leftarrow \mathbf{x} - \mathbf{x}^{p_i}$

        $\mathbf{x}^*, \hat{y}^* \leftarrow$ Algorithm 1 $(f, J, \mathbf{x}, y, \mathbf{x}^l, \boldsymbol{\alpha}, \mu, T)$

        break

    end if

end while

return $\mathbf{x}^*, \hat{y}^*$

---

3   Time Complexity of Our Approach

To summarize computational costs, we take a look at the costs of Algorithms 1 and 2. Algorithm 1 is a variation of the FGSM [10] algorithm in essence, so defining T as the number of iterations of the while loop, the dominant computation factor is $\eta \leftarrow \nabla_{\mathbf{x}^p} J(\boldsymbol{\theta}, \mathbf{x}_t^{p*}, y^p)$. This is typically $\mathcal{O}(n)$, with n the number of parameters of the network. We define

- U_p: as the number of pixels in the subset x^p
- C_f as the cost of one forward pass through the model f
- U as the total number of pixels in the image
- T as the max number of iterations
- d as the number of features (e.g. superpixels)
- I: max number of outer loop iterations of Algorithm 2

This means that the normalization operation costs $\mathcal{O}(U_p)$, which is smaller than the reconstruction time of the full image from subset and complement $\mathcal{O}(U)$. In total, the runtime of Algorithm 1 is $\mathcal{O}(T \cdot (n + C_f + U))$. Algorithm 2 employs

LIME, and considering that it performs linear regression the runtime cost of LIME is $\mathcal{O}(N \cdot C_f + N \cdot d^2 + d^3)$. This gives us the overall cost of

$$\mathcal{O}(N \cdot C_f + N \cdot d^2 + d^3 + I \cdot T(n + C_f + U))$$

In practice, the attack takes less than a second on an NVIDIA A100-SXM4-80GB GPU.

4   Evaluation of the Attack

We compare the performance of our attack against some of the most non-targeted popular white box strategies in the literature, namely FGSM [10], BIM [14], and Deepfool [19]. Beforehand, the pre-trained neural network architectures Resnet18, VGG19, and Inceptionv3 from Pytorch [26] are fine-tuned using the FER13 [9], JAFFE [17] and Imagenet [29] datasets. We used three different convolutional neural network architectures to evaluate CIFAR10 [12]. We fine-tuned the models for the highest validation accuracy on the various network architectures with a learning rate of 0.001 and momentum of 0.9, resulting in twelve models. The authors of [7] demonstrated that for 2 to 10 iterations, momentum iterative FGSM does not present high variations concerning the success rate. Thus, the number of iterations is set to $p0 = 10$. The neighbourhood factor for BIM was chosen heuristically, considering the findings of [14], making the perturbation imperceptible. The number of neighbours for LIME for the proposed approach is based on experiments testing how the number of neighbours affects the attack's success, optimized for the highest success rate. The experiments aim to compare the success rate, confidence, quality of the images generated, and transferability rate of the attacks of FGSM, BIM, Deepfool, and the suggested approach, observing in which of these criteria the attacks fail. We want to highlight not just the success rate, confidence, or any other single metric. The benefits of our attack strategy are summarized in Table 1.

Table 1. Characteristics analysed

Characteristic	FGSM	BIM	Deepfool	Our approach
Success rate	Low	High	High	**High**
Confidence	High	High	Low	**High**
Transferability	High	Low	High	**High**
Image Quality	High	Low	High	**High**

Table 2 presents the parameters used in each strategy. These remain unchanged across all target models and data sets exposed before in order to provide a comparable baseline: We wanted to measure how each attack performs if the perturbation size remains the same. The pixels of the images take values from 0 to 1. Furthermore, the adversarial attacks are only applied to images correctly labelled by the models to guarantee that the assignment of a label different from the actual class is due to the effectiveness of the attacks. [7] demonstrated that for 2 to 10 iterations, momentum iterative FGSM does not present high variations concerning the success rate, which is why we compared the attacks at 10 iterations- we allowed Deepfool to have even more iterations in order to find adversarials successfully. The perturbation choice is a common choice in literature for α [3].

4.1   Confidence of Adversaries

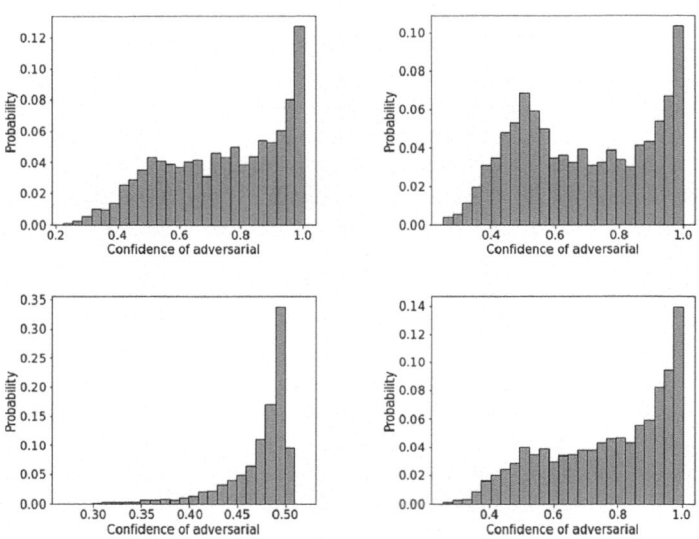

Fig. 1. confidence of adversaries generated using as target Resnet18-FER13 of BIM (top left), the proposed strategy (top right), Deepfool (bottom left), and FGSM (bottom right).

Comparing the confidence of the adversaries that successfully fooled the models, it is possible to recognize in Fig. 1 that in the case of the model Resnet18-FER13, the success rate of Deepfool is 100%. Still, the adversaries generated have a maximum confidence of 50%. This occurs given that Deepfool projects the original instance to the nearest hyperplane until a new label is predicted, and the algorithm is stopped once this has been attained; thus, the instances generated tend to reside close enough to the original images. In contrast, the new labels of the adversaries produced by BIM have an average probability of occurrence of 74%, and above 50% of them have a confidence of 77%. Considering that the termination criterion of the Basic Iterative Method algorithm is based on the completion of all iterations, the instances are continuously pushed toward the direction of maximum change, even though, most likely, a different label is already achieved on previous iterations. Regarding the proposed attack, changing only a subset of the feature space from the original image, i.e., the regions describing the samples' final class, obtains a distribution of probabilities with similar behaviour to FGSM and BIM, with an average confidence of the new predictions equal to $\sim70\%$. They all follow a common pattern: low confidence rates for Deepfool and a more even distribution for all other three strategies.

Table 2. Parameters used in adversarial strategies

Parameter	FGSM	BIM	Deepfool	Our approach
Iterations	-	10	12	10
Perturbation size α	0.001	0.001	-	0.001 to 0.01
Decay factor μ	-	-	-	0.006
Neighborhood factor ϵ	-	0.006	-	-
Number of neighbours	-	-	-	100 (FER), 70 (JAFFE)

The confidence and success rate experiments suggest that although Deepfool has the highest success rate, it cannot produce adversaries with high confidence values for the winning class. In contrast, FGSM has a low success rate, particularly for the models trained on FER13, but it produces perturbed images with the highest probabilities among the strategies studied. The proposed approach offers a trade-off between confidence and success rate, obtaining satisfactory results for both elements. This is also achieved using BIM. But, BIM does not offer high transferability or high image quality, which we will discuss in the following. It is self-explanatory why high image quality would benefit an attacker.

4.2 Transferability

In real-world scenarios, the attacker is unlikely to know the target model's architecture. Therefore, the transferability of adversaries is a desired property, guaranteeing that perturbations generated with a different model from the original thread model are still capable of fooling this. White box strategies can be employed on surrogate models simulating the predictive pattern of the target model either by querying the system multiple times or having *apriori* information about the distribution of the training set and the tuples (x,y). The worst-case scenario occurs when the attacker cannot query the model and does not have information about the training data distribution but only knows a portion of the pair of clean instances used during training and their respective labels [5]. Goodfellow et al. [35] suggest that adversaries can transfer even if another classifier differing in architecture from the target is used to create perturbations. Hence, using a white box strategy on a surrogate target classifier that might differ in architecture from the original target model but is trained on the tuples (x,y) is equivalent to a restricted black box setup. The transferability rate is calculated as:

Table 3. Transferability of attacks

Dataset	f_1	f_2	Success FGSM	Success BIM	Success Deepfool	Success New attack
FER13	Resnet18	VGG19	56.2%	25.3%	56.2%	**58.4%**
		InceptionV3	58.4%	33.7%	58.5%	**59.8%**
	VGG19	Resnet18	66.9%	48.3%	69.5%	**69.9%**
		InceptionV3	60.2%	44.1%	61.8%	**62.0%**
	InceptionV3	Resnet18	68.3%	36.9%	68.3%	**68.9%**
		VGG19	58.0%	25.1%	57.9%	**59.6%**
JAFFE	Resnet18	VGG19	**82.47%**	31.44%	**82.47%**	81.95%
		InceptionV3	**82.47%**	31.44%	**82.47%**	81.95%
	VGG19	Resnet18	**85.78%**	39.21%	**85.78%**	**85.78%**
		InceptionV3	83.33%	**100%**	83.33%	**100%**
	InceptionV3	Resnet18	**85.78%**	39.21%	**85.78%**	85.29%
		VGG19	83.33%	**100%**	83.33%	**100%**
CIFAR10	ConvNet1	ConvNet2	**76.9%**	**76.9%**	**77.0%**	**76.9%**
		ConvNet3	**98.7%**	**98.7%**	**98.7%**	**98.8%**
	ConvNet2	ConvNet1	**71.8%**	**71.8%**	**71.8%**	**71.8%**
		ConvNet3	**99.4%**	**99.4%**	**99.4%**	**99.4%**
Imagenet	Resnet18	VGG19	**100%**	**100%**	**100%**	**100%**
		InceptionV3	**100%**	**100%**	**100%**	**100%**

$$T^{f_1}_{f_2} = \frac{\sum_{\forall \mathbf{x}^*} \mathbb{1}(f_2(\mathbf{x}^*) \neq y_{true})}{N}, \qquad (2)$$

where $\mathbf{x}^*$ represents the adversarial crafted with the "substitute" model f_1 and $f_2(\mathbf{x}^*)$ is the classification obtained for the adversarial $\mathbf{x}^*$ using the target model f_2. It calculates the percentage of perturbed images originally crafted by f_1 and successfully misclassified using f_2 concerning their original label. Table 3 shows how adversaries generated from different strategies using the neural network architecture exposed at the beginning of transfer between each other. The lowest transferability rate is observed using the Basic Iterative Method. In contrast, FGSM, Deepfool, and the proposed method have similar success rates across all experiments, showing an advantage for the proposed approach with comparable averages. This indicates that the regions determined as the most critical features for the original classification of the instances are similar across all models, disregarding their architecture, implying that the property of implementation invariance is fulfilled [34]. Either the transferability was higher for our attack, or the networks were too close to one another to detect meaningful distances. If it performed slightly worse, it was within the statistical error rate.

4.3   Quality of the Images Produced

To quantify the indistinguishability of the crafted perturbation on the original image, three distance measures are calculated between the original and manipulated data points for each model and data set. The Root Mean Squared Error (RMSE) [4] measures the squared intensity differences between both images at a pixel level, averaging this over the total amount of pixels. Given the lack of representation of the perceived visual quality of an image [38] by using only the RMSE, the Structure Similarity Measure (SSIM) [38] and Visual Information Fidelity (VIF) [31] are as well analyzed. SSIM provides the normalized mean of the structural information of the two images, taking into account the luminescence, contrast, and structure of these. An SSIM equal to one is equivalent to the structure similarity being perfect. Visual Information Fidelity quantifies the loss of information in the distortion process of an image and how this relates to its visual quality. This is a non-negative measure for which values lower than 1 represent a worse quality of the perturbed image, a VIF equal to one suggests no loss of information in the distortion process, and values above mean that the contrast of the manipulated image is higher than in the clean one. Table 4 shows the values of these metrics for the FER13, JAFFE CIFAR-10, and Imagenet data set on Resnet18, VGG19, InceptionV3, and three Convolutional Neural Networks. It is visible that for the three metrics, the worst performance is obtained by far by the Basic iterative Method. FGSM, Deepfool, and the proposed approach have small values for the RMSE. This indicates minimal pixel differences between the perturbed and original instances. Regarding the Visual Information Fidelity of the adversaries, the Basic iterative method has across all experiments values greater than 1, indicating that the contrast of the regions in the images is higher, and the images have better quality [31].

However, in this context, it is desirable that the VIF of each pair of images is close to 1 so that the manipulation appears imperceptible. BIM performs significantly worse. FGSM, Deepfool, and the proposed approach have values close to 1, indicating that the loss of information in the perturbation process is relatively manageable for the strategies' output images. The Structure Similarity Measure for the images crafted by the Basic Iterative Method is, on average, much lower. At the same time, on the remaining strategies, the values are proximate to 1, disclosing again the strong distortion effect produced by the Basic Iterative Method (Table 5).

Table 4. Average similarity measure for each strategy

Dataset	Model	FGSM			BIM			Deepfool			Proposed Approach		
		RMSE	VIF	SSIM	RMSE	VIF	SSIM	RMSE	VIF	SSIM	RMSE	VIF	SSIM
FER13	Resnet18	5.3×10^{-4}	~1.00	0.99	0.56	1.89	0.59	3.8×10^{-4}	~1.00	0.99	3.8×10^{-3}	0.99	0.98
	VGG19	5.4×10^{-4}	0.99	0.99	0.56	1.90	0.59	2.2×10^{-4}	0.99	0.99	4.4×10^{-3}	0.99	0.97
	InceptionV3	5.3×10^{-4}	~1.00	0.99	0.56	1.88	0.59	2.8×10^{-4}	0.99	0.99	5.8×10^{-3}	0.99	0.96
JAFFE	Resnet18	6.1×10^{-4}	~1.00	0.99	0.82	3.22	0.49	3.6×10^{-4}	0.99	0.99	1.9×10^{-3}	0.99	0.98
	VGG19	6.1×10^{-4}	~1.00	0.99	0.81	3.21	0.49	3.5×10^{-4}	~1.00	0.99	2.2×10^{-3}	0.99	0.98
	InceptionV3	6.1×10^{-4}	~1.00	0.99	0.81	3.21	0.49	3.4×10^{-4}	~1.00	0.99	2.2×10^{-3}	0.99	0.98
CIFAR-10	ConvNet1	3.6×10^{-4}	~1.00	~1.00	0.59	4.10	0.33	5.7×10^{-4}	~1.00	~1.00	2.7×10^{-3}	0.99	0.99
	ConvNet2	3.6×10^{-4}	~1.00	~1.00	0.62	4.80	0.32	8.0×10^{-3}	0.99	0.99	0.001	~1.00	0.99
	ConvNet3	3.6×10^{-4}	~1.00	~1.00	0.59	4.10	0.33	3.8×10^{-4}	~1.00	0.99	0.001	0.99	0.99
Imagenet	Resnet18	4.6×10^{-4}	~1.00	0.99	0.81	2.92	0.37	3.8×10^{-4}	~1.00	0.99	4.8×10^{-3}	0.99	0.93
	VGG19	4.7×10^{-4}	~1.00	~1.00	0.76	2.63	0.39	7.6×10^{-4}	~1.00	~1.00	1.5×10^{-4}	~1.00	~1.00
	InceptionV3	4.6×10^{-4}	~1.00	~1.00	0.79	2.91	0.33	3.8×10^{-4}	~1.00	0.99	0.001	0.99	0.99

Table 5. Success rate of different strategies

Dataset	Images Tested	Model	FGSM	BIM	Deepfool	Our Approach
FER13	5000	Resnet18	68.4%	95.0%	100%	95.9%
		VGG19	64.3%	97.3%	100%	94.9%
		InceptionV3	59.3%	95.0%	100%	81.3%
JAFFE	213	Resnet18	86.5%	100%	100%	99.4%
		VGG19	83.3%	100%	100%	100%
		InceptionV3	83.3%	100%	100%	100%
CIFAR10	60000	ConvNet1	77.4%	16.4%	100%	94.0%
		ConvNet2	71.1%	16.5%	100%	91.8%
		ConvNet3	63.7%	30.3%	100%	87.3%
Imagenet	50000	Resnet18	94.4%	66.6%	100%	100%
		VGG19	90.2%	56.9%	100%	100%
		InceptionV3	40.0%	58.1%	100%	83.6%

4.4   Success Rate

An adversarial strategy should be able to fool a target model continuously to incur possible integrity violations given the inability of the model to provide an accurate prediction in the presence of small perturbations on natural data samples. The success rate of an adversarial strategy is measured as the percentage of times the target model mistakenly predicts a perturbed sample from an originally correctly classified image. Throughout all the experiments, FGSM is the strategy with the lowest capacity of fooling the models, obtaining an average success rate of 73.5%, suggesting that in some cases, the resulting one-step distortion becomes insufficient to generate a different classification by the model. The proposed strategy is a strong competitor against BIM and Deepfool, obtaining a success rate above 90% in almost all cases. It should be noted again here that the success rate alone is not the best for our strategy- instead, a combination of factors is shown in Table 1.

5 Conclusion and Ethical Statement

The proposed approach unifies transferability, high image quality, confidence and success rate in one method. To do so, we used explainable A.I. to craft a minimal manipulation to produce a new label. To identify the advantages of implementing this strategy, the proposed approach was compared with three of the most popular full-dimension techniques for finding adversaries in a non-targeted setup: FGSM, Basic Iterative Method, and Deepfool. The properties tested of each method were the success rate on a particular target model, the confidence of the adversaries generated, the visual quality of the final image compared to the original one, and the transferability rate. The experiments revealed that the proposed method had a similar success rate as the Basic Iterative Method and Deepfool, reaching a success rate above 90% in most cases. This was similar to the results found in the literature, suggesting that iterative adversarial attacks achieve a higher success rate than one-step attacks like FGSM. However, by analyzing the confidence of the adversaries produced, FGSM generated the adversaries with the highest confidence, followed by the proposed method and the Basic iterative method. Deepfool produced adversaries with a classification probability as high as 50%, showing its inability to generate high confidence perturbations. To assess the image quality of the adversaries produced, the Root Mean Squared Error, the Structural Similarity Index, and the Visual Information Fidelity measure were calculated on the pictures produced by each strategy compared to the clean instances. The results showed that FGSM, Deepfool, and the proposed approach were able to compute imperceptible perturbations that were smoother on the pixels and did not affect the structural information or visual quality of an image. In contrast, the Basic Iterative Method reduced the quality of the original images dramatically, meaning that this strategy computes more aggressive perturbations. Only our proposed method had high performance regarding image quality, success rate, confidence, and transferability. We argue that this result is especially ethically important: It shows us that features often proposed as key for the detection of adversarial examples can be made inconspicuous with the help of explanation methods. It is especially important to publish information regarding the ongoing back and forth between attacks and their mitigation and detection, especially due to the ease of implementing such aids.

Acknowledgment. This work was funded by the Research Center Trustworthy Data Science and Security.

References

1. Antorán, J., Bhatt, U., Adel, T., Weller, A., Hernández-Lobato, J.M.: Getting a clue: a method for explaining uncertainty estimates. arXiv preprint arXiv:2006.06848 (2020)
2. Bykov, K., Höhne, M.M.C., Müller, K.R., Nakajima, S., Kloft, M.: How much can I trust you?–quantifying uncertainties in explaining neural networks. arXiv preprint arXiv:2006.09000 (2020)
3. Carlini, N., Wagner, D.: Towards evaluating the robustness of neural networks. In: 2017 IEEE Symposium on Security and Privacy (SP), pp. 39–57. IEEE (2017)
4. Chai, T., Draxler, R.R.: Root mean square error (RMSE) or mean absolute error (MAE). Geoscientific Model Dev. Discuss. $7(1)$, 1525–1534 (2014)
5. Chakraborty, A., Alam, M., Dey, V., Chattopadhyay, A., Mukhopadhyay, D.: A survey on adversarial attacks and defences. CAAI Trans. Intell. Technol. $6(1)$, 25–45 (2021)
6. Chapman-Rounds, M., Bhatt, U., Pazos, E., Schulz, M.A., Georgatzis, K.: Fimap: feature importance by minimal adversarial perturbation. In: Proceedings of the AAAI Conference on Artificial Intelligence, vol. 35, pp. 11433–11441 (2021)
7. Dong, Y., et al.: Boosting adversarial attacks with momentum. In: Proceedings of the IEEE Conference on Computer Vision and Pattern Recognition, pp. 9185–9193 (2018)
8. Fidel, G., Bitton, R., Shabtai, A.: When explainability meets adversarial learning: detecting adversarial examples using shap signatures. In: 2020 International Joint Conference on Neural Networks (IJCNN), pp. 1–8. IEEE (2020)
9. Goodfellow, I.J., et al.: Challenges in representation learning: a report on three machine learning contests. In: Lee, M., Hirose, A., Hou, Z.-G., Kil, R.M. (eds.) ICONIP 2013. LNCS, vol. 8228, pp. 117–124. Springer, Heidelberg (2013). https://doi.org/10.1007/978-3-642-42051-1_16
10. Goodfellow, I.J., Shlens, J., Szegedy, C.: Explaining and harnessing adversarial examples. arXiv preprint arXiv:1412.6572 (2014)
11. Khalil, M., Kimelfeld, B.: The complexity of the Shapley value for regular path queries. arXiv preprint arXiv:2212.07720 (2022)
12. Krizhevsky, A., Hinton, G., et al.: Learning multiple layers of features from tiny images (2009)
13. Kurakin, A., Goodfellow, I., Bengio, S.: Adversarial machine learning at scale. arXiv preprint arXiv:1611.01236 (2016)
14. Kurakin, A., Goodfellow, I.J., Bengio, S.: Adversarial examples in the physical world. In: Artificial Intelligence Safety and Security, pp. 99–112. Chapman and Hall/CRC (2018)
15. Liu, Y., Chen, X., Liu, C., Song, D.: Delving into transferable adversarial examples and black-box attacks. arXiv preprint arXiv:1611.02770 (2016)
16. Lundberg, S.M., Lee, S.I.: A unified approach to interpreting model predictions. In: Advances in Neural Information Processing Systems, vol. 30 (2017)
17. Lyons, M., Akamatsu, S., Kamachi, M., Gyoba, J.: Coding facial expressions with gabor wavelets. In: Proceedings Third IEEE International Conference on Automatic Face and Gesture Recognition, pp. 200–205. IEEE (1998)
18. Moosavi-Dezfooli, S.M., Fawzi, A., Fawzi, O., Frossard, P.: Universal adversarial perturbations. In: Proceedings of the IEEE Conference on Computer Vision and Pattern Recognition, pp. 1765–1773 (2017)

19. Moosavi-Dezfooli, S.M., Fawzi, A., Frossard, P.: Deepfool: a simple and accurate method to fool deep neural networks. In: Proceedings of the IEEE Conference on Computer Vision and Pattern Recognition, pp. 2574–2582 (2016)
20. Newen, C., Müller, E.: Unsupervised deepview: global explainability of uncertainties for high dimensional data. In: 2022 IEEE International Conference on Knowledge Graph (ICKG), pp. 196–202. IEEE (2022)
21. Newen, C., Müller, E.: Unsupervised deepview: global uncertainty visualization for high dimensional data. In: 2022 IEEE International Conference on Data Mining Workshops (ICDMW), pp. 1–8. IEEE (2022)
22. Newen, C., Müller, E.: On the independence of adversarial transferability to topological changes in the dataset. In: 2023 IEEE 10th International Conference on Data Science and Advanced Analytics (DSAA), pp. 1–8. IEEE (2023)
23. Oquab, M., Bottou, L., Laptev, I., Sivic, J.: Is object localization for free?-weakly-supervised learning with convolutional neural networks. In: Proceedings of the IEEE Conference on Computer Vision and Pattern Recognition, pp. 685–694 (2015)
24. Papernot, N., McDaniel, P., Jha, S., Fredrikson, M., Celik, Z.B., Swami, A.: The limitations of deep learning in adversarial settings. In: 2016 IEEE European symposium on security and privacy (EuroS&P), pp. 372–387. IEEE (2016)
25. Papernot, N., McDaniel, P., Wu, X., Jha, S., Swami, A.: Distillation as a defense to adversarial perturbations against deep neural networks. In: 2016 IEEE Symposium on Security and Privacy (SP), pp. 582–597. IEEE (2016)
26. Paszke, A., et al.: Pytorch: an imperative style, high-performance deep learning library. In: Advances in Neural Information Processing Systems, vol. 32 (2019)
27. Ribeiro, M.T., Singh, S., Guestrin, C.: "Why should I trust you?" explaining the predictions of any classifier. In: Proceedings of the 22nd ACM SIGKDD International Conference on Knowledge Discovery and Data Mining, pp. 1135–1144 (2016)
28. Ribeiro, M.T., Singh, S., Guestrin, C.: Anchors: high-precision model-agnostic explanations. In: Proceedings of the AAAI Conference on Artificial Intelligence, vol. 32 (2018)
29. Russakovsky, O., et al.: Imagenet large scale visual recognition challenge. Int. J. Comput. Vis. **115**, 211–252 (2015)
30. Selvaraju, R.R., Cogswell, M., Das, A., Vedantam, R., Parikh, D., Batra, D.: Grad-cam: visual explanations from deep networks via gradient-based localization. In: Proceedings of the IEEE International Conference on Computer Vision, pp. 618–626 (2017)
31. Sheikh, H.R., Bovik, A.C.: Image information and visual quality. IEEE Trans. Image Process. **15**(2), 430–444 (2006)
32. Slack, D., Hilgard, S., Jia, E., Singh, S., Lakkaraju, H.: Fooling lime and shap: adversarial attacks on post hoc explanation methods. In: Proceedings of the AAAI/ACM Conference on AI, Ethics, and Society, pp. 180–186 (2020)
33. Smith, L., Gal, Y.: Understanding measures of uncertainty for adversarial example detection. arXiv preprint arXiv:1803.08533 (2018)
34. Sundararajan, M., Taly, A., Yan, Q.: Axiomatic attribution for deep networks. In: International Conference on Machine Learning, pp. 3319–3328. PMLR (2017)
35. Szegedy, C., et al.: Intriguing properties of neural networks. arXiv preprint arXiv:1312.6199 (2013)
36. Vilone, G., Longo, L.: Explainable artificial intelligence: a systematic review. arXiv preprint arXiv:2006.00093 (2020)
37. Wang, Z., Guo, H., Zhang, Z., Liu, W., Qin, Z., Ren, K.: Feature importance-aware transferable adversarial attacks. In: Proceedings of the IEEE/CVF International Conference on Computer Vision, pp. 7639–7648 (2021)

38. Wang, Z., Bovik, A.C., Sheikh, H.R., Simoncelli, E.P.: Image quality assessment: from error visibility to structural similarity. IEEE Trans. Image Process. **13**(4), 600–612 (2004)

39. Wiyatno, R., Xu, A.: Maximal Jacobian-based saliency map attack. arXiv preprint arXiv:1808.07945 (2018)

40. Wu, L., Zhu, Z.: Towards understanding and improving the transferability of adversarial examples in deep neural networks. In: Asian Conference on Machine Learning, pp. 837–850. PMLR (2020)

41. Wu, L., Zhu, Z., Tai, C., et al.: Understanding and enhancing the transferability of adversarial examples. arXiv preprint arXiv:1802.09707 (2018)

42. Xie, C., et al.: Improving transferability of adversarial examples with input diversity. In: Proceedings of the IEEE/CVF Conference on Computer Vision and Pattern Recognition, pp. 2730–2739 (2019)

43. Yao, Z., Gholami, A., Xu, P., Keutzer, K., Mahoney, M.W.: Trust region based adversarial attack on neural networks. In: Proceedings of the IEEE/CVF Conference on Computer Vision and Pattern Recognition, pp. 11350–11359 (2019)

Integrating Bitcoin Transactions into Relational Databases for IoT: Challenges and Solutions

Rebeca Tonu$^{(\boxtimes)}$, Otilia Muntean , and Ciprian Pungilă

Faculty of Mathematics and Informatics, West University of Timișoara,
Timișoara, Romania
{rebeca.tonu,otilia.muntean97,ciprian.pungila}@e-uvt.ro
https://info.uvt.ro

Abstract. Extracting and processing Bitcoin transaction data into relational database structures involves numerous technical and conceptual challenges, primarily due to the decentralized and append-only nature of blockchain. This project examines the complex challenges encountered in the process of finding the best methods for processing and potentially integrating data from the Bitcoin blockchain into a relational database suitable for IoT purposes. We perform empirical experiments in order to determine the validity of our findings, by comparing common queryable datasets using both our proposed relational database schema and optimizations, and blockchain-driven methods. We observed large performance gaps between our proposed schema and the alternative Bitcoin queryable databases, and we discuss their impact in real-world scenarios.

Keywords: Bitcoin blockchain · big data · relational database · data extraction tools · heterogeneous computing · fraud detection · money laundering

1 Introduction

The growing integration of blockchain, particularly Bitcoin, into IoT enables devices to act as autonomous agents—handling tasks like delivery, charging, or maintenance while earning and spending Bitcoin. Its decentralized nature suits logging sensor data in supply chains, enhancing transparency. Devices like printers or parking meters can use Bitcoin for seamless per-use payments. However, integrating Bitcoin data into relational databases poses challenges due to the blockchain's complexity, size, and the limited performance and compatibility of existing tools. This research evaluates current solutions, identifies gaps, and highlights areas for improvement. An extended version of this paper is present at [1].

© The Author(s), under exclusive license to Springer Nature Switzerland AG 2026
C. K. Leung et al. (Eds.): DaWaK 2025, LNCS 16048, pp. 225–230, 2026.
https://doi.org/10.1007/978-3-032-02215-8_16

2   State of the Art

Blockchain-IoT integration improves supply chains by cutting admin costs and boosting efficiency via smart contracts and real-time data sharing [2]. Hannon et al. [3] propose a scalable, low-resource Bitcoin payment channel for IoT. Miraz et al. [4] emphasize blockchain's role in enhancing IoT security. Fernández-Caramés et al. [5] address authentication and privacy using decentralized ledgers and suggest alternative consensus mechanisms. Degala et al. [6] automate agriculture with smart contracts, while Peker et al. [7] reduce storage costs in supply chains. Other uses include secure storage [8], micropayments [9], data marketplaces [10], and health monitoring privacy [11].

For blockchain analysis, Schnoering et al. [19] created a Bitcoin transaction graph with over 252M nodes. Yue et al. [20] assessed SQL-based Bitcoin queries; Xuanwu Yue et al. [21] built BitExTract for visual analytics. Spagnuolo et al. [22] and Mun et al. [23] explored RDBMS approaches, and Nathan et al. [24] proposed a decentralized PostgreSQL-based system.

3   Implementation

Bitcoin data processing remains difficult due to its size and outdated tools. By Dec. 2024, syncing 620 GB with Bitcoin Core took 10 days. Most parsers are unmaintained, hard-coded, or lack key features. A Septâ€“Dec 2024 search found few viable tools. Only functional, goal-aligned tools were used in our tests. Tools like znort-block-parser [13], bitcoin-blockchain-parser [16], and Bitcoin-DatabaseGenerator [17] had compatibility issues. Others, like blockchain-parser [14] and python-bitcoin-blockchain-parser [15], require heavy pre-processing or fail to run with Bitcoin Core. BlockSci [18] is capable but unsupported and resource-intensive. RPC [25] is reliable but complex and bandwidth-limited. Relational databases show promise, but tool support is scarce. Only Bitcoin-DatabaseGenerator offers parallelism, yet it's outdated. Can a new relational schema outperform these tools for querying Bitcoin transactions in IoT use cases?

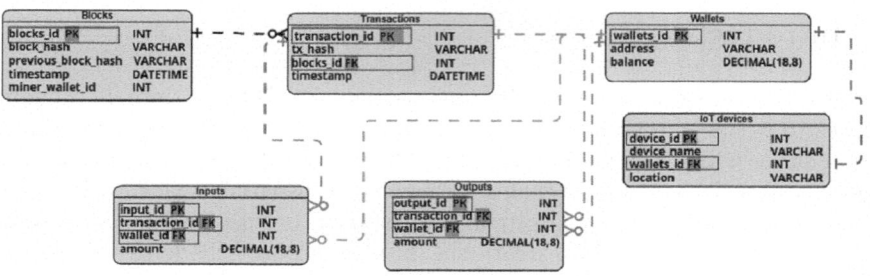

Fig. 1. Proposed database structure.

Our public GitHub repo [12] lists tool attributes and classifies them by deployment difficulty, dependencies, access method, development status, and functionality. Pre-experiment setup required adapting tools with varied languages and structures. Google Cloud's BigQuery [26] was tested but abandoned due to high costs ($750 for <50 queries). No tool supported fast parallel processing natively, so all operations were manually implemented.

Figure 1 shows a scalable, normalized blockchain schema with six linked tables: blocks, transactions, inputs, outputs, wallets, and IoT devices. Transactions connect blocks to fund flows; wallets link many-to-many with transactions, and each IoT device maps to one wallet for payments and authentication. Foreign keys and indexing enable fast queries. Designed for real-time IoT micropayments, it supports ongoing optimization. Based on this schema, three scenarios are presented: (1) BTC spending by EV chargers over a week, (2) most active devices in 24 h, and (3) a toll system flagging payments over 0.5 BTC (Fig. 2).

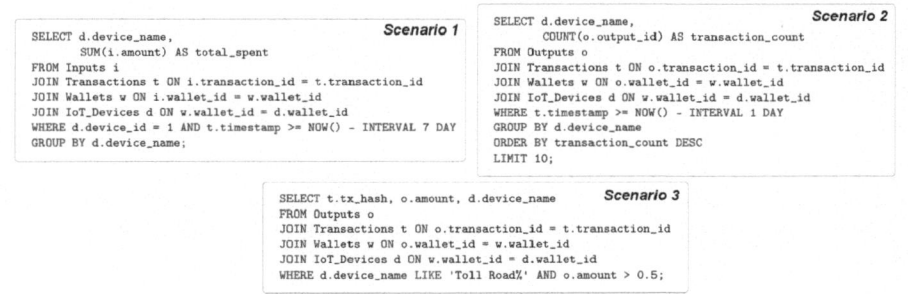

Fig. 2. Scenario queries used.

4   Results

We evaluated performance using Bitcoin Core to download 620 GB of blockchain data. Tests ran on a Windows 10 machine with an i7-12700H CPU, 16 GB RAM, and a VM with 6 GB RAM on Ubuntu 22.04. Parsing bandwidth (Mbps) was used as the main efficiency metric. We analyzed 100 BLK files (153MB, 29.6M transactions across 275,000 blocks), measuring extraction speed to ensure consistent tool comparison and highlight analysis scale.

We benchmarked several blockchain parsers on 100 raw Bitcoin blocks (99.84 MB), but all were outdated or inadequate [1]. We then tested our proposed IoT-focused Bitcoin relational database (approach 1) using MySQL, averaging query times over three runs for the three use-case scenarios. In parallel, we implemented the same queries programmatically (approach 2) by directly querying the Bitcoin Core daemon for wallets, transaction hashes, and I/O data. Data was stored using hashtables and binary-searchable lookup tables for efficient access. Figure 3 shows results; Fig. 4 compares retrieval times between the two approaches.

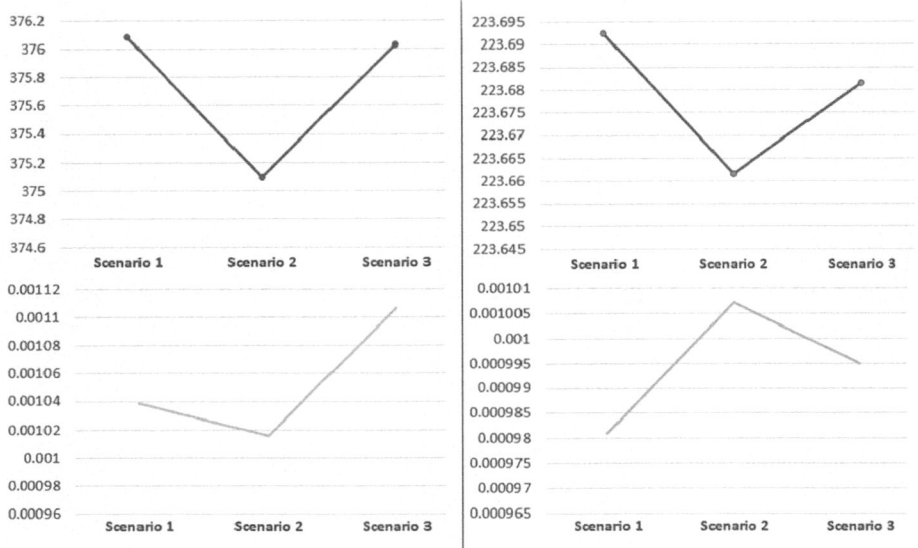

Fig. 3. Execution time performance (in seconds) for test A (left side) and test B (right side), for both approaches, i.e. programmatic implementations (top, with blue color) and relational database queries (bottom, with green color). (Color figure online)

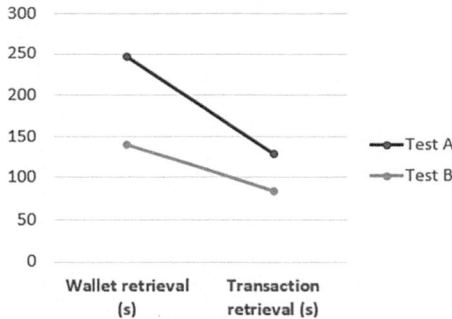

Fig. 4. Comparison of programmatic Bitcoin Core data retrieval methods.

We conducted two tests. Test A used 50,000 entries each for wallets, transactions, inputs, outputs, and IoT devices to simulate a detailed environment. Test B used 50,000 wallets, 100,000 transactions, 40,000 inputs, 30,000 outputs, and 1,000 IoT devices to reflect real-world conditions with more transactions per wallet and fewer devices. Results showed minimal variation: MySQL differed by <8.2% in test A and <3% in test B; the programmatic approach showed <1% variation in both. However, direct Bitcoin database queries were much slower, making them unsuitable for real-time or repeated queries.

5   Conclusions

This study examines the challenges of using blockchain storage for IoT, focusing on the inefficiencies of current Bitcoin-to-database parsing tools. Due to limited solutions, it explores faster, more efficient methods for retrieving and querying blockchain data. Results show most tools lack scalability and adaptability. Our proposed schema addresses these issues by optimizing key queries, validated through programmatic and relational database real-world tests. Future work includes exploration of NoSQL databases, and the development of memory-efficient, high-throughput heterogeneous architectures for processing big data as part of blockchain integration for IoT devices.

This work has been partially supported by (1) the project virtuaLedger [27], (2) the project RoNaQCI, part of EuroQCI, DIGITAL-2021-QCI-01-DEPLOY-NATIONAL, 101091562, (3) the project "Romanian Hub for Artificial Intelligence - HRIA", Smart Growth, Digitization and Financial Instruments Program, 2021–2027, MySMIS no. 334906.

References

1. Tonu, R., Muntean, O., Pungila, C.: Exploring the Integration of Bitcoin Transactions with Relational Databases in IoT Environments: Issues and Mitigation Strategies. https://www.academia.edu/129920080/Exploring_the_Integration_of_Bitcoin_Transactions_with_Relational_Databases_in_IoT_Environments_Issues_and_Mitigation_Strategies. Accessed 12 June 2025

2. Unlocking the Potential of IoT and Blockchain: Harnessing the Power of Decentralized Data and Smart Devices. https://www.pwc.de/en/strategy-organisation-processes-systems/viewpoint-unlocking-the-potential-of-iot-and-blockchain.pdf. Accessed 29 Dec 2024

3. Hannon, C., Jin, D.: Bitcoin payment-channels for resource limited IoT devices. In: Proceedings of the International Conference on Omni-Layer Intelligent Systems (COINS 2019), pp. 50–57. Association for Computing Machinery, New York (2019). https://doi.org/10.1145/3312614.3312629

4. Miraz, M.H., Ali, M.: Blockchain enabled enhanced IoT ecosystem security. In: Miraz, M.H., Excell, P., Ware, A., Soomro, S., Ali, M. (eds.) iCETiC 2018. LNICST, vol. 200, pp. 38–46. Springer, Cham (2018). https://doi.org/10.1007/978-3-319-95450-9_3

5. Fernández-Caramés, T.M., Fraga-Lamas, P.: A review on the use of blockchain for the internet of things. IEEE Access 6, 32979–33001 (2018). https://doi.org/10.1109/ACCESS.2018.2842685

6. Degala, S., Sarvani, K.L., Jwalitha, D., Latha, I.S.: Integration of blockchain and IoT sensor networks for enhanced transparency and efficiency in agricultural supply chains. Int. Res. J. Eng. Technol. (IRJET) 10(11), 268–273 (2023)

7. Peker, Y.K., Rodriguez, X., Ericsson, J., Lee, S.J., Perez, A.J.: A cost analysis of internet of things sensor data storage on blockchain via smart contracts. Electronics 9(2), 244 (2020). https://doi.org/10.3390/electronics9020244

8. Parmar, M., Kaur, H., Rinku: Blockchain-based secured data transmission of IoT sensors using ThingSpeak. In: Book Title (if available), pp. 77–86 (2021). https://doi.org/10.1201/9781003150664-9

9. Mercan, S., Kurt, A., Akkaya, K., Erdin, E.: Cryptocurrency solutions to enable micropayments in consumer IoT. IEEE Consum. Electron. Mag. **11**(2), 97–103 (2022). https://doi.org/10.1109/MCE.2021.3060720

10. Sober, M., Scaffino, G., Schulte, S., Kanhere, S.S.: A blockchain-based IoT data marketplace. Clust. Comput. **26**(6), 3523–3545 (2023). https://doi.org/10.1007/s10586-022-03745-6

11. Makka, S., Arora, G., Mopuru, B.: IoT based health monitoring and record management using distributed ledger. In: Journal of Physics: Conference Series, vol. 2089, no. 1, p. 012030. IOP Publishing (2021). https://doi.org/10.1088/1742-6596/2089/1/012030

12. Extracted Attributes from Blockchain Parsing Tools. https://github.com/RebecaTonu/btcrelationaldb/blob/master/README.md. Accessed 03 Mar 2024

13. Blockparser. https://github.com/znort987/blockparser. Accessed 12 Dec 2024

14. Blockchain-parser. https://github.com/ragestack/blockchain-parser. Accessed 12 Dec 2024

15. Python-bitcoin-blockchain-parser. https://github.com/alecalve/python-bitcoin-blockchain-parser. Accessed 10 Dec 2024

16. Blockchain-parser. https://github.com/rmull/blockchain-parser. Accessed 10 Dec 2024

17. BitcoinDatabaseGenerator. https://github.com/ladimolnar/BitcoinDatabaseGenerator. Accessed 10 Dec 2024

18. BlockSci. https://github.com/citp/BlockSci. Accessed 08 Feb 2025

19. Schnoering, H., Vazirgiannis, M.: Bitcoin research with a transaction graph dataset. arXiv preprint arXiv:2411.10325 (2024)

20. Yue, K.-B., Chandrasekar, K., Gullapalli, H.: Storing and querying bitcoin blockchain using SQL databases. Inf. Syst. Educ. J. **17**(4), 24–41 (2019)

21. Yue, X., et al.: BitExTract: interactive visualization for extracting bitcoin exchange intelligence. IEEE Trans. Vis. Comput. Graph. **25**(1), 162–171 (2019). https://doi.org/10.1109/TVCG.2018.2864814

22. Spagnuolo, M., Maggi, F., Zanero, S.: BitIodine: extracting intelligence from the bitcoin network. In: Christin, N., Safavi-Naini, R. (eds.) FC 2014. LNCS, vol. 8437, pp. 457–468. Springer, Heidelberg (2014). https://doi.org/10.1007/978-3-662-45472-5_29

23. Mun, H., Lee, Y.: BitSQL: a SQL-based bitcoin analysis system. In: 2022 IEEE International Conference on Blockchain and Cryptocurrency (ICBC), pp. 1–8. IEEE (2023). https://doi.org/10.1109/ICBC54727.2022.9805551

24. Nathan, S., Govindarajan, C., Saraf, A., Sethi, M., Jayachandran, P.: Blockchain meets database: design and implementation of a blockchain relational database. arXiv preprint arXiv:1903.01919 (2019)

25. Bitcoin Developer. https://developer.bitcoin.org/reference/rpc/. Accessed 02 Dec 2024

26. Bitcoin Cryptocurrency. https://console.cloud.google.com/marketplace/product/bitcoin/crypto-bitcoin?project=cvdfd-32441. Accessed 02 Dec 2024

27. The VirtuaLedger project. https://virtualedger.com. Accessed 13 Dec 2024

Effects of Response Length on User Search Experience in Spoken Conversational Search

Ken Tobioka[ID], Takehiro Yamamoto[✉][ID], and Hiroaki Ohshima[ID]

University of Hyogo, Kobe, Japan
af24h009@guh.u-hyogo.ac.jp, t.yamamoto@sis.u-hyogo.ac.jp,
ohshima@ai.u-hyogo.ac.jp

Abstract. Response design plays a critical role in user experience in spoken conversational search (SCS). This study examines how two response styles—long response presentation (LRP) and short response presentation (SRP)—affect users' cognitive load and search behavior under different task complexities. A user study ($N = 139$) was conducted using an LLM-based voice agent across low and high complexity search tasks. Results show that SRP reduced cognitive load and improved satisfaction, especially for low complexity tasks. However, in high complexity tasks, SRP led to more query issuance. This may indicate that users were actively engaging with the fragmented information to better understand the task or fill in missing details. These findings suggest that no single response style is optimal across all situations.

Keywords: SCS · exploratory search · user interface

1 Introduction

Voice-based conversational agents are gaining traction as natural language processing advances. With the rise of large language models (LLMs), such agents are expected to handle more complex tasks, including exploratory search tasks.

Long voice responses from LLMs often exceed users' memory capacity, especially under high task complexity, making information processing difficult.

In this study, we compare two response presentation styles in spoken conversational search (SCS): **long response presentation (LRP)**, which presents information more comprehensively, and **short response presentation (SRP)**, which provides concise and focused responses. Our goal is to investigate how these styles affect users' search experience.

Although prior work has examined response length preferences or compared visual and voice modalities [2,5,6], few studies have systematically investigated how response length in voice-only settings affects cognitive load, users' subjective experience and search behavior.

To address this, we developed a desktop-based SCS system. We conducted a user study with 139 participants to examine the following research questions:

© The Author(s), under exclusive license to Springer Nature Switzerland AG 2026
C. K. Leung et al. (Eds.): DaWaK 2025, LNCS 16048, pp. 231–236, 2026.
https://doi.org/10.1007/978-3-032-02215-8_17

RQ1: Is the short response presentation method effective in spoken conversational search ?

RQ2: Do task complexity and response length affect users' search behavior ?

Findings show that SRP reduced cognitive load and increased satisfaction in low complexity tasks, but triggered more queries in high complexity tasks. These results suggest that response strategies should adapt to task demands.

2   Related Work

Prior studies show that reading textual search results aloud can negatively affect the user experience [6]. Trippas et al. [5] found that users often prefer shorter summaries in voice-based interfaces, though task-dependent trade-offs exist between conciseness and informativeness.

For routine tasks, shorter responses improve usability [1], but this may not extend to exploratory search. Also, most prior work uses multimodal interfaces and does not isolate the effects of response length in voice-only settings [2].

Our study extends this line of research by systematically examining how response presentations affect user experience and behavior under varying task complexities.

3   Methodology

3.1   Task Topics and Cognitive Complexity Levels

Two search scenarios were designed to reflect different levels of cognitive complexity, based on Trippas et al. [4].

– **Low complexity (Remember):** Where does cinnamon come from?
– **High complexity (Analyze):** Comparison of per capita alcohol consumption

3.2   Experimental Conditions

We defined two response styles:

– **LRP:** Comprehensive answers with multiple details, often in long sentences or bulleted form.
– **SRP:** Concise, one-to-two sentence summaries directly addressing the query.

Both styles used the gpt-4o-2024-08-06 model. Prompts were adjusted to encourage longer or shorter output.

Our system provided fully spoken interaction using OpenAI's Whisper for speech-to-text conversion and a text-to-speech engine for audio output. Users initiated voice input by pressing an on-screen microphone button. Notably, system responses were presented exclusively via speech, without any textual display.

3.3   Flow of the User Study

After informed consent and a tutorial video[1], participants completed a training task. Then they performed two main search tasks, each followed by questionnaires. The user study was approved by the research ethics committee at our university (No. UHIS-EC-2024-016).

3.4   Questionnaire

The questionnaire consisted of three parts: (1) a free-description item asking participants to summarize what they had understood through the search task, (2) the NASA Task Load Index (NASA-TLX) to measure cognitive load, and (3) participants rated the perceived effectiveness of response length in supporting information retrieval using a 5-point Likert scale (1: Strongly Disagree – 5: Strongly Agree).

3.5   Participants

We recruited 156 Japanese participants via Lancers.jp[2]; 139 completed the study successfully. Participants (aged 20–70 s): 61.2% male, 38.8% female; 65.5% had prior experience with voice agents. Participants were compensated with approximately 1,100 JPY.

4   Results

We employed GLM to examine the main effects of response length (LRP vs. SRP) and topic complexity (low vs. high), as well as the interaction between these factors, on each NASA-TLX subscale, questionnaire (Q1–Q6) and number of issued queries and query issuance time.

4.1   Impact on Effectiveness (RQ1)

To investigate the effectiveness of SRP in SCS, we analyzed NASA-TLX scores as an indicator of cognitive load and participants' responses to Q1–Q6 as indicators of satisfaction with their search behavior. Table 1 shows the results.

Cognitive Load: Mental demand showed significant main effects for response length ($F(1, 204) = 6.00$, $p < 0.05$) and task complexity ($F(1, 204) = 4.49$, $p < 0.05$). Overall workload showed significant main effects for both response length and task complexity, with a significant interaction ($F(1, 204) = 4.08$, $p < 0.05$). Post-hoc tests revealed SRP reduced workload especially under low complexity (mean: 54.34 for LRP vs. 44.55 for SRP, $t(121.9) = 3.10$, $p < 0.01$). This suggests that the benefit of **SRP** is more evident in **low complexity**

---

[1] https://youtu.be/DSRzbFkI9f8.

[2] https://www.lancers.jp/.

Table 1. Participant workload (NASA-TLX) and subjective ratings (Q1–Q6) across response length (LRP vs. SRP) and task complexity (low vs. high). Significance levels: * $p < 0.05$, ** $p < 0.01$, *** $p < 0.001$. Bold indicates better scores.

Measure		LRP		SRP		Response Length	Task complexity	Interaction
		Low	High	Low	High			
Cognitive load	Mental	44.17	45.82	**32.62**	43.21	*	*	0.07
	Temporal	22.93	25.54	**15.63**	23.87	0.09	*	0.47
	Frustration	24.51	28.25	**17.62**	27.49	0.17	*	0.31
	Overall	54.34	57.46	**44.55**	55.52	**	***	*
User satisfaction	Q1 (response length)	3.45	3.36	**4.04**	3.87	***	0.27	0.85
	Q2 (info helpfulness)	3.76	3.61	**4.06**	3.99	**	0.31	0.58
	Q3 (Info diversity)	**4.15**	3.87	3.87	4.01	0.51	0.51	*
	Q4 (Info coverage)	**4.07**	3.51	4.06	3.58	0.79	***	0.56
	Q5 (Opinion formation)	**4.08**	3.57	4.06	3.56	0.86	***	0.83
	Q6 (Info retention)	3.07	2.99	**2.44**	2.79	**	0.32	0.13

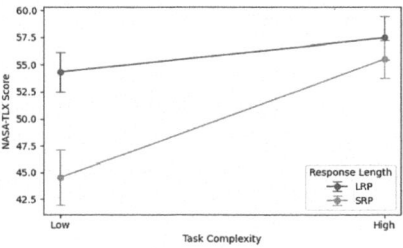

Fig. 1. Interaction effect between response length (LRP vs. SRP) and task complexity (low vs. high) on overall workload (error bars indicate the standard error of the mean).

tasks, while its advantage becomes less pronounced in high complexity tasks (see Fig. 1).

User Satisfaction: For Q1 and Q2, significant main effects of response length were observed, with **SRP** rated more appropriate ($F(1, 204) = 22.66, p < 0.001$) and helpful ($F(1, 204) = 10.06, p < 0.01$). Q3 showed an interaction effect ($F(1, 204) = 5.32, p < 0.05$), but follow-up t-tests showed only a weak trend favoring **LRP** in low complexity tasks. For Q6, **SRP** reduced the feeling of forgetting important information ($F(1, 204) = 9.67, p < 0.01$).

4.2   Impact on Search Behavior (RQ2)

To explore how the combination of response length and task complexity affects search behavior, we analyzed system interaction metrics including the number of issued queries and query issuance time. Table 2 shows the results.

Number of Issued Queries: We observed a significant interaction between response length and task complexity ($F(1, 204) = 5.54, p < 0.05$). Post-hoc t-tests confirmed the difference (mean: 5.55 for high vs. 3.85 for low, $t(134.33) =$

Table 2. Participant subjective ratings across response length (LRP vs. SRP) and task complexity (low vs. high). Significance levels: * $p < 0.05$, ** $p < 0.01$, *** $p < 0.001$.

Measure		LRP		SRP		Response length	Task complexity	Interaction
		Low	High	Low	High			
Search behavior	Number of issued queries	2.96	3.28	3.80	5.60	***	***	*
	Query issuance Time (s)	11.89	9.19	13.90	13.44	*	0.38	0.30

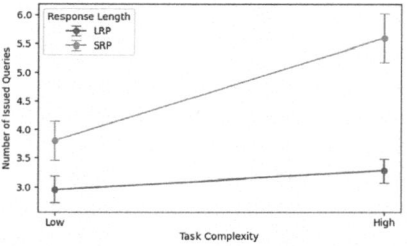

Fig. 2. Interaction between response length (LRP vs. SRP) and task complexity (low vs. high) on the number of issued queries (error bars indicate the standard error of the mean).

-3.12, $p < 0.01$). This suggests that **SRP** prompts more iterative search in **high complexity tasks** (see Fig. 2).

Query Issuance Time: We observed a significant main effect of response length ($F(1, 141) = 4.21$, $p < 0.05$). This suggests that **LRP** may have provided more informative content, which helped users quickly identify follow-up questions and issue queries more efficiently.

5   Discussion

RQ1: SRP reduced cognitive load, particularly mental demand. Users found SRP responses appropriately concise (Q1), helpful for information acquisition (Q2), and easier to remember (Q6). A participant noted, *"The agent provided concise and direct answers, which was very helpful. When the responses were long, I tended to forget parts of the information."*

However, a significant interaction in overall workload showed **SRP** was more effective only in **low complexity tasks**. In high complexity tasks, SRP's brevity may have fragmented information, potentially increasing cognitive burden.

Q3 revealed a preference for **LRP** in low complexity tasks for acquiring diverse information, though some users found lengthy outputs excessive.

These findings suggest the need for *adaptive strategies*—e.g., providing LRP followed by a summary. One user suggested, *" It would be nice if the output adjusted to whether I want detailed or concise responses."*

RQ2: Response length and task complexity affected behavior. **SRP** led to more queries, especially in **high complexity tasks**, suggesting users compensated for limited information. In contrast, **LRP** seemed to help users issue queries more efficiently—possibly due to richer content or added planning time during longer responses. Systems might benefit from a dynamic approach—starting with LRP to aid exploration, then shifting to SRP to reduce overload.

Limitations: We did not assess the *quality* of acquired information (e.g., correctness or relevance). Future work could evaluate user summaries using information nuggets [3] and compare coverage with system outputs.

6   Conclusion

This study examined how response length and task complexity affect user experience in SCS. The main findings are twofold. First, SRP significantly reduced cognitive load and increased user satisfaction, particularly in low complexity tasks. SRP is likely effective in mitigating the memory burden inherent in SCS. Second, when the task complexity increased, SRP condition led to a significantly higher number of issued queries, reflecting more active search behavior. However, this may also increase cognitive strain, as users need to reconstruct fragmented information across multiple interactions. This suggests the need for task-adaptive rather than uniform response strategies.

Acknowledgments. This work was supported by JSPS KAKENHI Grant Numbers JP24K03228, JP22H03905, JP21H03554, and JP21H03775.

References

1. Haas, G., Rietzler, M., Jones, M., Rukzio, E.: Keep it short: a comparison of voice assistants' response behavior. In: Proceedings of the 2022 CHI Conference on Human Factors in Computing Systems, pp. 1–12 (2022)
2. Kaushik, A., Jones, G.: Examining the potential for conversational exploratory search using a smart speaker digital assistant. In: Proceedings of the 7th International Conference on Human Computer Interaction Theory and Applications, pp. 305–317 (2023)
3. Lin, J., Demner-Fushman, D.: Will pyramids built of nuggets topple over? In: Proceedings of the Human Language Technology Conference of the NAACL, Main Conference, pp. 383–390. Association for Computational Linguistics (2006)
4. Trippas, J.R., Spina, D., Cavedon, L., Sanderson, M.: How do people interact in conversational speech-only search tasks: a preliminary analysis. In: Proceedings of the 2017 Conference on Human Information Interaction and Retrieval, pp. 325–328 (2017)
5. Trippas, J.R., Spina, D., Sanderson, M., Cavedon, L.: Towards understanding the impact of length in web search result summaries over a speech-only communication channel. In: Proceedings of the 38th International ACM SIGIR Conference on Research and Development in Information Retrieval, pp. 991–994 (2015)
6. Vtyurina, A., Clarke, C.L.A., Law, E., Trippas, J.R., Bota, H.: A mixed-method analysis of text and audio search interfaces with varying task complexity. In: Proceedings of the 2020 ACM SIGIR on International Conference on Theory of Information Retrieval, pp. 61–68 (2020)

Fair Proportional Top-k Ranking

Nina A. Liebrand(✉) [iD], Manh Khoi Duong [iD], and Stefan Conrad [iD]

Heinrich Heine University, Universitätsstraße 1, 40225 Düsseldorf, Germany
{nina.liebrand,manh.khoi.duong,stefan.conrad}@hhu.de

Abstract. Selecting the k most relevant candidates from a larger set is known as top-k ranking. Traditional ranking methods prioritize candidates based on their relevance, which can lead to discrimination. Due to the AI Act, fair top-k ranking has recently gained attention. We introduce a new positional fairness metric that considers the ranking positions of groups in the top-k ranking. Secondly, we propose a novel algorithm, *FairNormRank*, that optimally fulfills the three fair top-k ranking criteria of proportional fairness, maximum relevance, and ordering consistency and accounts for positional fairness. Our method works for non-binary and intersectional groups, therefore enhancing its applicability in realistic scenarios. An evaluation on a real-world dataset shows that we outperform existing methods in terms of fulfilling the fairness criteria.

Keywords: Fair Ranking · Positional Fairness · Proportional Representation · Intersectional Fairness

1 Introduction

Top-k ranking is a widely used method in information retrieval, where the goal is to select the most relevant candidates from a larger set [11]. This technique is increasingly applied, for example, in search engine results, college admissions, or hiring. Traditional ranking methods prioritize candidates solely based on relevance, leading to discrimination of protected groups. For instance, in university admissions, ranking applicants based on SAT scores disadvantages female candidates, whose scores tend to be lower despite comparable academic performance [9]. *Fair top-k ranking* addresses the challenge of selecting the most relevant candidates while ensuring equal representation across social groups.

Fairness in ranking has only been studied rarely [8]. While some studies focused on measuring fairness in rankings [1,4,6,10], other studies introduced algorithms for fair ranking. For instance, Asudeh et al. [3] directly adjust the weights of the scoring function to create fair relevance scores. Feldman et al. [5] proposed a preprocessing technique that aligns the relevance score distribution of each group. Zehlike et al. [11] introduced a fairness-aware ranking approach, which iteratively selects candidates for the top-k set based on their relevance scores and several fairness criteria. In their algorithm, a statistical test based on

---

N.A. Liebrand and M.K. Duong—Equal contribution.

© The Author(s), under exclusive license to Springer Nature Switzerland AG 2026
C. K. Leung et al. (Eds.): DaWaK 2025, LNCS 16048, pp. 237–243, 2026.
https://doi.org/10.1007/978-3-032-02215-8_18

the difference between observed and expected proportions of protected groups is used. As a result, they do not fully fulfill the fairness criteria.

Existing algorithms mainly focused on ensuring proportional representation without addressing positional fairness. Positional unfairness occurs, for example, in search engine results, where fairness-aware ranking algorithms may ensure proportional representation of marginalized groups but still position their links predominantly at the bottom of the results page [4]. Zehlike et al. [12] later extended their algorithm to handle multiple protected groups, as well as positional fairness by ensuring that protected groups are proportionally represented in subsets of the top-k ranking. However, we state that this approach alone does not fully capture positional fairness. For instance, the ranking where a privileged group with members in better positions can still be considered fair if the same group has members in worse positions as well.

This work's contributions mainly include the introduction of a novel positional fairness metric, as well as the proposal of an algorithm, *FairNormRank*, that optimally fulfills existing fair ranking criteria and accounts for positional fairness. This algorithm can handle non-binary and intersectional groups and is available on GitHub (https://github.com/NinaL-ai/FairRanking/).

2   Preliminaries

This section introduces the notation used throughout the paper. In classical ranking only the relevance score of a document is considered. In contrast, fair ranking deals with ranking human candidates and additionally considers their protected attributes. Hence, we denote $\mathcal{C}$ as a set of candidates, each associated with a *relevance score* and belonging to one or more *social groups*.

Definition 1 (Relevance Score). *Each candidate $c \in \mathcal{C}$ is assigned a relevance score $q(c)$. The function $q : \mathcal{C} \to \mathbb{R}$ maps each candidate to a real-valued score, where higher values correspond to better qualifications.*

Definition 2 (Intersectional Groups). *Let $\mathcal{Z}_1, \ldots, \mathcal{Z}_m$ be a set of discrete protected attributes. Each candidate $c \in \mathcal{C}$ belongs to a group g defined as a combination of attribute values $(z_1, \ldots, z_m)$, where $z_i \in \mathcal{Z}_i$. The set of all available intersectional groups in a dataset is denoted as $\mathcal{G}$. We note that $\mathcal{G} \subseteq \mathcal{Z}_1 \times \ldots \times \mathcal{Z}_m$. We do not consider combinations of groups that are not present in the dataset, i.e., $\mathcal{Z}_1 \times \ldots \times \mathcal{Z}_m \setminus \mathcal{G}$.*

Definition 3 (Group Membership). *The group membership function $M : \mathcal{C} \to \mathcal{G}$ returns the intersectional group $g \in \mathcal{G}$ to which each candidate $c \in \mathcal{C}$ belongs to:*

$$M(c) = g = (z_1, \ldots, z_m) \in \mathcal{G},$$

where $z_i \in \mathcal{Z}_i$ for $i = 1, \ldots, m$.

Definition 4 (Candidate Ranking). *A ranking function $R : \mathcal{C} \to \{1, \ldots, |\mathcal{C}|\}$ is a bijective function that assigns a unique rank to each candidate, where a lower rank indicates a better position in the ranking.*

Definition 5 (Top-k Candidates). *Given a ranking $R : \mathcal{C} \to \{1, \ldots, |\mathcal{C}|\}$, the top-k candidates are defined as $\tau_k = \{c \in \mathcal{C} \mid R(c) \leq k\}$, i.e., candidates that have a rank lower or equal to $k \in \mathbb{N}$.*

Definition 6 (Group Proportion in Top-k Ranking). *We define group proportion as the fraction of candidates in a group g in the top-k ranking and denote it with:*

$$P_k(g) := \frac{|\{c \in \tau_k \mid M(c) = g\}|}{k}.$$

3   Fairness-Aware Ranking Criteria

Zehlike et al. [11] introduced three criteria that a fair ranking should satisfy. We modify these criteria and introduce a new criterion for positional fairness.

3.1   Maximum Relevance

Intuitively, we want our top-k ranking set to include the most relevant candidates. But due to proportional fairness, we cannot rank candidates solely based on their relevance scores. Hence, we introduce a criteria that ensures that each candidate in τ_k has a higher relevance score than any other candidate not in τ_k, if they are coming from the same group. Formally, this can be expressed with:

$$\forall g \in \mathcal{G}, \forall c \in \tau_k, \forall c' \in \mathcal{C} \setminus \tau_k : M(c) = M(c') = g \implies q(c) \geq q(c'). \tag{1}$$

3.2   Ordering Consistency

The ranking should preserve the order of relevance scores, meaning that candidates with higher relevance scores should receive better ranks than those with lower scores. A ranking satisfies *ordering consistency* if [11]:

$$q(c) > q(c') \implies R(c) < R(c'), \tag{2}$$

for all $c, c' \in \tau_k$. This ensures that while fairness constraints may introduce adjustments, they do not completely worsen the quality of the ranking.

3.3   Proportional Fairness

In proportional fairness, we aim to ensure that each group is represented in the top-k candidate set equally. This prevents systematic underrepresentation of certain groups and thus ensures a more fair ranking.

The group proportion for each group g in the top-k ranking can be set to achieve a desired fraction t_g, where $\sum_{g \in \mathcal{G}} t_g = 1$. For equal representation, one could set each t_g to the same value, i.e., $t_g = \frac{1}{|\mathcal{G}|}$. In practice, the desired proportions may not be achievable due to the limited number of candidates or the choice of k.

Definition 7 (Maximum Proportional Disparity (MPD)). *To measure proportional fairness for* τ_k, *the largest deviation among all group from their desired target proportions is reported. This score is normalized by the maximum possible disparity:*

$$MPD(\tau_k) := \max_{g \in \mathcal{G}} \frac{|P_k(g) - t_g|}{\max(1 - t_g, t_g)}.$$

3.4 Positional Fairness

Suppose we have two groups g and g' with equal representation in the top-k candidate set, but every candidate in g is ranked better than g', i.e., $R(c) < R(c')$ for all $c, c' \in \tau_k$ with $M(c) = g$ and $M(c') = g'$. This contradicts the intuition of a fair ranking.

Definition 8 (Positional Score). *The positional score of a group* $g \in \mathcal{G}$ *in a ranking* τ_k *is defined as:*

$$PS(g, \tau_k) := \sum_{c \in \tau_k \wedge M(c) = g} W(R(c)),$$

where positions are weighted using a logarithmic discount $W(i) = \frac{1}{\log_2(i+1)}$, *reflecting the greater importance of better ranked positions.*

Definition 9 (Maximum Positional Disparity (MPoD)). *To measure the positional fairness for* τ_k, *we report the largest deviation among all groups, normalized by the highest possible positional score:*

$$MPoD(\tau_k) := \frac{\max_{i \in \mathcal{G}} PS(i, \tau_k) - \min_{j \in \mathcal{G}} PS(j, \tau_k)}{\sum_{r=1}^{k} W(r)}.$$

4 Fair Top-k Ranking Algorithm

The FairNormRank algorithm begins with a preprocessing step by partitioning candidates into group specific sets C_g, and appling a min-max normalization to the relevance scores within each group. Both steps require $\mathcal{O}(|\mathcal{C}|)$ operations. The normalization is controlled by $\alpha \in [0, 1]$, where $\alpha = 0$ preserves the original scores and $\alpha = 1$ applies full normalization. Each group's candidates are then sorted by normalized relevance, costing $\mathcal{O}(|\mathcal{C}| \log |\mathcal{C}|)$. For each group, the top desired number of candidates $\lfloor t_g \cdot k \rfloor$ are selected, forming the set S. To compensate for rounding, the next best candidate from each group is stored in a residual set X. These selections are $\mathcal{O}(1)$ as the sets C_g are pre-sorted. Here, we assume that each group has sufficient candidates. The remaining top $k - |S|$ candidates are chosen from X, which is sorted in $\mathcal{O}(|\mathcal{G}| \log |\mathcal{G}|)$ time. This ensures that the total number of candidates reaches k. Finally, the top-k ranking τ_k is sorted by the normalized relevance scores, requiring $\mathcal{O}(k \log k)$ time. Since $k, |\mathcal{G}| \ll |\mathcal{C}|$,

Algorithm 1. Fair Normalized Top-k Ranking Algorithm (FairNormRank)

Require: Candidate set $\mathcal{C}$, relevance $q : \mathcal{C} \rightarrow \mathbb{R}$, group membership $M : \mathcal{C} \rightarrow \mathcal{G}$, target proportions t_g for each $g \in \mathcal{G}$, $k \in \mathbb{N}$, and $\alpha \in [0, 1]$.

Ensure: Fair top-k candidate set τ_k

1: Initialize $S, X \leftarrow \emptyset$
2: $\mathcal{C}_g \leftarrow \{c \in \mathcal{C} \mid M(c) = g\}$ for all $g \in \mathcal{G}$
3: $q_{\text{norm}}(c_g) \leftarrow (1 - \alpha)q(c_g) + \alpha \frac{q(c_g) - \min_{c \in \mathcal{C}_g} q(c)}{\max_{c \in \mathcal{C}_g} q(c) - \min_{c \in \mathcal{C}_g} q(c)}$ $\forall c_g \in \mathcal{C}_g, \forall g \in \mathcal{G}$
4: Sort each $\mathcal{C}_g$ in descending order by relevance $q_{\text{norm}}(c_g)$
5: **for** each group $g \in \mathcal{G}$ **do**
6:     $n_g \leftarrow \lfloor t_g \cdot k \rfloor$
7:     $S_g \leftarrow$ First n_g candidates from $\mathcal{C}_g$
8:     $S \leftarrow S \cup S_g$
9:     $X \leftarrow X \cup \{\text{candidate at position } n_g + 1 \text{ in } \mathcal{C}_g\}$
10: **end for**
11: $X \leftarrow$ First $(k - |S|)$ elements from X, sorted descending by $q_{\text{norm}}(c)$
12: $\tau_k \leftarrow (S \cup X)$ sorted descending by $q_{\text{norm}}(c)$
13: **return** τ_k

the runtime is dominated by the initial sorting. Hence, the overall complexity remains $\mathcal{O}(|\mathcal{C}| \log |\mathcal{C}|)$. □

5   Evaluation

We evaluate the effectiveness of the FairNormRank algorithm on the real-world dataset COMPAS [7], which is a risk assessment tool used in the US criminal justice system to predict the likelihood of recidivism. This scoring system has been shown to be biased against African-American defendants [2].

Firstly, we evaluate the effect of the normalization factor on the fair ranking criteria by creating top-100 rankings of the defendants that are least likely to recidivate using FairNormRank and a colorblind baseline that ranks candidates purely based on relevance. The target proportions of each ethnicity are set equally to $t_g = \frac{1}{|\mathcal{G}|} = \frac{1}{9}$. As expected, independent of the normalization factor, FairNormRank fulfills the proportional fairness, maximum relevance, and ordering consistency criteria. We measured the corresponding metrics, MPD, selection utility loss [11], and NDCG loss [6], which always achieved optimal values. In contrast, the baseline clearly violates the proportional fairness criterion, as seen in the MPD score (Fig. 1a). Our method shows a better positional fairness than the baseline, which further improves as the normalization factor increases, reaching the highest score at a normalization factor of 1.0 (Fig. 1b). In contrast, Kendall's τ decreases with an increasing normalization factor, indicating that the rankings correlate less with the original ranking (Fig. 1c).

Secondly, we compare our results with the algorithm from Zehlike et al. [11]. On the COMPAS dataset, their algorithm violates all four criteria (MPD = 0.539, selection utility loss = 0.718, NDCG loss = 0.090, MPoD = 0.599), as it only ensures proportional representation of the protected groups as a whole, rather than for individual groups, leading to four missing ethnicities.

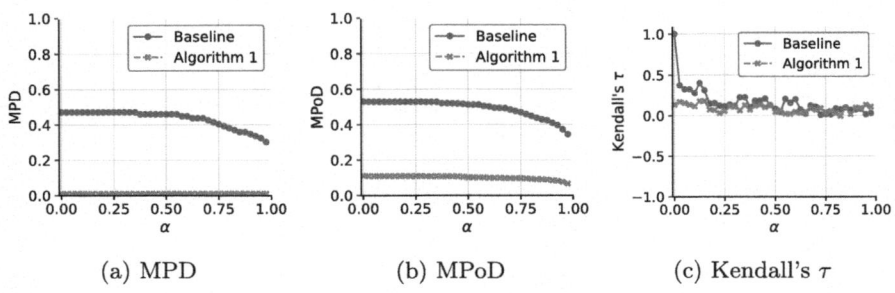

(a) MPD   (b) MPoD   (c) Kendall's τ

Fig. 1. Proportional fairness, positional fairness and Kendall's τ across different normalization factors α.

6   Conclusion

We proposed a fair ranking algorithm, *FairNormRank*, that optimally satisfies proportional fairness, maximum relevance, and ordering consistency while supporting non-binary and intersectional groups. Our experiments on the COMPAS dataset showed that our introduced normalization technique is able to improve positional fairness. These findings highlight the effectiveness of our approach in mitigating bias of ranks while maintaining quality according to ranking criteria.

References

1. Agarwal, A., Zaitsev, I., Wang, X., Li, C., Najork, M., Joachims, T.: Estimating position bias without intrusive interventions. In: WSDM, pp. 474–482 (2019)
2. Angwin, J., Larson, J., Mattu, S., Kirchner, L.: Machine bias. In: Ethics of Data and Analytics, pp. 254–264 (2022)
3. Asudeh, A., Jagadish, H.V., Stoyanovich, J., Das, G.: Designing fair ranking schemes. In: MOD, pp. 1259–1276 (2019)
4. Celis, L.E., Straszak, D., Vishnoi, N.K.: Ranking with fairness constraints. In: ICALP (2018)
5. Feldman, M., Friedler, S.A., Moeller, J., Scheidegger, C., Venkatasubramanian, S.: Certifying and removing disparate impact. In: KDD, pp. 259–268 (2015)
6. Järvelin, K., Kekäläinen, J.: Cumulated gain-based evaluation of IR techniques. ACM Trans. Inf. Syst. (TOIS) **20**(4), 422–446 (2002)
7. Larson, J., Angwin, J., Mattu, S., Kirchner, L.: Machine bias (2016)
8. Patro, G.K., Porcaro, L., Mitchell, L., Zhang, Q., Zehlike, M., Garg, N.: Fair ranking: a critical review, challenges, and future directions. In: FAcct, pp. 1929–1942 (2022)
9. The College Board: Sat percentile ranks (2014)

10. Yang, K., Stoyanovich, J.: Measuring fairness in ranked outputs. In: SSDBM (2017)
11. Zehlike, M., Bonchi, F., Castillo, C., Hajian, S., Megahed, M., Baeza-Yates, R.: Fa*ir: A fair top-k ranking algorithm. In: CIKM, pp. 1569–1578 (2017)
12. Zehlike, M., Sühr, T., Baeza-Yates, R., Bonchi, F., Castillo, C., Hajian, S.: Fair top-k ranking with multiple protected groups. Inf. Process. Manag. **59**(1), 102707 (2022)

PAID: Power-Efficient AI-Optimized Databases

Ayoub Bouhatous[1], Ladjel Bellatreche[2(✉)], El Hassan Abdelwahed[1],
and Carlos Ordonez[3]

[1] Cadi Ayyad University, Marrakech, Morocco
[2] LIAS/ISAE-ENSMA, Poitiers, France
bellatreche@ensma.fr
[3] Department of Computer Science, University of Houston, Houston, USA

Abstract. Traditionally, query processors (QPs) have been designed
to optimize response time, but not energy consumption. Heeding this
new optimization goal, during the last decade, the database community
turned its attention to enhancing the energy efficiency (EE) of QPs, but
there are still major challenges. By analyzing recent work on EE of QPs, we
observe that even though Machine Learning (ML) models can accurately
predict energy consumption, they do not modify the core QP functionality
(software) nor dynamically adjust CPU configuration parameters (hardware: # of cores and clock frequency), to actually save energy for a query.
To address this gap, we introduce PAID, a subsystem integrated with the
query optimizer that combines old AI with new AI: a Genetic Algorithm
(GA) with a Neural Network (NN). The NN model predicts query energy
consumption, whereas GA determines the optimal CPU configuration,
deciding clock speed frequency and core allocation for each query. The
GA configuration is then fed back into the NN model to tune prediction
accuracy. Our experiments, conducted on the TPC-H benchmark with
PostgreSQL, show that PAID effectively finds CPU configurations that
exceed the performance of default settings, achieving significant energy
savings.

1 Introduction

The development of efficient query processors (QP) remains an active research
area. As data science and AI applications proliferate, QPs have become the backbone for data preparation and data exploration phases, enabling AI model computations. In parallel, the ICT industry is estimated to be responsible for 1.8% -
2.8% of the global carbon footprint. Therefore, growing environmental concerns
will require policymakers, industry stakeholders and researchers to prioritize
energy efficiency (EE) in the design of future computing systems. Thus, improving the EE of QPs will remain an important problem for green computing.

Aligned with this motivation, the database community has shown notable
interest over the past two decades in developing various initiatives to address
the challenge of EE in databases. These efforts include surveys [2, 7, 15], methodic

© The Author(s), under exclusive license to Springer Nature Switzerland AG 2026
C. K. Leung et al. (Eds.): DaWaK 2025, LNCS 16048, pp. 244–250, 2026.
https://doi.org/10.1007/978-3-032-02215-8_19

studies for facilitating research on this topic [1], and prediction models for assessing the energy consumption of traditional DBMS QPs [2–4,6,8,9,13,16].

By conducting an in-depth analysis of these research initiatives, we observe that they introduce both hardware and software tactics to enhance EE [1]. Hardware manufacturers have made significant strides in developing high-performance CPUs [14], GPUs [10], and specialized hardware accelerators [12]. Dynamic Voltage and Frequency Scaling (DVFS) is recognized as one of the most effective techniques for reducing power consumption in both CPUs and GPUs. By dynamically adjusting the voltage and frequency based on workload demands, DVFS helps optimize EE without significantly compromising performance [5,11]. Software tactics include, among others, analytical cost models designed to predict the query energy consumption. They extend conventional query optimizers by incorporating parameters such as IO, CPU usage, and memory costs. Machine learning (ML) techniques use these parameters to predict energy consumption of queries [2]. More recently, these models have been further refined by incorporating hardware-related parameters, such as the number of CPU cores and frequency [3]. However, they primarily rely on basic predictive methods such as linear regression, multiple regression, and random forest [4,6,13].

The analysis of the current state of the art reveals two important findings (F1 and F2) that need to be consolidated, as well as two main limitations (L1 and L2). **F1** The pivotal role of ML techniques in aiding the development of environmentally-friendly QP. **F2** The importance of ML solutions that integrate both software and hardware parameters to accurately predict energy consumption. **L1** The existing energy consumption prediction models rely on simple ML techniques. **L2** Despite the CPU's dominance in energy consumption, existing ML-based solutions often overlook the importance of optimizing CPU configurations—specifically, the number of cores and CPU frequency. Instead, they rely on the default settings predefined by the host machine of the target DBMS, potentially missing opportunities for improvedEE.

To overcome the above two limitations, we propose a novel optimization system PAID that integrates Genetic Algorithm (GA) and Neural Networks (NN). GA aims at selecting the optimal number of CPU cores and a CPU frequency for a given query. The configuration chosen by the GA is subsequently incorporated into the NN model to enhance the accuracy of the prediction. Thereafter, NN are utilized to predict both energy consumption and response time of queries.

2   Integrating Old and New AI: GA and NN

In this section, we describe the two components of PAID system: GA and NN.

2.1   NN Model

Our NN model aims to estimate the energy consumption (output) considering critical CPU settings, including the number of cores and CPU frequency. According to the state-of-the-art, estimating the energy consumption for a given query

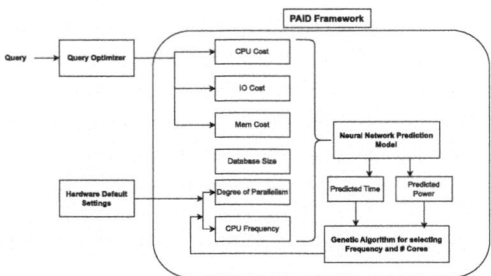

Fig. 1. The PAID Subsystem.

Q_i requires considering the following key features (that can be extracted from query optimizers): CPU cost, I/O cost, memory cost, and database size. Therefore, from an energy consumption perspective, Q_i can be represented by the following feature vector $\overrightarrow{Q_i}$, which serves as input to our NN model:
$\overrightarrow{Q_i} = (COST_{CPU_i}, COST_{IO_i}, COST_{Memory_i}, DB_{Size_i})$, where $COST_{CPU_i}$, $COST_{IO_i}$, $COST_{Memory_i}$, and DB_{Size_i} represent respectively the number of instructions executed by the CPU, the number of pages read/written from secondary storage (persistent storage), the number of pages accessed in main memory when executing Q_i, and the size of the target database. A CPU configuration used for executing a query Q_i is represented by the following vector:
$\overrightarrow{CPU_i} = (FRQ_i, CORE_i)$, where FRQ_i, and $CORE_i$ represents respectively the CPU frequency and the number of cores used during Q_i execution. *It is important to highlight that, in existing studies, all CPU vectors associated with queries are fixed. That is, the CPU configuration remains fixed across all queries, rather than being dynamically optimized based on individual query characteristics.* Based on these two vectors, a query Q_i is then represented by a vector obtained by concatenating the query feature vector $\overrightarrow{Q_i}$ and the CPU configuration vector $\overrightarrow{CPU_i}$. The query vector is initially passed through a set of fully connected layers of monotonically decreasing size.

2.2 Selecting an Optimal CPU Configuration: A Genetic Algorithm

Recall that our GA aims at selecting the best configuration of CPU for a given query Q_i. Let us first formalize the problem of CPU configuration selection. Given the extended query (Q_i) vector $\overrightarrow{EQ_i} = (\overrightarrow{Q_i}, \overrightarrow{CPU_i})$ and our NN model that predicts the energy consumption of queries under a given CPU configuration, our problem consists in setting the best CPU configuration that minimizes both the power consumption (F_{Power}) and execution time F_{Time}:
$minimize_{(j,k)} \quad F_{power}(\overrightarrow{Q_i}, \overrightarrow{CPU_{i,j,k}}) \times F_{time}(\overrightarrow{Q_i}, \overrightarrow{CPU_{i,j,k}})$
where $min\_frequency \leq j \leq max\_frequency$ and $min\_number\_cores \leq k \leq max\_number\_cores$.

Our NN model predicts energy consumption and response time for a query based on considered the features, without varying them. In contrast, the GA selects the optimal CPU configuration for a query. By sending predicted energy from the NN to theGA, PAID enables NN to get the optimal CPU configuration. Figure 1 illustrates the connection between NN and GA, where the fitness function used by the GA is provided by the NN model.

3   Experimental Study

We present and discuss experimental results obtained using a Dell Precision Tower 3620 server equipped with a Core i7-6700 CPU (4 cores, 8 threads), 16 GB DDR4 RAM, and a 256 GB SSD. Our experimental setup comprises three main components: a client machine (monitor), a PostgreSQL DBMS (version 14.1) running on Ubuntu 20.04 (kernel 20.04.4 LTS), and an external power meter for energy measurement called Yocto-Watt[1] at a frequency of 1 Hz. It is directly placed between the database server and the electrical power supply and it is linked using a USB cable to the client machine for data collection. Our server is installed with We utilize the TPC-H benchmark to train and evaluate our models. Three databases are generated with sizes of 10 GB, 30 GB, and 50 GB (used to validate our proposal). In addition to the benchmark's original 22 queries, we generate 70 additional queries randomly. For each database instance, we collected query execution plans and measured energy consumption using our power meter. Each query was executed multiple times while varying the number of CPU cores (parallelism degree) from 1 to 4 and adjusting CPU frequency configurations between 0.8 GHz and 3.4 GHz. Before conducting our experiments, we deactivated unnecessary background tasks and cleared both the operating system and PostgreSQL buffers before each query execution. To assess the effectiveness of our NN model, we selected Random Forest Regression (RFR) as a baseline. The input layer of our NN model consists of 6 neurons, corresponding to the number of features in the input data. The subsequent three layers are densely connected, with each neuron in a layer being connected to every neuron in the previous layer. We use the rectified linear unit (ReLU) as the activation function for the hidden layers, which introduces non-linearity to the network. The output layer consists of two neurons, each producing a single output value, indicating the two outputs predicted by the network. The activation function for these output neurons is linear, which allows for continuous predictions of power consumption and query execution time. The model is compiled using the Adam optimizer with a learning rate of 0.0125, and the loss function employed during training is mean absolute error (MAE).

NN model vs. RFR Table 1 presents the evaluation results of Random Forest Regression (RFR) and our NN model, showcasing their performance metrics for predicting power consumption and query execution time. The NN model generally outperforms the RFR model in terms of both MAE and R^2 for predicting

---

[1] https://www.yoctopuce.com/FR/products/yocto-watt.

Table 1. Evaluation results of RFR and NN in predicting power consumption and time of queries.

Model's Output	Evaluation Metrics	RFR	NN
Power	MAE	2.3	2.5
	R^2	63.82	64.17
Time	MAE	54.72	13.85
	R^2	47.51	83.64

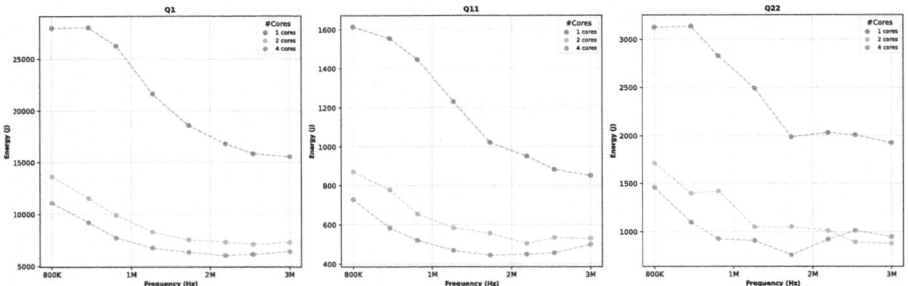

Fig. 2. Impact of variation of CPU configuration on queries Q1, Q11, and Q22

power and time. This indicates that the NN model provides superior accuracy in estimating the energy consumption and execution time of queries, making it a more reliable choice for predicting these metrics.

Impact of CPU Configuration. To evaluate the impact of CPU configuration, we varied the number of cores and measured response time and energy consumption for 22 queries. As expected, increasing the number of cores generally improved response time, except for simple queries (with less joins and sorting). Surprisingly, energy consumption was found to correlate with elapsed time. Using all available cores reduced energy consumption for most queries, except some queries with high number of joins and sorting using two attributes, energy usage remained high despite full core utilization. This suggests that for complex queries, the overhead of parallelism may offset energy savings, especially when managing numerous joins and sort operations. Due to the large number of queries and the wide spectrum of values, plotting all results together was impractical. Therefore, we chose to present representative results for three queries (Fig. 2): Q1 (0 join, 4 sums, 3 avg, 1 sort on two attributes), Q11 (4 joins, 2 nested queries, 3 sums, 1 sort), and Q22 (4 nested queries, 2 joins, 2, one sort).

4   Conclusion

We introduced PAID, a subsystem that integrates a GA with a NN to optimize energy consumption in query processors (QPs). PAID addresses the complexity of selecting an optimal CPU configuration for a given query due to the large number of possible combinations. On the other hand, going beyond existing linear regression models, we explained NN can learn a more accurate non-linear model to predict energy and reduce energy consumption based on a query workload. Therefore, we gave evidence both techniques need to be integrated, complementing each other. Our results on the TPC-H Benchmark show that energy consumption can be reduced by up to 30%, often accompanied by improved query performance or a minor decrease. Interestingly, increasing the number of cores led to reductions in both energy use and response time. In contrast, the effect of CPU frequency varied across query types, contradicting earlier work aiming for a single default configuration.

We are exploring other optimization algorithms, like greedy algorithms, and dynamic programming, but we believe they will not produce a drastic difference. With the ongoing migration from on-premise to cloud-based database systems, it is essential to reconsider all aspects of system design—including virtual CPUs, elastic computing resources, and the extension of cloud database architectures—to incorporate not only scalability and cost, but also energy efficiency.

References

1. Bellatreche, L., Djellali, F., Macyna, W., Ordonez, C.: Energy-aware query processing: a case study on join reordering. In: IEEE Big Data, pp. 3743–3752 (2023)
2. Binglei, G., Jiong, Y., Dexian, Y., Hongyong, L., Bin, L.: Energy-Efficient Database Systems: A Systematic Survey. ACM Computing Survey (2022)
3. Bouhatous, A., Bellatreche, L., Abdelwahed, E.H., Ordonez, C.: The impact of multicore cpus on eco-friendly query processors in big data warehouses. In: IEEE Big Data, pp. 4463–4472 (2022)
4. Dembele, S.P., Bellatreche, L., Ordonez, C., Roukh, A.: Think big, start small: a good initiative to design green query optimizers. Clust. Comput. **23**(3), 2323–2345 (2020)
5. Etinski, M., Corbalán, J., Labarta, J., Valero, M.: Understanding the future of energy-performance trade-off via DVFS in HPC environments. J. Parallel Distrib. Comput. **72**(4), 579–590 (2012)
6. Guo, B., Yu, J., Liao, B., Yang, D., Lu, L.: A green framework for DBMS based on energy-aware query optimization and energy-efficient query processing. J. Netw. Comput. Appl. **84**, 118–130 (2017)
7. Harizopoulos, S., Shah, M.A., Meza, J., Ranganathan, P.: Energy efficiency: the new holy grail of data management systems research. In: CIDR (2009)
8. Kunjir, M., Birwa, P.K., Haritsa, J.R.: Peak power plays in database engines. In: EDBT, pp. 444–455 (2012)
9. Lang, W., Kandhan, R., Patel, J.M.: Rethinking query processing for energy efficiency: slowing down to win the race. IEEE Data Eng. Bull. **34**(1), 12–23 (2011)
10. Mittal, S., Vetter, J.S.: A survey of methods for analyzing and improving GPU energy efficiency. ACM Comput. Surv. **47**(2), pp. 19:1–19:23 (2014)
11. Psaroudakis, I., et al.: Dynamic fine-grained scheduling for energy-efficient main-memory queries. In: DaMoN Workshop, pp. 1:1–1:7 (2014)
12. Qasaimeh, M., Zambreno, J., Jones, P.H., Denolf, K., Lo, J., Vissers, K.A.: Analyzing the energy-efficiency of vision kernels on embedded cpu, GPU and FPGA platforms. In: FCCM, p. 336 (2019)

13. Roukh, A., Bellatreche, L., Bouarar, S., Boukorca, A.: Eco-physic: eco-physical design initiative for very large databases. Inf. Syst. **68**, 44–63 (2017)
14. Tsirogiannis, D., Harizopoulos, S., Shah, M.A.: Analyzing the energy efficiency of a database server. In: ACM SIGMOD, pp. 231–242 (2010)
15. Wang, J., Feng, L., Xue, W., Song, Z.: A survey on energy-efficient data management. SIGMOD Rec. **40**(2), pp. 17–23 (2011)
16. Xu, Z., Tu, Y.C., Wang, X.: Dynamic energy estimation of query plans in database systems. In: ICDCS, pp. 83–92 (2013)

On the Costs and Benefits of Learned Indexing for Dynamic High-Dimensional Data

Terézia Slanináková[1,2] , Jaroslav Olha[1] , David Procházka[1] ,
Matej Antol[1,2] , and Vlastislav Dohnal[1(✉)]

[1] Faculty of Informatics, Masaryk University, Brno, Czechia
`dohnal@fi.muni.cz`
[2] Institute of Computer Science, Masaryk University, Brno, Czechia

Abstract. One of the main challenges within the growing research area of learned indexing is the lack of adaptability to dynamically expanding datasets . This paper explores the dynamization of a static learned index for complex data through operations such as node splitting and broadening, enabling efficient adaptation to new data. Furthermore, we evaluate the trade-offs between static and dynamic approaches by introducing an amortized cost model to assess query performance in tandem with the build costs of the index structure, enabling experimental determination of when a dynamic learned index outperforms its static counterpart. We apply the dynamization method to a static learned index and demonstrate that its superior scaling quickly surpasses the static implementation in terms of overall costs as the database grows.

Keywords: Learned indexing · Dynamization · Dynamic datasets · k-NN search · ANN search

1 Introduction

The problem of adapting to dynamically expanding datasets remains a challenge in many indexing approaches. For instance, many recent advances in indexing involve machine learning models, leading to the emergence of a new specialized field of study called *learned indexing*. Once trained, machine learning models typically cannot be updated with new data or classification categories without losing prior knowledge, often requiring a full retraining instead.

Despite this limitation, learned indexing has proven successful at indexing structured, low-dimensional datasets [4,7,14], and continues to gain traction in complex data indexing and retrieval [1,5,6,8].

We propose a dynamization method capable of transforming a static learned indexing structure for complex, high-dimensional data into a dynamic one. This is achieved through generalized node-splitting and node-broadening operations, along with a set of simple rules for their application. Even though it was developed with a learned indexing use case in mind, this approach is general enough

© The Author(s), under exclusive license to Springer Nature Switzerland AG 2026
C. K. Leung et al. (Eds.): DaWaK 2025, LNCS 16048, pp. 251–258, 2026.
https://doi.org/10.1007/978-3-032-02215-8_20

to be applicable to any partition-based indexing technique that struggles with dynamization.

A key challenge, however, is the evaluation of the costs and benefits of such a dynamization, since it can be difficult to compare an index with high upfront construction costs to one where those costs are distributed more evenly over the lifetime of the database.

Thus, we compare the static and dynamic approaches in terms of their amortized costs. We do this by establishing several indexing scenarios, and defining them in a way that allows us to quantify the amortized build costs of an average query, in addition to its search costs. This results in a single amortized cost metric that can be used to evaluate, in a very clear and straightforward manner, when dynamization is appropriate and when it might be counter-productive.

2   Related Work

The central theme of this paper is dynamization through the use of index-agnostic operators. The goal of dynamization, as proposed in the original Bentley-Saxe method [3], is to abstract away from the details of particular indexing solutions by transforming a single index into a series of indexes of progressively larger sizes. Although applied in numerous studies [10,12,15], they struggle with deletion performance, limited query type support and a lack of generality. Recent work [13] addressed the shortcomings of Bentley-Saxe by proposing a more general dynamization framework.

As opposed to [13], the dynamization method presented in this paper is not an extensive ready-to-use framework. Instead, we introduce a minimalistic approach consisting of two extension operations and a basic set of restructuring policies that can be implemented into any partitioning-based index.

3   Methods

The static index we have chosen for dynamization is a hierarchical learned index called the Learned Metric Index (LMI) [1,11]. The model was first conceptualized as a tree structure composed of learned models, where the root node is a model trained on all the data, with a pre-defined number of classification categories corresponding to its children. The child nodes are either leaf nodes, i.e., buckets containing the given subset of the data, or inner nodes, which correspond to another learned model – this model once again partitions its given subset of data into a pre-defined number of classes.

3.1   Dynamized Learned Index

For a tree-like structure to become dynamic, we need to define mechanisms for adaptive expansion of its nodes: node split (deepening) and node expansion (broadening). *Deepening* triggers vertical growth of the index. Once a leaf node reaches maximum capacity, the node is transformed into an inner node, and its objects are dispersed into newly created child nodes. In a learned index, deepening would be equivalent to creating a new model and training on the node's objects with a given target number of child nodes. *Broadening* extends the index horizontally – it is defined as the re-creation of a node (either inner or leaf) from scratch with its current objects. In learned indexing, this involves re-partitioning and retraining of the learned model with all the relevant objects (potentially also objects on the grandchildren's level). The *shortening* operation addresses the removal of severely underpopulated nodes. This involves deleting the node and reinserting its objects into the index by removing the corresponding output neuron and its connections from the learned model (MLP). Unlike adding a neuron – which requires global retraining and risks catastrophic forgetting – this localized operation allows well-populated categories to redistribute the obsolete category's data points. All three defined operations are visualized in Fig. 1.

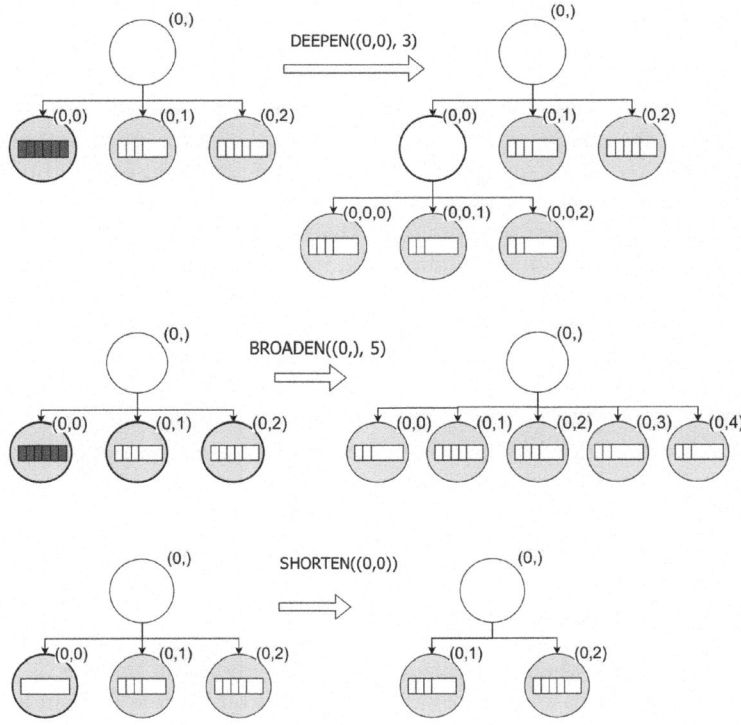

Fig. 1. Overview of the deepening, broadening and shortening operations.

The next step is to establish policies to invoke these operations. First, we define minimum and maximum bounds on the leaf node capacity to impose limits on the costs of sequential search. Similarly, these bounds are defined on the inner nodes for optimal discriminative power of a single model. We implement a policy of detecting and resolving any violations of these bounds – in case of a leaf node having fewer than 5 objects (*underflow*), we invoke the *shorten* operation. To manage *overflow*, the structure ensures that the average occupancy of child nodes remains below 1 000. When this bound is violated, we invoke node extension, alternating between deepen and broaden operations to maintain a shallow index. Specifically, a node will be deepened (split) until the index reaches a maximum depth of two levels. After this depth is reached, the broaden operation is invoked on the appropriate node to accommodate further growth, re-partitioning, and retraining of the learned model.

3.2   Evaluation Baseline

A dynamized index is not necessarily an improvement in terms of total costs – while it should theoretically scale better with growing datasets, it introduces additional overhead not present in the static version. To determine the conditions under which dynamization is worthwhile, the dynamized index should be evaluated against its static counterpart.

One way to do this is to simply keep inserting objects into the static index and compare it with the growing adaptive version. This will necessarily lead to deterioration of the quality of the query results in the static index, but the static version will only require a single build, likely leading to lower overall build costs. In the experimental evaluation, we will refer to this as the **No rebuild** baseline.

Another way is to take advantage of the fact that any static index can be made somewhat dynamic by rebuilding it in its entirety after a certain threshold of new objects, thus allowing it to adapt to new data at the cost of additional build costs – we will refer to this as the **Naive rebuild** baseline. Implementing the Naive rebuild method involves the selection of a single parameter which determines how often the index is discarded and rebuilt from scratch. Let us call this parameter the *rebuild interval*, and define it in terms of the number of new objects that can be added before the index is rebuilt (e.g., a rebuild interval of 10 000 means that after we build the index, the next 9 999 objects are simply added to the structure as is, and the 10 000th object triggers a full rebuild). In the next section, we will discuss the selection and optimization of this parameter.

3.3   Amortized Cost Model

The typical problem when evaluating static and dynamic indexes is the fact that their costs are distributed very differently – if build time is included in the evaluation, a static index may be greatly disadvantaged at the beginning, as it takes a long time to even answer the first query. If the build time is excluded, however, various adaptive methods may be punished for not taking the time to build the perfect structure ahead of time. An ideal evaluation metric should combine the search costs and build costs into a single objective that can be compared on a per-query basis regardless of how dynamic the method is.

To unify static and dynamic evaluation, let us first define a new property of the system: the relative number of queries per inserted object. We will call this the *querying frequency*, and define it as $\frac{\#queries}{\#new\_objects}$. We will assume querying frequency to be an inherent property of the given indexing scenario, which does not change during the lifetime of the database. This allows us to relate the number of new objects added to the database (which dictates the rebuild costs of a dynamic structure) with the number of queries performed on the database (which determines the amortization).

The only other factor that determines whether structural adjustments are worth their costs is how accurate the queries need to be – the *target recall*. If lower query recall is sufficient, deterioration of the structure is less problematic, and rebuilds are not as worthwhile. If high recall is needed, the index needs to be maintained in peak condition via more frequent rebuilds – deterioration of query quality is more punishing, so even more costly rebuilds are justified.

Once we determine the querying frequency of our indexing scenario and the desired (average) recall of our queries, we can infer the amortized search cost as:

$$\texttt{AC} = \texttt{SC} + \frac{\texttt{BC}}{\texttt{RI} * \texttt{QF}}$$

where $\texttt{AC}$ is the amortized cost, $\texttt{SC}$ is the search cost of a single query (i.e., how many seconds it takes for an average query to achieve the target recall), $\texttt{RI}$ is the rebuild interval of the given index (in terms of the average number of new objects that trigger a rebuild), $\texttt{BC}$ is its build cost (in seconds), and $\texttt{QF}$ is the querying frequency of the given indexing scenario (in terms of queries per insert).

In other words, if an index is rebuilt after adding 1K new objects, and the querying frequency is 100 queries per new object, then one build of the index will last for 100K queries. Therefore, the amortized cost of a query should include $\frac{1}{100000}$ of the index build costs in addition to the immediate search costs. This allows methods with infrequent but costly rebuilds to be directly compared with methods that perform gradual, less extensive updates.

Optimal Rebuild Interval. The concept of amortized cost can also be used to optimize the *Naive rebuild* baseline method described earlier – if we can determine the indexing scenario ahead of time, we can optimize when a full rebuild of the structure should be triggered so as to balance build costs with structural deterioration.

By definition, the rebuild interval can only be optimized for a specific scenario, that is, a single combination of query frequency and target recall. The penalty for a suboptimal selection of the rebuild interval parameter will be evaluated in the next section.

4   Experiments

For our experimental evaluation, we use SIFT descriptors [9], one of the datasets used in ANN-benchmarks [2]. This dataset consists of 1 million objects of 128 dimensions with the Euclidean distance metric, and additionally contains 10K queries in a 30-NN setup for experimental evaluation.

Since the amortized cost metric depends on two variables – the number of *queries per inserted object* (QPI) and *target recall* (TR) – the evaluation considers scenarios with two extreme settings of either variable, for a total of 4 scenarios. We set high querying frequency at 100 queries per insert (corresponding, for instance, to a typical social media feed) and low frequency at 1 query per insert (e.g., a monitoring service or a messaging system). For target recall, we use extremes of 0.9 (high) and 0.5 (low) – in a 30-NN query, this means the database must return, on average, at least 27 objects at high recall and 15 at low recall.

As described in Sect. 3, we compare the dynamized version of the learned index against two baselines – the *Naive rebuild* baseline, which rebuilds the entire index periodically based on a pre-defined rebuild interval, and the *No rebuild* baseline, which simply stores incoming objects without adjusting its structure until it runs out of experimental data to process. We performed the experiment with various database sizes, ranging from 100K to 900K. For the baselines, this means that the database was first built using the given number of initial objects, and then new objects were added – in the case of the *Naive rebuild* baseline, the objects were added until a rebuild was triggered, in the *No rebuild* baseline, they were added until all 1M experimental data objects were indexed. The dynamized index was built gradually starting with no objects, and its amortized performance was simply evaluated after every 100K objects as the index grew and adapted.

Since the *Naive rebuild* baseline is sensitive to the selection of rebuild interval, we optimized this parameter for each of the 4 experimental scenarios, resulting in 4 different performance curves of this method. As a result, in each evaluated scenario, one of the Naive rebuild baselines shows the optimal solution (since the rebuild frequency is set up perfectly for that scenario), and the other Naive rebuild baselines show the penalty for parameterizing the method incorrectly.

As for the *No rebuild* baseline, since the index is only built once (using the initial number of objects), the build costs are as low as possible for the given initial database size, but the quality of queries will keep deteriorating towards exhaustive search in the limit. Being limited by a finite experimental dataset (1M objects), we can only simulate smaller and smaller increments of this deterioration as we use more objects to build the initial database and thus have fewer objects left to insert. Thus, when plotting the amortized cost, the results of this baseline keep converging towards the dynamized methods, not due to improved performance of the method, but because the fixed size of the experimental dataset limits the ability to model larger-scale growth.

Note that the static index used in the *Naive rebuild* and *No rebuild* baselines is a single-level structure, implemented as a single MLP, and is parameterized to hold an average of 1K objects per bucket. Additionally, note that only insertion of new objects is considered – deletion and re-balancing can also be implemented in a straightforward manner (as shown in Sect. 3.1), but the experimental evaluation of such operations is beyond the scope of this paper.

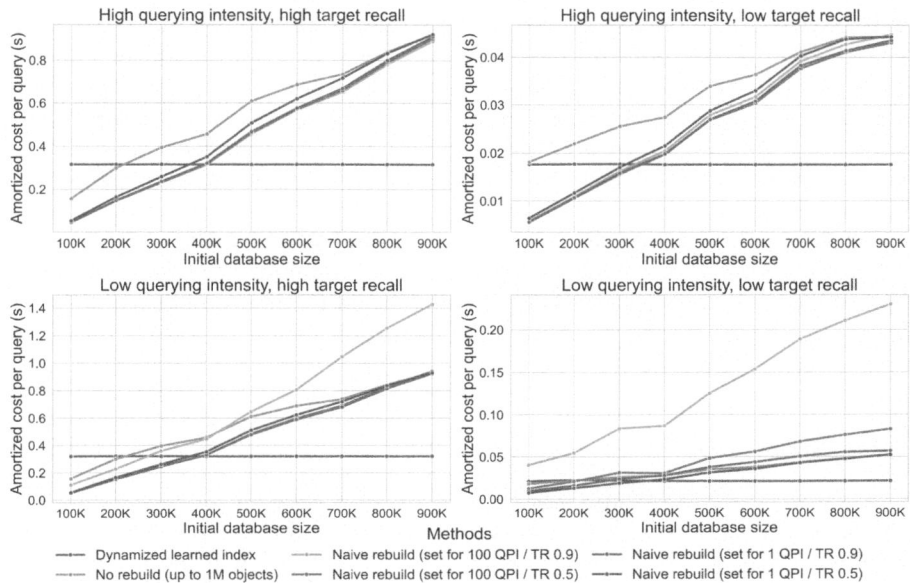

Fig. 2. Amortized costs of the dynamized index and various baselines in four experimental scenarios – querying intensity: 100 queries per insert (high) vs. 1 query per insert (low); target recall: 0.9 (high) vs. 0.5 (low).

Figure 2 shows how amortized costs scale with the database size for each combination of high/low querying frequency and high/low target recall. Keep in mind that the "Initial database size" label only holds true for the baseline methods – the dynamized index always has an initial database size of 0, and it is merely evaluated at various size thresholds as it grows.

As expected, the amortized costs of the *Naive rebuild* and *No rebuild* baselines keep increasing with database size – as a result, the dynamized method inevitably performs better on larger datasets.

In scenarios with higher querying intensity (100 queries per object), the various rebuild intervals of the *Naive rebuild* method do not seem to make much of a difference. In lower intensity scenarios, however, the method with the most frequent rebuilds (parameterized for high intensity scenarios) is severely punished, usually performing worse than a method that performs no rebuilds at all. Aside from this extreme case, the choice between the naive and dynamized approaches is usually much more important than the choice of the rebuild interval between different *Naive rebuild* variants.

5   In Conclusion

We introduced a method for dynamizing a static index, using a streamlined set of policies and operations to enable adaptive expansion. This approach was used to transform a previously static learned index for complex datasets into a dynamic one. In addition, we introduced an amortized cost model to evaluate the benefits of the dynamic version against its static counterpart in various scenarios. As expected, the results show that the dynamized index scales very well with database size in all scenarios, and while its overhead may not be justified for smaller databases, the dynamization quickly proves advantageous as the database grows.

Taken together, our dynamization approach offers a straightforward way to transition from static to dynamic learned indexes, significantly broadening their applicability in complex data, while the evaluation method provides a practical way to assess the trade-offs of dynamization and identify scenarios where its benefits outweigh its costs.

References

1. Antol, M., Olha, J., Slanináková, T., Dohnal, V.: Learned metric index - proposition of learned indexing for unstructured data. Inf. Systems **100**, 101774 (2021)
2. Aumüller, M., Bernhardsson, E., Faithfull, A.: Ann-benchmarks: a benchmarking tool for approximate nearest neighbor algorithms. Inf. Syst. **87**, 101374 (2020)
3. Bentley, J.L., Saxe, J.B.: Decomposable searching problems i. static-to-dynamic transformation. J. Algorithms **1**(4), pp. 301–358 (1980)
4. Ding, J., et al.: Alex: an updatable adaptive learned index. In: Proc. ACM SIGMOD 2020, pp. 969–984 (2020)
5. Dong, Y., Indyk, P., Razenshteyn, I.P., Wagner, T.: Learning space partitions for nearest neighbor search. ICLR (2020)
6. Gupta, G., Medini, T., Shrivastava, A., Smola, A.J.: Bliss: a billion scale index using iterative re-partitioning. In: Proc. ACM SIGKDD 2022, pp. 486–495 (2022)
7. Kraska, T., Beutel, A., Chi, E.H., Dean, J., Polyzotis, N.: The case for learned index structures. In: Proc. ACM SIGMOD 2018, pp. 489–504 (2018)
8. Li, L., Han, A., Cui, X., Wu, B.: Flex: a fast and light-weight learned index for KNN search in high-dimensional space. Inf. Sci. **669**, 120546 (2024)
9. Lowe, D.G.: Distinctive image features from scale-invariant keypoints. Int. J. Comput. Vision **60**, 91–110 (2004)
10. Naidan, B., Hetland, M.L.: Static-to-dynamic transformation for metric indexing structures. In: International Conference on Similarity Search and Applications, pp. 101–115. Springer (2012)
11. Procházka, D., Slanináková, T., Čerňanský, J., Olha, J., Antol, M., Dohnal, V.: Scaling learned metric index to 100m datasets. In: International Conference on Similarity Search and Applications, pp. 266–273. Springer (2025)
12. Rumbaugh, D.B., Xie, D.: Practical dynamic extension for sampling indexes. Proc. ACM Manag. Data **1**(4), 1–26 (2023)
13. Rumbaugh, D.B., Xie, D., Zhao, Z.: Towards systematic index dynamization. Proc. VLDB Endow. **17**(11), 2867–2879 (2024)
14. Wu, J., Zhang, Y., Chen, S., Wang, J., Chen, Y., Xing, C.: Updatable learned index with precise positions. Proc. VLDB Endow. **14**(8), 1276–1288 (2021)
15. Xie, D., Phillips, J.M., Matheny, M., Li, F.: Spatial independent range sampling. In: Proceedings of the 2021 International Conference on Management of Data, pp. 2023–2035 (2021)

A Bayesian Reinforcement Learning Framework for Online Index Tuning

Md Rakibul Hasan[1], Xiaoying Wu[2], and Dimitri Theodoratos[1]([⊠])

[1] New Jersey Institute of Technology, Newark, NJ, USA
{mh629,dth}@njit.edu
[2] School of Computer Science, Wuhan University, Wuhan, China
xiaoying.wu@whu.edu.cn

Abstract. Efficient index tuning is critical to maintain high query performance in database systems with dynamic workloads. Traditional offline and heuristic-driven tuning methods often incur high overhead due to frequent reconfiguration and fail to adapt to evolving workloads. To overcome these limitations, we address online index selection and formulate it as a sequential decision-making problem under uncertainty. We propose a Bayesian Reinforcement Learning framework that adaptively tunes index configurations based solely on the observed workload history, without requiring prior knowledge of the workload. Our framework leverages Q-learning with Thompson Sampling, a posterior distribution sampling method, to adaptively maintain and refine index configurations over time. The probabilistic mechanism of our approach enables the Q-learning agent to effectively balance exploration and exploitation, allowing it to traverse the vast exponential index configuration space. Our comprehensive experimental evaluation demonstrates that our algorithm excels at online index tuning across a diverse range of workloads on a standard benchmark dataset and outperforms other index tuning algorithms which are based on alternative learning methods.

1 Introduction

Indexing plays a crucial role in optimizing database query performance, yet selecting an optimal set of indexes remains a challenging task. Traditionally, Database Administrators (DBAs) manually manage index selection, either proactively during schema design or reactively based on query execution patterns. Furthermore, workload patterns in real-world applications are highly dynamic, such as anomaly detection and exploratory analytics, require indexing solutions that adapt in real-time to evolving query patterns. That's why making static, offline indexing strategies ineffective because it fails to capture changing access patterns, leading to suboptimal indexing configurations.

Unlike static tuning, online index selection adapts indexing in response to workload history, ensuring sustained performance. Online methods fall into learning-based and non-learning-based approaches: learning-based methods [1,9–11] adaptively optimize indexes from data, while non-learning heuristics [6,8,12]

© The Author(s), under exclusive license to Springer Nature Switzerland AG 2026
C. K. Leung et al. (Eds.): DaWaK 2025, LNCS 16048, pp. 259–267, 2026.
https://doi.org/10.1007/978-3-032-02215-8_21

lack flexibility for complex workloads. This paper proposes a reinforcement learning framework with Bayesian exploration—Reward-Based Learning Algorithm (RBLA)—which autonomously adapts to workload shifts. By integrating Thompson Sampling, RBLA balances exploration and exploitation, optimizing both query performance and index maintenance for evolving databases. The contributions of this work include the following:

- We introduce *RBLA*, an adaptive indexing architecture based on Q-learning, which autonomously learns to tune indexes online in response to workload changes, overcoming static tuning limitations.
- RBLA jointly considers query performance and maintenance overhead, making data-driven decisions to create, retain, or drop indexes, thus reducing total operational cost.
- We integrate Thompson Sampling with Q-learning, enabling efficient exploration and exploitation across a large configuration space and preventing premature convergence.
- Extensive experiments on benchmark datasets show that RBLA consistently outperforms state-of-the-art learning-based approaches in both query execution efficiency and transition cost.

2   Online Index Selection (OIS) Problem

We formulate online index selection as a sequential decision-making problem: the database dynamically adjusts its index configuration to optimize query performance while controlling overhead. At each step k, the active configuration $s_k \subseteq I$ (where I is the set of possible indexes) is used to execute query q_k, with transitions between configurations incurring a cost. The benefit of adding index i to configuration s_{k-1} for query q_k is $b(q_k, i, s_{k-1}) = g_{\text{exec}}(q_k, s_{k-1} \setminus \{i\}) - g_{\text{exec}}(q_k, s_{k-1} \cup \{i\})$, where g_{exec} denotes the (estimated) query execution cost under a given configuration. The cumulative benefit for a configuration is additive over its indexes:

$$B_k(s_k) = \sum_{i \in s_k} B_k(i) \tag{1}$$

with each $B_k(i)$ updated online as new queries are processed. Changing from s_{k-1} to s_k incurs a transition cost, denoted *transition-cost*(s_{k-1}, s_k). The objective is to find an optimal sequence of configurations $S(s_0, s_1, \ldots, s_N)$ that minimizes the total expected cost. For a policy π starting at s_0, the cost includes execution time, transition overhead, and recommendation cost:

$$J_\pi(s_0) = \sum_{k=1}^{N} \mathbb{E} \left[g_{\text{exec}}(q_k, s_{k-1}) + \textit{transition-cost}(s_{k-1}, s_k) + g_{\text{rec}}(s_k) \right] \tag{2}$$

Thus, the goal is to determine the optimal policy π^* that minimizes this cost:

$$J_{\pi^*}(s_0) = \min_{\pi \in \Pi} J_\pi(s_0) \tag{3}$$

3   Q-Learning with Thompson Sampling for OIS

Q-learning [14] is a value-based reinforcement learning method for learning the expected cumulative reward of actions in a Markov Decision Process (MDP), defined by the tuple $(\mathcal{S}, \mathcal{A}, P, R, \gamma)$. The agent iteratively updates the Q-function as:

$$Q(s_t, a_t) \leftarrow Q(s_t, a_t) + \alpha \left[r_t + \gamma \max_{a'} Q(s_{t+1}, a') - Q(s_t, a_t) \right] \qquad (4)$$

where α is the learning rate and γ the discount factor. The optimal policy is derived as $\pi^*(s) = \arg\max_a Q(s, a)$. To balance exploration and exploitation, Q-learning can be combined with Thompson Sampling (TS), a Bayesian approach that samples actions according to their posterior probability of being optimal [3]. In this framework, each per-index Q-value is modeled by a Gaussian posterior $Q(s, x) \sim \mathcal{N}(\mu_x, \sigma_x^2)$ and sampled at each decision step. The agent selects indexes with the highest samples, then updates the posteriors with observed rewards using TD learning and Bayesian inference.

OIS problem naturally aligns with a Q-learning formulation, where the database system serves as the *environment* and the indexing mechanism operates as the *agent*. At each decision step k, corresponding to the arrival of a new query, the agent evaluates and potentially modifies the current index configuration to optimize query execution performance while minimizing associated costs. Formally, the key components of this reinforcement learning approach are follows:

- **State Space ($\mathcal{S}$):** The state $s_{k-1} \in \mathcal{S}$ at step k implicitly encapsulates the current set of active indexes and relevant workload features—primarily the set of columns appearing in the incoming query workload. Rather than explicitly enumerating all possible states, the agent identifies the state implicitly based on active indexes and workload characteristics.
- **Action Space ($\mathcal{A}$):** Actions in our formulation represent atomic index modifications. Specifically, each action corresponds to adding or removing a single candidate index $x \in I$. Each index x thus maintains its own Q-value estimate $Q(s, x)$, reflecting the incremental expected benefit or utility of including or excluding the index from the current configuration.
- **Reward Function (R):** The immediate reward $r_k(x)$ for an index x at step k directly quantifies the performance gain achieved by modifying the current configuration. This reward is calculated as the reduction in query execution cost resulting from adding index x, from $r_k(x) = g_{\text{exe}}(w_k, s_{k-1}) - g_{\text{exe}}(w_k, s_{k-1} \cup \{x\})$
- **Transition Dynamics (P):** State transitions occur deterministically with respect to index modifications (creation or deletion), but are inherently stochastic concerning future query arrivals. Incoming queries are assumed to be drawn from an unknown and evolving distribution, reflecting the dynamic nature of real-world database workloads.

4   Reward Based Learning Algorithm(RBLA) for OIS

4.1   Algorithm Overview

The Reward-Based Learning Algorithm (RBLA, Algorithm 1) adaptively selects index configurations to minimize total workload cost. At each query step, candidate indexes are identified from the query's columns, defining the current state. RBLA maintains Gaussian posteriors for each index's Q-value, modeling both benefit (mean) and uncertainty (variance). The configuration (state) value is given by

$$V(s) = \sum_{x \in s} Q(s, x) \tag{5}$$

For each candidate, Thompson Sampling is applied by drawing a Q-value sample from its posterior. Marginal per-index rewards $r_k(x)$ update the posteriors. The top-K sampled indexes form the next configuration. After executing this configuration, the aggregate reward R_k is used in a TD update of the state value, and the resulting TD error δ_k is evenly distributed among the configuration's Q-values. This separates fast per-index reward updates from slower configuration-level policy learning.

To avoid excessive changes, RBLA only adopts a new configuration if its expected value gain exceeds transition cost by a threshold. This iterative, sample-driven approach enables RBLA to adaptively optimize index configurations in dynamic workloads while controlling tuning overhead.

4.2   RBLA-Agent Decision Process

In this subsection, we describe the detailed decision-making process of the RBLA-agent underlying RBLA for online index selection. The agent continuously interacts with the environment, observing queries and adapting the index configuration over time. The process is divided into the following key stages:

Agent Initialization and State Sensing. At startup, the RBLA-agent sets all Q-values $Q(s, x)$ to zero and initializes Gaussian posteriors for each index x with mean $\mu_x = 0$ and variance $\sigma_x^2 = \sigma_0^2$, reflecting initial uncertainty. At each step, the current state s_{k-1} is defined by the active configuration and the columns appearing in query w_k, which specify the candidate indexes to evaluate.

Action Sampling via Thompson Sampling. To balance exploration and exploitation, RBLA maintains a Gaussian posterior over each index's Q-value, with mean μ_x and variance σ_x^2. At each step, the agent computes the marginal per-index reward $r_k(x)$, updates the posterior (μ_x, σ_x^2) for each x, and then samples θ_x from these posteriors. These samples guide the selection of indexes, encouraging exploration for uncertain candidates (high variance) and exploitation of those with high expected benefit.

Configuration Selection and Transition Decision. After sampling, RBLA selects the top-K indexes by θ_x to form configuration s_k. The value of this configuration, $V(s_k)$, aggregates per-index Q-values as in (5), estimating its expected utility. The agent computes net benefit $B = V(s_k) - \sum_{x \in s_k \setminus s_{k-1}} g_{\text{trans}}(s_{k-1}, s_k)$, where g_{trans} is the index creation cost. Comparing the benefit ratio $B/V(s_{k-1})$ to threshold β, the new configuration is adopted only if the gain justifies the transition; otherwise, the agent retains s_{k-1}. This mechanism controls unnecessary index churn and balances execution and transition costs.

Reward Computation and Temporal Difference Update. After selecting s_k, the agent computes the configuration reward R_k as the execution cost reduction by $R_k = g_{\text{exe}}(w_k, s_{k-1}) - g_{\text{exe}}(w_k, s_k)$. This drives a Temporal Difference (TD) update at the configuration level. The TD error is calculated between the new TD target and the previous configuration value, then distributed evenly to update Q-values of all indexes in s_{k-1}. This hybrid approach ensures uncertainty-aware exploration and efficient convergence to optimal configurations.

Algorithm 1. *RBLA (Reward-Based Learning Algorithm)*

Input: $W = (w_1, w_2, \ldots, w_N)$: query workload, I: index catalog, α: learning rate, γ: discount factor, β: transition threshold, K: maximum number of indexes, s_0: initial configuration, σ_0^2: initial variance;

Output: Sequence $S = (s_0, \ldots, s_N)$

1: Initialize $Q(s, x) = 0$, $\mu_x = 0$, $\sigma_x^2 = \sigma_0^2$ for $x \in I$
2: **for** $k = 1, \ldots, N$ **do**
3:     Observe w_k and s_{k-1}; $V(s_{k-1}) \leftarrow \sum_{x \in s_{k-1}} Q(s_{k-1}, x)$
4:     **for** each $x \in I$ **do**
5:         $r_k(x) \leftarrow g_{exe}(w_k, s_{k-1}) - g_{exe}(w_k, s_{k-1} \cup \{x\})$ if x relevant; else 0
6:         Update (μ_x, σ_x^2) using $r_k(x)$; sample $\theta_x \sim \mathcal{N}(\mu_x, \sigma_x^2)$
7:     **end for**
8:     $s_k \leftarrow$ top-K indices by θ_x; $V(s_k) \leftarrow \sum_{x \in s_k} Q(s_k, x)$
9:     $B \leftarrow V(s_k) - \sum_{x \in s_k \setminus s_{k-1}} \text{cost}(x)$; $B' \leftarrow V(s_{k-1})$
10:    **if** $B/B' < \beta$ **then**
11:        $s_k \leftarrow s_{k-1}$, $V(s_k) \leftarrow V(s_{k-1})$
12:    **end if**
13:    $R_k \leftarrow g_{exe}(w_k, s_{k-1}) - g_{exe}(w_k, s_k)$
14:    $\text{TD}_{target} \leftarrow R_k + \gamma V(s_k)$; $\delta_k \leftarrow \text{TD}_{target} - V(s_{k-1})$
15:    Update $Q(s_{k-1}, x) \mathrel{+}= \frac{\alpha \delta_k}{|s_{k-1}|}$ for $x \in s_{k-1}$
16:    Append s_k to S
17: **end for**
18: **return** S

Iteration and Policy Evolution. At each step, the RBLA-agent appends the selected configuration s_k to the sequence S. The process—observing queries, updating index beliefs, sampling actions, and refining configuration values—repeats for each workload step. Over time, this iterative loop enables the agent to adaptively improve its policy, balancing exploration and exploitation to minimize cumulative cost. Continuous learning from observed rewards ensures that RBLA efficiently tunes indexes for dynamic workloads, without requiring prior workload knowledge.

5   Experimental Evaluation

We conducted comprehensive experiments to evaluate the effectiveness and efficiency of our proposed RBLA for online index selection using the TPC-H benchmark. We compared RBLA with state-of-the-art index tuning algorithms.

5.1   Algorithms for Comparison

We evaluated our approach (RBLA) against the following methods:

- **Anytime** [4] Anytime is a state-of-the-art offline tuning algorithm. It is used in Database Tuning Advisor of Microsoft SQL Server.
- **SmartIX** [7]: A DRL-based offline index tuning algorithm. We adapted it for online use by training a deep model to handle dynamic workloads.
- **LiteSelect** [15]: A recent learning-based online index selection method that employs exponential smoothing for real-time tuning.
- **C2UCB** [9]: An online tuning technique based on MAB-UCB principles that uses advanced exploration strategies in dynamic environments.

We also include a **NoIndex** baseline, which runs the workload using only the primary and foreign key indexes automatically created, to quantify the maximum possible benefit from index tuning.

5.2   Experimental Setup and Evaluation Metrics

We evaluate our approach using the TPC-H benchmark (scale factor 1), which contains 8 tables, 45 indexable columns, and 22 query templates featuring complex analytical queries. Workloads include: (1) **Shifting** -four query groups executed sequentially over 80 rounds, (2) **Noisy** - shifting plus 10% injected noise queries, totaling 96 rounds, and (3) **Random** - randomized query grouping, 25 rounds, capturing diverse and dynamic query behaviors. Query costs are estimated using the What-if optimizer [5].

RBLA parameters were tuned for performance: learning rate $\alpha = 0.1$, discount factor $\gamma = 0.9$, maximum indexes $K = 10$, what-if call ratio 0.5, and index width fixed at 1. Comparative algorithms used their default or published settings (e.g., SmartIX was trained with 20K TPC-H queries and a matching learning rate).

Performance is measured by index recommendation time, transition time (for creating/dropping indexes), and query execution time. Queries exceeding a 30-second timeout are capped at 30 s for fairness. SmartIX and other baselines are evaluated with these metrics for direct comparison. All experiments ran on an Intel Core i5 system with 8 GB RAM, Ubuntu, and PostgreSQL-16. This setup assesses RBLA's ability to minimize costs and adapt to evolving workloads in a realistic environment.

5.3   Performance Analysis

Recommendation Time Performance. Figure 1a shows that C2UCB incurs the highest recommendation times, while RBLA is 23x faster than both Lite and C2UCB across all workloads. Lite and Anytime show moderate overhead due to their more conservative strategies. All methods take slightly longer under noisy conditions. SmartIX is omitted since its caching mechanism makes its times not directly comparable. These results highlight RBLA's efficiency in rapidly recommending effective index configurations.

Transition Time Performance. Figure 1b shows that C2UCB incurs the highest transition times due to frequent index changes, while Lite maintains the lowest overhead across workloads. RBLA achieves significantly lower transition times than Anytime (about 2x improvement) and dramatically outperforms C2UCB (7–8x lower), with values steadily dropping from 34.75 s (shifting) to 10.74 s (random). These results demonstrate that RBLA's learning-driven approach efficiently balances adaptation and overhead, minimizing costly index modifications compared to exploration-heavy methods.

Execution Time Performance. Figure 1c compares execution times for all algorithms under shifting, noisy, and random TPC-H workloads. RBLA consistently achieves the lowest execution times (1421 s, 1764 s, 2308 s), outperforming NoIndex, Anytime, SmartIX, C2UCB, and Lite by up to 62% on shifting and following a similar trend in other scenarios. Lite and C2UCB yield intermediate results, with C2UCB's aggressive exploration leading to higher costs. SmartIX and Anytime lag further due to lack of adaptation. The NoIndex baseline performs worst. These results confirm that RBLA minimizes execution cost by adaptively converging to high-quality index configurations across varying workload dynamics.

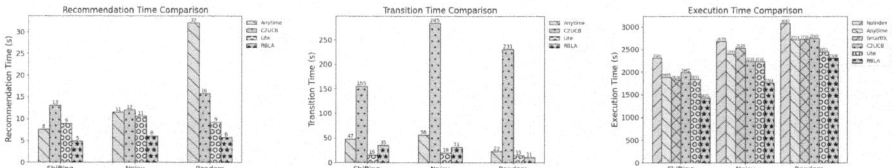

(a) Index Recommendation   (b) Index Transition Time   (c) Query Execution Time
Time

Fig. 1. Performance metrics comparison for RBLA, Lite, C2UCB, SmartIX, Anytime, and NoIndex on the TPC-H shifting, noisy, and random workloads. (a) Total Index Recommendation Time, (b) Total Index Transition Time, (c) Total Query Execution Time.

6 Related Work

Prior research on online index selection has focused on both traditional heuristic methods and more recent learning-based approaches. Conventional techniques, as described in [2,4,8,12], assume that future queries will resemble recent ones and rely on fixed heuristics to iteratively select an index configuration based on historical workload performance. However, these static heuristics lack adaptability and cannot refine their decisions as new data becomes available.

Several auto-indexing methods employ deep reinforcement learning (DRL) to learn cost models [1,13]. For instance, one approach uses a policy iteration algorithm to build and update value functions, but it suffers from high recommendation costs [15]. Similarly, SmartIX [7] adopts a Q-learning-based epsilon-greedy strategy that, despite achieving low latency in index recommendation, requires significant computational resources for training and exhibits slower adaptation due to its static deep model. Another work [9] proposes a sequence-based tuning framework that exploits query order to adjust index configurations. However, its fixed heuristic exploration can incur substantial computational overhead as the candidate index space expands. In contrast, RBLA employs a Q-learning approach enhanced with Thompson Sampling to enable fast, online updates with low computational cost. By continuously refining its action-value estimates in real time, RBLA effectively balances exploration and exploitation, making it highly suitable for dynamic workload environments.

7 Conclusion

In this paper, we introduced RBLA, an automated database indexing architecture that leverages Q-learning with Thompson Sampling for online index selection. By framing index selection as a sequential decision-making problem, RBLA adapts to evolving workloads and effectively balances exploration and exploitation. The use of Thompson Sampling enables our agent to sample from posterior distributions, efficiently navigating the configuration space to avoid premature convergence on suboptimal policies. Moreover, RBLA continuously monitors query performance and update overhead to minimize operational costs associated with index changes. Extensive experiments demonstrate that RBLA outperforms recent learning-based solutions in both efficiency and effectiveness. We are currently working on how the RBLA framework can be extended to automate other knob tuning for performance optimization in DBMSs.

References

1. Basu, D., et al.: Cost-model oblivious database tuning with reinforcement learning. In: DEXA (2015)
2. Bruno, N., Chaudhuri, S.: An online approach to physical design tuning. In: ICDE, pp. 826–835 (2007)

3. Chapelle, O., Li, L.: An empirical evaluation of Thompson sampling. In: Neural Information Processing Systems (2011)
4. Chaudhuri, S., Narasayya, V.: Anytime algorithm of database tuning advisor for Microsoft SQL server (2020)
5. Chaudhuri, S., Narasayya, V.R.: Autoadmin "what-if" index analysis utility. In: SIGMOD. ACM Press (1998)
6. Chaudhuri, S., Narasayya, V.R.: Self-tuning database systems: a decade of progress. In: VLDB, pp. 3–14 (2007)
7. Licks, G.P., Couto, J.M.C., de Fátima Miehe, P., Paris,R.D., Ruiz, D.D.A., Meneguzzi, F.: SmartIX: a database indexing agent based on reinforcement learning. Appl. Intell. **50**(8) (2020)
8. Lühring, M., Sattler, K., Schmidt, K., Schallehn, E.: Autonomous management of soft indexes. In: ICDEW, pp. 450–458 (2007)
9. Perera, R.M., Oetomo, B., Rubinstein, B.I.P., Borovica-Gajic, R.: HMAB: self-driving hierarchy of bandits for integrated physical database design tuning. Proc. VLDB Endow. **16**(2), 216–229 (2022)
10. Powell, W.B.: Reinforcement Learning and Stochastic Optimization: A Unified Framework for Sequential Decisions. Wiley, Hoboken (2022)
11. Sadri, Z., Gruenwald, L., Leal, E.: DRLindex: deep reinforcement learning index advisor for a cluster database. In: Proceedings of the 24th Symposium on International Database Engineering & Applications (2020)
12. Schnaitter, K., Abiteboul, S., Milo, T., Polyzotis, N.: COLT: continuous on-line tuning. In: SIGMOD (2006)
13. Wang, Z., Liu, H., Lin, C., Bao, Z., Li, G., Wang, T.: Leveraging dynamic and heterogeneous workload knowledge to boost the performance of index advisors. Proc. VLDB Endow. **17**, 1642–1654 (2024)
14. Watkins, C., Dayan, P.: Q-learning. Mach. Learn. **8**, 279–292 (1992)
15. Wu, X., Wang, S., Liu, X., Theodoratos, D., Hasan, M.R.: LiteSelect: a lightweight adaptive learning algorithm for online index selection. In: International Conference on Data Warehousing and Knowledge Discovery (2024)

Large Language Models (LLMs)

Explaining Recovery Trajectories of Older Adults Post Lower-Limb Fracture Using Modality-Wise Multiview Clustering and Large Language Models

Shehroz S. Khan[1,2(✉)] , Ali Abedi[1,3] , and Charlene H. Chu[3]

[1] KITE Research Institute, Toronto Rehabilitation Institute,
University Health Network, Toronto, Canada
{shehroz.khan,ali.abedi}@uhn.ca
[2] College of Engineering and Technology, American University of the Middle East,
Egaila, Kuwait City, Kuwait
[3] Lawrence Bloomberg Faculty of Nursing, University of Toronto, Toronto, Canada
charlene.chu@utoronto.ca

Abstract. Interpreting large volumes of high-dimensional, unlabeled data in a manner that is comprehensible to humans remains a significant challenge across various domains. In unsupervised healthcare data analysis, interpreting clustered data can offer meaningful insights into patients' health outcomes, which hold direct implications for healthcare providers. This paper addresses the problem of interpreting clustered sensor data collected from older adult patients recovering from lower-limb fractures in the community. A total of 560 days of multimodal sensor data, including acceleration, step count, ambient motion, GPS location, heart rate, and sleep, alongside clinical scores, were remotely collected from patients at home. Clustering was first carried out separately for each data modality to assess the impact of feature sets extracted from each modality on patients' recovery trajectories. Then, using context-aware prompting, a large language model was employed to infer meaningful cluster labels for the clusters derived from each modality. The quality of these clusters and their corresponding labels was validated through rigorous statistical testing and visualization against clinical scores collected alongside the multimodal sensor data. The results demonstrated the statistical significance of most modality-specific cluster labels generated by the large language model with respect to clinical scores, confirming the efficacy of the proposed method for interpreting sensor data in an unsupervised manner. This unsupervised data analysis approach, relying solely on sensor data, enables clinicians to identify at-risk patients and take timely measures to improve health outcomes.

Keywords: multiview clustering · large language models · context-aware prompting · multimodal data · older adults

ⓒ The Author(s), under exclusive license to Springer Nature Switzerland AG 2026
C. K. Leung et al. (Eds.): DaWaK 2025, LNCS 16048, pp. 271–285, 2026.
https://doi.org/10.1007/978-3-032-02215-8_22

1   Introduction

Clustering is the process of grouping data objects such that the objects within the same clusters are more similar to each other than to the objects in other clusters [1]. The majority of clustering algorithms use some notion of similarity to group different objects into distinct clusters. Based on the data type (numerical, categorical, or mixed), similarity or dissimilarity can be calculated using many available metrics, such as Euclidean distance, Hamming distance, Jaccard coefficient, and other specialized metrics [2]. Several types of clustering algorithms exist, including partitional, hierarchical, density-based, spectral, subspace, and others [2,3]. In multiview clustering approaches [4], the data are clustered based on 'views' or subspaces of the data, where a view contains a specific aspect of information (e.g., data collected from different sensors makes different views of information). Then, the data are clustered using those views and eventually merged to give final clusters using various approaches [5].

Irrespective of the clustering approach, the overall clustering process is objective, data-driven, and unsupervised, i.e., it groups data objects using some notion of (dis)similarity, mostly without any context, which results in cluster labels that are largely arbitrary. Human involvement is needed to understand the contents of these clusters to determine what each cluster represents. In a healthcare setting, these clusters could represent healthy, non-healthy, or at-risk patients. This would be vital information for artificial intelligence (AI)-driven healthcare decision support systems, enabling clinicians to intervene promptly and improve patient outcomes.

It is well known that in high-dimensional multimodal data, different data modalities or features extracted from different data modalities interact with each other in complex ways, making it difficult to explain the clusters obtained through a clustering algorithm. Therefore, interpreting clusters is a challenging problem. Recent developments in Large Language Models (LLMs), e.g., ChatGPT [6], DeepSeek [7], provide an opportunity to analyze and query multimodal datasets following clustering. These models can assist in generating human-interpretable explanations for cluster structures that are otherwise difficult to interpret [8–13]. This paper presents a novel multiview clustering pipeline in which data are clustered based on specific data modalities, and cluster labels are inferred through context-aware LLM prompting. To evaluate the proposed method, a multimodal sensor dataset (MAISON-LLF) [14] was used, comprising 560 days' worth of data, including acceleration, step count, ambient motion, GPS location, heart rate, and sleep, collected from ten community-dwelling older adult patients recovering from lower-limb fractures. The data were continuously collected using a smartwatch, smartphone, and sleep and motion sensors. Using the proposed multiview clustering–LLM pipeline, the contribution of each modality to the recovery trajectory was contextualized. The MAISON-LLF dataset [14] is also accompanied by a range of clinical scores, such as social isolation and other metrics related to functional recovery. Through rigorous statistical analysis and visualization, it was shown that most of the derived cluster labels are in agreement with clinical scores, thereby reaffirming the validity of the proposed approach. The low statistical significance of a few clusters underscores the importance of improved feature extraction methods to more effectively capture and explain patient health outcomes, suggesting that domain-informed feature engineering or representation learning techniques may further enhance the interpretability and clinical relevance of clustering results.

The paper is structured as follows: Sect. 2 reviews related LLM-based cluster interpretation methods. Section 3 details the proposed methodology. Section 4 presents experimental settings and results. Section 5 concludes and outlines future work.

2   Literature Review

This section reviews related work on unsupervised sensor data analysis through clustering and the use of LLMs. The primary rationale behind these approaches is to enable interpretation of sensor data without manual effort. LLMs can serve as a valuable tool for labeling clustered sensor data. Typically, the sensor data is first clustered algorithmically, and then an LLM is prompted to assign labels or meaningful descriptions to the clusters or their centers. These LLM-generated labels are subsequently propagated to the data samples within each cluster for further analysis.

Hasasneh et al. [8] proposed an unsupervised learning framework using smartwatch sensor data, heart rate, heart rate variability, and step count, for COVID-19 detection and monitoring. By applying clustering techniques like K-means and DBSCAN, the framework identifies health states based on deviations from baseline physiology. An explainability layer using Decision Trees and GPT-3 Davinci [15] provides interpretable summaries. Evaluated on 28 participants (11 COVID-positive), the approach effectively detects infection-related changes, correlating with symptom onset and severity. Results show that unsupervised clustering outperforms human-labeled data, reducing the need for supervised annotations in early infection detection.

Gao et al. [9] proposed LLMs and Iterative Evolution for Unsupervised Human Activity Recognition (LLMIE-UHAR), an unsupervised human activity recognition framework that combines LLMs with iterative evolution. The method begins by applying K-means clustering to extract key representative sensor samples, which are then converted into textual prompts enriched with contextual and semantic information. LLMs use these prompts to annotate activities without manual labeling. A CNN model is trained on the annotated samples and refined iteratively. Evaluated on the Activity Recognition with Ambient Sensing dataset [16] using ambient sensor data, the method achieved 96% accuracy across nine selected human activities, demonstrating its effectiveness in unsupervised activity recognition using structured LLM-based interpretation.

Hota et al. [10] investigated the feasibility of using LLMs as virtual annotators for time-series physical sensing data, specifically for human activity recognition, to eliminate the need for traditional human-in-the-loop annotation. They introduced a self-supervised learning-based encoding approach, where raw inertial measurement unit data is transformed into feature-rich embeddings, so-called clusters, using contrastive learning techniques before being fed into the LLM for annotation. These embeddings enhance the LLM's ability to classify human activities without fine-tuning or complex prompt engineering. The method was evaluated on four benchmark human activity recognition datasets. Results show that LLM-based annotation significantly improves human activity recognition accuracy when combined with contrastive learning-based embeddings, demonstrating that while LLMs alone are insufficient for direct sensor data analysis, integrating contrastive learning bridges this gap, making LLMs effective virtual annotators.

Natarajan et al. [11], along with Khare and Kumar [17], introduced a novel approach to interpreting clustered data in different domains using LLMs. The method derives key attributes from clusters by analyzing statistical properties such as cluster size, variance, density, spatial positioning, and inter-cluster relationships. These extracted keywords serve as structured inputs for the cluster interpreter, which transforms them into coherent linguistic descriptions of the clusters. The model is trained using T5 [18] and GPT [15] architectures, both of which are fine-tuned to produce fluent, non-monotonous, and contextually rich cluster interpretations and summaries across various domains. The approach is evaluated using multiple real-world datasets, including the Iris dataset, airline dataset, and credit card dataset, where clustering algorithms such as K-means, DBSCAN, and single-linkage hierarchical clustering are applied. The study's results highlight that LLMs enhance the interpretability of clustering outcomes, making data-driven insights more accessible to domain non-experts.

Paraschou [12] introduced a method that combines clustering, explainable AI, and LLMs to improve the interpretability of a complex dataset. The framework begins with clustering algorithms such as K-means, DBSCAN, and hierarchical clustering to group similar data points. Explainability techniques such as Shapley Additive Explanations (SHAP) [19] are then applied to identify influential features within each cluster. Finally, LLMs such as GPT-4 [20] and Mistral 7B [21] generate human-readable summaries that translate numerical patterns into intuitive insights. The approach was evaluated on the LifeSnaps dataset [22], a multimodal time-series dataset containing data modalities such as physical activity, sleep, heart rate variability, stress, and personality traits. Results demonstrate that LLM-based summaries significantly enhance the interpretation of sensor data related to physical and mental health, offering more accessible and meaningful analysis compared to traditional methods in healthcare and well-being monitoring.

Ke et al. [13] proposed Integrate Vision-Language Semantic Graphs in multiview Clustering (IVSGMV), a novel method that integrates vision-language semantic graphs with multiview clustering to enhance performance on high-dimensional image datasets. Using Contrastive Language-Image Pretraining (CLIP) [23], the method builds bipartite graphs linking image embeddings to text concepts from WordNet [24]. An adaptive hybrid graph filter then balances homophily- and heterophily-based information to refine clustering. Evaluated on large-scale multiview datasets, IVSGMV outperforms traditional baselines and effectively leverages zero-shot capabilities of pre-trained models, achieving improved clustering without the need for labeled data.

Despite existing evidence in the literature highlighting the applicability and effectiveness of LLMs for labeling and interpreting clustered sensor data [8–13,17], particularly in the context of unsupervised remote patient monitoring [8,12], none of the previous studies have specifically focused on understanding social isolation and functional decline among older adult patients discharged home after lower-limb fracture surgery. Moreover, no prior research has combined multimodal sensor data with corresponding multiview clustering and LLM-driven interpretation to provide a holistic view of behavioral patterns in this population. Given the unique and often varied patterns of daily activities, mobility, and lifestyle commonly observed among older adults, a clear research gap remains, especially regarding the integration of contextual information about this specific problem into LLM frameworks. To address this gap, this paper presents a novel approach that leverages multimodal sensor data with multiview clustering to identify patterns associated with the daily behaviors of older adults recovering from lower-limb fractures. This methodology is further enhanced by the use of problem-specific context-aware prompts, enabling LLMs to automatically label and explain the resulting clusters, supporting an unsupervised and comprehensive understanding of social isolation and functional decline. This approach minimizes dependence on subjective clinical data or patient self-reports, offering objective, scalable insights into the health and behavioral patterns of older adults. The quality of the inferred cluster labels was further validated using visualizations and statistical analysis of the corresponding clinical scales.

3 Multiview Clustering and LLM

This section presents the idea of explainable multiview clustering aimed at interpreting multimodal data in an unsupervised manner. The dataset under consideration is called MAISON-LLF [25], which contains 560 days of sensor data across multimodal features collected from various sensors deployed in the homes of patients recovering from lower-limb fractures (see more details in Sect. 3.1). A single-view clustering approach could result in multiple clusters across many features that may be difficult to interpret. To address this, the proposed method clusters the data by each modality and then uses context-aware prompting to generate explainable labels describing each modality-specific cluster. This approach facilitates understanding the role of each modality in mapping the recovery trajectory of patients once they return to the community after being discharged following a lower-limb fracture. Finally, statistical analysis is conducted to determine whether the clusters formed through the proposed explainable multiview clustering represent statistically significant clinical information. The overall pipeline is shown in Fig. 1.

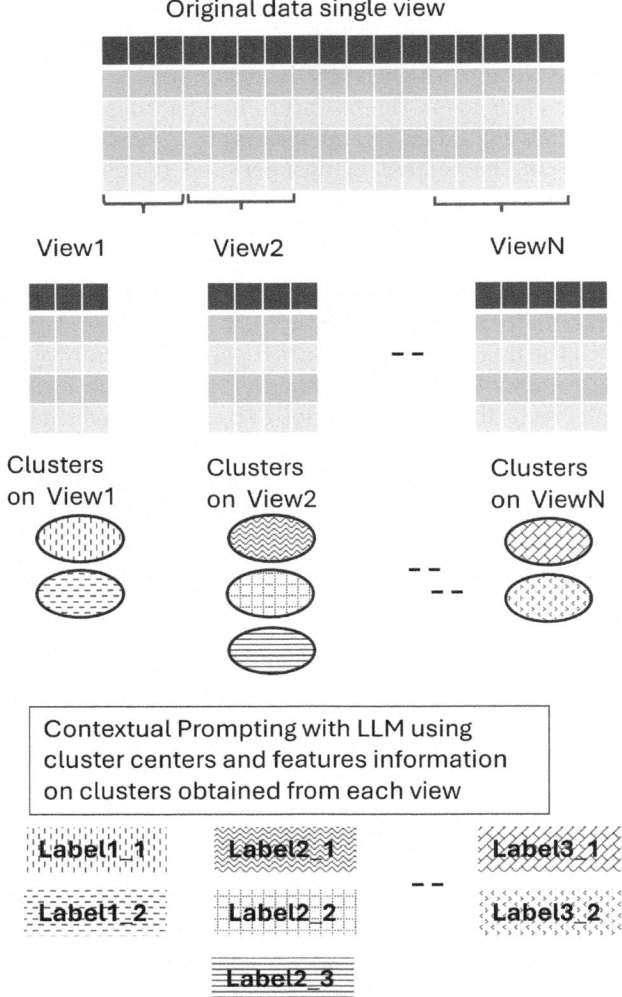

Fig. 1. Generation of cluster labels on multimodal data using multiview clustering and context-aware prompts provided to LLMs.

3.1 MAISON-LLF Dataset

The Multimodal AI-based Sensor platform for Older iNdividuals with Lower-Limb Fractures (MAISON-LLF) dataset is a publicly available, multimodal sensor dataset collected from older adults recovering from lower-limb fractures in community settings [25]. It includes data from smartphone and smartwatch sensors, motion detectors, sleep-tracking mattresses, and clinical questionnaires related to social isolation and functional decline. The dataset was collected from 10 older adults living alone for eight weeks each, totaling 560 days of continuous, 24-hour monitoring. It spans six sensor modalities: acceleration, heart rate, step, GPS location, motion, and sleep, with a range of features extracted from each. Additionally, participants completed biweekly clinical assessments, including the Social Isolation Scale (SIS) [26], Oxford Hip and Knee Scores (OHS, OKS) [27,28], the Timed Up and Go (TUG) test [29], and the 30-Second Chair Stand Test [30], providing ground-truth measures of social and functional health. The data collection process began shortly after discharge from the hospital. The dataset is publicly accessible via Zenodo [14], offering raw sensor data, extracted features, and detailed documentation in multiple formats.

3.2   Data Processing

The version of the MAISON-LLF dataset used in this paper is available in tabular format [25]. It contains 560 rows, each representing a day of data, and 80 columns comprising daily clinical information and features extracted from multimodal sensors. Clinical assessments were conducted bi-weekly; however, all 14 preceding days were assigned the same values for clinical assessments. Moreover, the clinical information should not be confused as a feature, but rather considered ground truth in this context to validate the clustering of days. For simplicity, we refer to clinical information as features. Therefore, out of the 80 features, three, 'participant', 'timestamp', and 'clinical-timestamp', were deemed unnecessary and removed from the analysis. Of the remaining 77 features, 42 were clinical or demographic features, while 35 were numerical features representing data extracted from six data modalities on a daily basis. For clustering purposes, these modalities are referred to as 'views', as they represent different types of features extracted from patients. The position view included 3 features, the motion sensor view included 5, the heart rate view included 4 features, the sleep view included 11 features, the step view included 5 features, and the acceleration view included 7 features. The clinical and demographic features were excluded from further analysis, and only numerical features were used. This was done to investigate the role of sensor data in understanding the recovery trajectory of patients. The clinical features were used to assess the quality of the clusters in accurately grouping patients.

3.3   Multiview Clustering and Cluster Labeling

K-means clustering was used to perform multiview clustering. The random state was set to a fixed value to ensure reproducibility of results. If the random state is changed to another value, the final clustering results (and cluster centers) could differ. All the data were normalized before clustering to prevent features with larger numerical ranges from disproportionately influencing the results. After clustering was completed, the cluster centers were computed and converted back to the original scale using reverse normalization. This facilitated the interpretation of features and the design of context-aware prompts for LLMs to infer relevant cluster labels.

4 Experiments and Results

Table 1. Two cluster centers obtained from the position view

Cluster	Features		
	position-count	position-duration	position-travelled-distance
1	41.218925	94.892652	3.445000
2	16.500000	21.000000	3632.845000

4.1 Multiview Clustering and Cluster Labeling

Each view described in Sect. 3.2 was clustered using the K-means clustering algorithm, varying the number of clusters from $K = 2$ to 15, and the best clusters were recorded for each data modality. The position, motion, step, and acceleration views each yielded 2 optimal clusters, whereas the heart rate view yielded 5, and the sleep view yielded 4 optimal clusters. As an example, the cluster centers generated by the position view are outlined in Table 1.

It can be observed that Cluster 1 represents days when patients traveled short distances but remained outside, whereas Cluster 2 represents days involving longer-distance travel. The following context-aware prompt was provided to the GPT-4o LLM to infer labels for each cluster, along with short descriptions:

I have clustered data into two clusters. Each cluster has three features, namely,

- position-count - The total count of position data in a day.
- position-duration - The duration (in hours) of being outside the home in a day.
- position-travelled-distance - The total distance (in kilometers) traveled outside the home in a day.

The two cluster centers are as follows:

- Cluster 1: 41.218925 94.892652 3.445000
- Cluster 2: 16.500000 21.000000 3632.845000

This data was collected from smartphones with GPS carried by different persons and features were extracted from it. These people are patients recovering from lower-limb fractures and were living at home. Please suggest cluster names for each cluster for better interpretation. Only provide a short cluster name and interpretation. Do not provide any additional analysis.

The LLM provided the following suggestions:

- Cluster 1: *Low Mobility with Extended Outdoor Time*- Patients spend a long duration outside but travel minimal distances, possibly engaging in sedentary or localized activities.
- Cluster 2: *High Mobility with Long-Distance Travel*- Patients travel significant distances, indicating higher mobility and potentially advanced recovery or external assistance for transportation.

Using a similar two-step process, cluster labels were generated for each sensor data modality as shown in Table 2.

Table 2. Cluster labels and their short description obtained for each cluster (for each sensor view or modality) after performing K-means clustering and context-aware prompting with the GPT-4o LLM.

View	Cluster Labels	Cluster Label Description
Motion	Active Home Movers	Higher motion count and spread-out activity suggest frequent movement within the living space
	Sedentary Home Dwellers	Lower motion count and reduced activity indicate a more stationary lifestyle with minimal movement
Heart Rate	Moderate Activity with Stable Heart Rate	Balanced heart rate values with low variability
	Low Activity with Anomalous Readings	Extremely low mean heart rate and missing minimum values
	Elevated Heart Rate with Variability	Higher heart rate with noticeable fluctuations
	Erratic Heart Rate Patterns	Very low mean heart rate with extreme variability
	Resting State with Mild Fluctuations	Lower heart rate with minimal variation
Sleep	Extended Sleep with Frequent Snoring	Long total sleep duration with notable snoring
	Short Sleep with Low Disturbance	Reduced total sleep with minimal wake-ups
	Light Sleep Dominant with High Wake-Ups	Predominantly light sleep with frequent wake-ups
	Elevated Heart Rate with Moderate Sleep	Moderate sleep duration with high heart rate
Step	Sedentary Routine	Lower step count and less frequent movement throughout the day
	Frequent Movers	Higher step count and more consistent movement patterns
Acceleration	Low Activity Movers	Lower acceleration count and variability, indicating limited movement
	High Activity Movers	Higher acceleration count and variability, suggesting more frequent or intense movement

4.2   Clinical Validation

As described in Subsect. 3.1, the MAISON-LLF dataset also contains bi-weekly clinical assessments for each patient. These assessments were applied to the dataset on a daily basis, such that a value collected on a given day was assigned to the preceding two weeks. The distribution of mean clinical scores for each cluster obtained by each sensor data modality is shown in Fig. 2. As an example, the mean SIS, OHS, OKS, and TUG values for the two clusters from the position view are $23.16, 27.67, 31.65, 21.98$ and $28.0, 36.0, 46.0, 9.0$, respectively (see Fig. 2a). Visually, these values differ for each clinical score. Referring back to the cluster labels obtained in the previous section, Clusters 1 and 2 were labeled as *Low Mobility with Extended Outdoor Time* and *High Mobility with Long-Distance Travel*, respectively. Thus, it can be anticipated that Cluster 1 could exhibit higher social isolation, lower hip function, lower knee function, and reduced overall body function compared to Cluster 2's mean clinical values. This confirms that the cluster labels obtained are potentially meaningful representations of the information contained in each cluster from the position view. However, this distinction may not be as visually evident in other cases, such as the heart rate view (see Fig. 2c). The heart rate cluster centers had abnormal values due to occasional zero heart rate readings, which are not realistic and were caused by hardware limitations of the smartwatch. Therefore, statistical tests were performed to determine whether the clinical scores corresponding to each cluster within a modality are significantly different from one another, as described in the following subsection.

4.3   Statistical Testing

Statistical tests were conducted to assess the significance of the distributions of SIS, OHS, OKS, and TUG clinical scores associated with the clusters obtained for each sensor data modality. If there were two clusters and both followed a normal distribution, a t-test was performed; otherwise, the Mannâ ĂŞWhitney U test (MWUT) was used. If there were more than two clusters and all followed a normal distribution, a repeated-measure ANOVA test was conducted; otherwise, the KruskalâĂŞWallis test (KWT) was used. Although the dataset included a limited number of participants, the number of samples per cluster was sufficient in most cases to meet the assumptions of the statistical tests. The null hypothesis was that there is no significant difference in clinical scores between clusters, i.e., the mean clinical scores of the clusters are the same. If the p-value was less than 0.05, the null hypothesis was rejected, indicating a statistically significant difference in clinical scores between clusters.

Table 3 shows the statistical testing results for each cluster corresponding to each sensor data modality. The columns Cluster1-5 shows the number of clusters per modality. A dash (-) means the cluster is absent, for e.g., Position view has only two clusters, so there is a dash for Cluster3-5. In each of the cluster1-5 column, a Yes, No or NaN means whether the samples in a cluster are normally distributed or not or insufficient data (less than three samples) following the

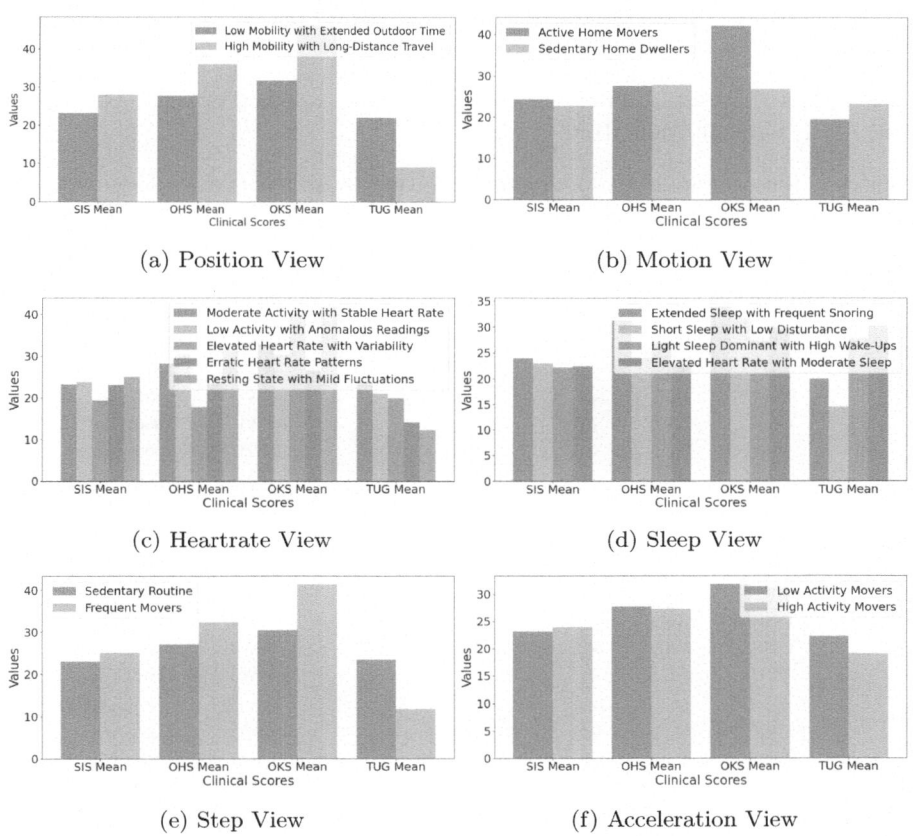

(a) Position View

(b) Motion View

(c) Heartrate View

(d) Sleep View

(e) Step View

(f) Acceleration View

Fig. 2. The mean distribution of various clinical scales for each clusters obtained from different views from the multimodal sensor data.

Shapiro-Wilk test. For clusters obtained from heart rate, sleep, and step data, there are statistically significant differences between clusters. This indicates that the sensor data was able to group samples into clusters with clinical values that were significantly different from one another. For the position clusters, OHS and OKS scores were not statistically significant, and for the motion clusters, OHS was not statistically significant. In the position view, the clusters were heavily skewed, with Cluster 1 and Cluster 2 containing 558 and 2 data points, respectively. As a result, the test could not compute reliable statistics (hence the NaN value in Table 3 corresponding to Cluster 1). Overall, statistical testing shows that the clusters contained samples that were statistically distinguishable based on the distribution of clinical scores. This further supports the validity of the cluster labels inferred in Subsect. 4.1.

The clusters obtained from the acceleration view were not found to be statistically significant with respect to any of the clinical scores. This was also visually

Table 3. Statistical testing results for each cluster corresponding to each sensor data modality. P-values less than 0.0001 are reported as "<0.0001". All other p-values are rounded to four decimal places. SIS: Social Isolation Scale, OHS: Oxford Hip Score, OKS: Oxford Knee Score, TUG: Timed Up and Go test, MWUT: Mann–Whitney U Test, KWT: Kruskal–Wallis Test.

View	Clinical Score	Cluster1	Cluster2	Cluster3	Cluster4	Cluster5	Test Used	p-value	Statistical Significance
Position	SIS	No	NaN	-	-	-	MWUT	0.0360	Yes
	OHS	No	NaN	-	-	-	MWUT	0.1581	No
	OKS	No	NaN	-	-	-	MWUT	0.0703	No
	TUG	No	NaN	-	-	-	MWUT	0.0499	Yes
Motion	SIS	No	No	-	-	-	MWUT	< 0.0001	Yes
	OHS	No	No	-	-	-	MWUT	0.8137	No
	OKS	No	No	-	-	-	MWUT	< 0.0001	Yes
	TUG	No	No	-	-	-	MWUT	0.0006	Yes
Heart Rate	SIS	No	No	No	No	No	KWT	< 0.0001	Yes
	OHS	No	No	No	No	No	KWT	< 0.0001	Yes
	OKS	No	No	No	No	No	KWT	< 0.0001	Yes
	TUG	No	No	No	No	Yes	KWT	0.0013	Yes
Sleep	SIS	No	No	No	No	-	KWT	< 0.0001	Yes
	OHS	No	No	No	No	-	KWT	< 0.0001	Yes
	OKS	No	No	No	No	-	KWT	< 0.0001	Yes
	TUG	No	No	No	No	-	KWT	< 0.0001	Yes
Step	SIS	No	No	-	-	-	MWUT	< 0.0001	Yes
	OHS	No	No	-	-	-	MWUT	< 0.0001	Yes
	OKS	No	No	-	-	-	MWUT	< 0.0001	Yes
	TUG	No	No	-	-	-	MWUT	< 0.0001	Yes
Acceleration	SIS	No	No	-	-	-	MWUT	0.1020	No
	OHS	No	No	-	-	-	MWUT	0.8241	No
	OKS	No	No	-	-	-	MWUT	0.6818	No
	TUG	No	No	-	-	-	MWUT	0.3892	No

observed in Fig. 2f, where most of the cluster center values appear similar for each clinical score. Although movement is known to be directly correlated with improved functional recovery and reduced social isolation [31] (as shown by the results from the motion sensor view, see Fig. 2b), this result appears counterintuitive. It is likely due to limitations in the sensitivity and resolution of the wearable device in capturing nuanced movement and inconsistencies in participant usage with the wearable device.

A python notebook containing all the experiments presented in this paper is available at https://github.com/shehrozskhan/InterpretMultiview.

4.4 Limitations

The results presented in this paper provide unique insights into the recovery trajectories of older adult patients following lower-limb fractures. However, the study has certain limitations:

- All clustering results were obtained using a fixed random seed. The optimal number of clusters and their interpretations may vary slightly with different random initializations, as this could lead to a different clustering outcome.
- Cluster centers average the features of all data points, so the impact of distant points or outliers within a cluster may not be well represented.
- LLMs can produce different cluster labels even when given the same prompts and cluster centers. However, it is expected that, despite variations in wording, the underlying meaning of the labels should remain consistent.
- The clusters obtained from the acceleration view were not found to be statistically significant, likely due to limitations in the study design that affected the consistency and richness of the data collected. The protocol may not have sufficiently accounted for variations in daily routines or provided enough structure to ensure uniform data capture across participants. As a result, the LLM was unable to detect reliable patterns, even with a context-aware prompt. This points to the importance of refining study protocols, such as incorporating clearer guidance, structured activity periods, or enhanced data monitoring to generate high quality data that can meaningfully inform the recovery process.

5   Conclusions and Future Work

This paper introduced the idea of clustering multimodal sensor data by each modality to improve understanding of their impact on the recovery of community-dwelling older adult patients after lower-limb fracture. Clustering by each modality breaks down a high-dimensional clustering problem into smaller and more manageable sub-problems that can be combined to cluster the overall dataset. Context-aware prompting and LLMs were used to infer cluster labels. Visualizations of the clinical scores corresponding to the clusters obtained per modality were then performed. Through statistical testing, it was shown that the clusters and their inferred labels are in agreement with each other and that their meanings can help in understanding the recovery trajectories of patients. It was also found that if a deeper understanding of the collected data is not provided to the LLM, the inferred cluster labels may not be useful from an interpretability perspective. This opens up new avenues to extract more informative and clinically relevant features per modality to improve the dataset's discriminative power.

This preliminary analysis leads to multiple future directions. The most intuitive direction is to combine the outcomes of multiple clusterings from each modality to perform a final clustering on the overall dataset. This will help interpret those clusters to group patients based on different sensing modalities. A comparison with single-view clustering will highlight the benefits of using multiview clustering in terms of explainability of the dataset. The clinical data could also be combined with sensor data to assess whether that improves clustering performance and contributes to explainability. The data have temporal and

284   S. S. Khan et al.

value-based dimensions; thus, spatio-temporal clustering algorithms [32] should be applied to understand the relationship between time and patterns in sensor data during patients' recovery. Lastly, this idea should be tested on other real-world sensor datasets, and its clinical implications must be further investigated.

Acknowledgment. The research work is supported through grants funded by the Natural Sciences and Engineering Research Council of Canada.

References

1. Khan, S.S., Ahmad, A.: Cluster center initialization algorithm for k-means clustering. Pattern Recogn. Lett. **25**(11), 1293–1302 (2004)
2. Xu, R., Wunsch, D.: Survey of clustering algorithms. IEEE Trans. Neural Netw. **16**(3), 645–678 (2005)
3. Xu, D., Tian, Y.: A comprehensive survey of clustering algorithms. Ann. Data Sci. **2**(2), 165–193 (2015)
4. Chao, G., Sun, S., Bi, J.: A survey on Multiview clustering. IEEE Trans. Artif. Intell. **2**(2), 146–168 (2021)
5. Yang, Y., Wang, H.: Multi-view clustering: a survey. Big Data Min. Anal. **1**(2), 83–107 (2018)
6. Mann, B., et al.: Language models are few-shot learners. arXiv preprint arXiv:2005.14165, vol. 1, p. 3 (2020)
7. Bi, X., et al.: Deepseek LLM: scaling open-source language models with longtermism. arXiv preprint arXiv:2401.02954 (2024)
8. Hasasneh, A., Hijazi, H., Talib, M.A., Afadar, Y., Nassif, A.B., Nasir, Q.: Wearable devices and explainable unsupervised learning for COVID-19 detection and monitoring. Diagnostics **13**(19), 3071 (2023)
9. Gao, J., Zhang, Y., Chen, Y., Zhang, T., Tang, B., Wang, X.: Unsupervised human activity recognition via large language models and iterative evolution. In: ICASSP 2024-2024 IEEE International Conference on Acoustics, Speech and Signal Processing (ICASSP), pp. 91–95. IEEE (2024)
10. Hota, A., Chatterjee, S., Chakraborty, S.: Evaluating large language models as virtual annotators for time-series physical sensing data. ACM Trans. Intell. Syst. Technol. (2024). https://doi.org/10.1145/3696461
11. Verma, S., Kumar, D., et al.: Enhanced interpretation of novel datasets by summarizing clustering results using deep-learning based linguistic models. Appl. Intell. **55**(5), 1–23 (2025)
12. Paraschou, E.: A human-centric and LLM-enhanced explainable framework for sensing data exploratory analysis. Ph.D. dissertation, Aristotle University of Thessaloniki (2024)
13. Ke, J., et al.: Integrating vision-language semantic graphs in multi-view clustering. In: Proceedings of the Thirty-Third International Joint Conference on Artificial Intelligence, pp. 4273–4281 (2024)
14. Abedi, A, Chu, C., Khan, S.: MAISON-LLF: multimodal AI-based sensor platform for older iNdividuals–lower limb fracture. dataset (2025). https://zenodo.org/records/14597613
15. Brown, T.B., et al.: Language models are few-shot learners. arXiv preprint arXiv:2005.14165 (2020)

16. Alemdar, H., Ertan, H., Incel, O.D., Ersoy, C.: Aras human activity datasets in multiple homes with multiple residents. In: 2013 7th International Conference on Pervasive Computing Technologies for Healthcare and Workshops, pp. 232–235. IEEE (2013)

17. Khare, A., Kumar, D.: Summarizing clustering results using sentence prototype based language models. In: 2023 3rd International Conference on Electrical, Computer, Communications and Mechatronics Engineering (ICECCME), pp. 1–6. IEEE (2023)

18. Raffel, C.: Exploring the limits of transfer learning with a unified text-to-text transformer. J. Mach. Learn. Res. **21**(140), 1–67 (2020)

19. Lundberg, S.M., Lee, S.I.: A unified approach to interpreting model predictions. In: Proceedings of the 31st International Conference on Neural Information Processing Systems (NeurIPS), pp. 4765–4774 (2017)

20. OpenAI. GPT-4 technical report (2023). https://openai.com/research/gpt-4

21. Jiang, A.Q., et al.: Mistral 7B (2023). https://mistral.ai/news/announcing-mistral-7b/

22. Yfantidou, S., et al.: Lifesnaps, a 4-month multi-modal dataset capturing unobtrusive snapshots of our lives in the wild. Sci. Data **9**(1), 663 (2022)

23. Radford, A., et al.: Learning transferable visual models from natural language supervision. arXiv preprint arXiv:2103.00020 (2021)

24. Fellbaum, C.: WordNet: An Electronic Lexical Database. MIT Press, Cambridge (1998)

25. Abedi, A., Chu, C.H., Khan, S.S.: Multimodal sensor dataset for monitoring older adults post lower-limb fractures in community settings. arXiv preprint arXiv:2501.13888 (2025)

26. Nicholson Jr., N.R., Feinn, R., Casey, E., Dixon, J.: Psychometric evaluation of the social isolation scale in older adults. Gerontologist **60**(7), e491–e501 (2020)

27. Wylde, V., Learmonth, I.D., Cavendish, V.J.: The oxford hip score: the patient's perspective. Health Qual. Life Outcomes **3**, 1–8 (2005)

28. Whitehouse, S.L., Blom, A.W., Taylor, A.H., Pattison, G.T., Bannister, G.C.: The oxford knee score; problems and pitfalls. Knee **12**(4), 287–291 (2005)

29. Podsiadlo, D., Richardson, S.: The timed "up & go": a test of basic functional mobility for frail elderly persons. J. Am. Geriatr. Soc. **39**(2), 142–148 (1991)

30. Chen, H.-T., Lin, C.-H., Yu, L.-H.: Normative physical fitness scores for community-dwelling older adults. J. Nurs. Res. **17**(1), 30–41 (2009)

31. Rosso, A.L., Taylor, J.A., Tabb, L.P., Michael, Y.L.: Mobility, disability, and social engagement in older adults. J. Aging Health **25**(4), 617–637 (2013)

32. Ansari, M.Y., Ahmad, A., Khan, S.S., Bhushan, G., Mainuddin: Spatiotemporal clustering: a review. Artif. Intell. Rev. **53**, 2381–2423 (2020)

Parameter Drift as a Signal for Membership Inference in Overfit-Tuned LLMs

Takuto Kitamura and Yu Suzuki[✉]

Gifu University, Gifu, Japan
kitamura.takuto.f6@s.gifu-u.ac.jp,kitamura.takuto.f6@f.gifu-u.ac.jp,
suzuki.yu.r4@s.gifu-u.ac.jp,suzuki.yu.r4@f.gifu-u.ac.jp

Abstract. We propose a novel white-box membership inference attack (MIA) for large language models (LLMs) that leverages internal parameter dynamics to determine whether a given text sample was included in a model's pre-training data. Prior MIA approaches rely primarily on input-output behavior and struggle to distinguish memorized samples from semantically similar but unseen ones due to the probabilistic nature of LLM generation. To address this challenge, we introduce a method based on *parameter drift*—defined as the Euclidean distance between the entire set of model parameters before and after continual pre-training on a single input. Our hypothesis is that continual pre-trained inputs induce minimal parameter changes, while unseen inputs require greater updates to the model. Notably, we find that even semantically similar inputs yield distinct drift magnitudes, enabling more precise membership inference. We validate our approach on multiple LLMs, including Pythia and LLaMA-2, and show that it consistently outperforms existing MIA baselines such as Min-K% Prob and SaMIA*zlib under various evaluation settings. Furthermore, we demonstrate that focusing on parameters with high drift further improves inference accuracy, achieving state-of-the-art results on benchmark datasets.

Keywords: Membership Inference Attacks · White-box Attacks · Parameter Drift · Large Language Models

1 Introduction

Large language models (LLMs), such as GPT-4 and LLaMA, have been shown to unintentionally memorize parts of their training data, including content that may not be publicly licensed or privacy-safe [5,7]. Since the training corpora of most LLMs are not disclosed, users cannot determine whether specific text was used during pre-training. This uncertainty raises concerns in terms of privacy, fairness, and intellectual property. Moreover, benchmark datasets may overlap with training data, potentially inflating evaluation scores if the model has memorized test samples [6,9]. These risks motivate the need for methods that can reliably verify whether a given sample was part of the training set of an LLM.

© The Author(s), under exclusive license to Springer Nature Switzerland AG 2026
C. K. Leung et al. (Eds.): DaWaK 2025, LNCS 16048, pp. 286–301, 2026.
https://doi.org/10.1007/978-3-032-02215-8_23

Membership Inference Attacks (MIA) aim to determine whether a specific data—e.g., an image or a sentence—was included in a model's training dataset [3]. Early work primarily targeted image classification models [8,11], while recent research has extended MIA to LLMs using their input-output behavior [4,10]. However, since LLMs generate outputs randomly from their learned distributions, distinguishing between memorized content and coincidentally similar outputs remains challenging. Even if a sentence was not seen during training, an LLM may generate it if it is semantically close to previously learned content. This makes black-box MIA based on textual similarity inherently unreliable.

Unlike these black-box MIA methods, which rely on surface-level model behavior, the internal parameters of neural networks encode rich information about the data seen during training. Prior work has shown that overfitting accentuates differences in model behavior between member and non-member samples [11]. Building on this insight, we propose a white-box MIA method that measures *parameter drift*—defined as the Euclidean distance between model parameters before and after continual pre-training on a single input. If the input was seen during pre-training, the model's parameters will change minimally; otherwise, the drift will be substantially larger. Notably, we find that even semantically similar non-member inputs induce measurably greater drift, enabling more precise membership inference.

We evaluate our method on several open-source LLMs, including Pythia, OPT, and LLaMA-2, and compare it against state-of-the-art MIA baselines such as Min-$K\%$ Prob [10] and SaMIA*zlib [4]. Our results show that the proposed method consistently outperforms existing approaches on certain models (e.g., Pythia) and performs comparably on others (e.g., LLaMA-2 and OPT). Furthermore, we find that selecting a subset of parameters with large drift leads to further gains in inference accuracy.

This paper makes the following contributions:

- We propose a novel white-box membership inference attack that measures internal parameter drift induced by continual pre-training of a single input.
- We show that selecting high-drift parameters significantly improves membership inference accuracy over using all parameters.
- We achieve state-of-the-art results on multiple LLMs and benchmark datasets, outperforming or matching existing black-box MIA baselines in several settings.

2   Related Work

Membership inference attacks (MIA) have been extensively studied in recent years. Hu [3] provides a comprehensive survey of MIA, categorizing attacks based on the adversary's level of access to the target model: (i) black-box attacks, where only the model's outputs (e.g., predictions or confidence scores) are observable, and (ii) white-box attacks, where internal model information such as parameters or gradients is accessible.

While black-box attacks are widely used due to their practicality and ease of deployment, white-box attacks provide access to richer internal signals and can enable more precise and fine-grained membership inference. In this work, we adopt the white-box setting to leverage internal parameter dynamics for more accurate inference.

Hu [3] also classifies MIA methods into two categories: classifier-based methods and metric-based methods. Classifier-based methods train a binary classifier to distinguish between member and non-member samples using features such as confidence scores or gradient norms [8,11]. In contrast, metric-based methods compute scores—such as prediction loss, log-likelihood, or entropy—without additional classifier training. Our proposed method belongs to the latter category: we introduce a new metric based on internal parameter drift observed after targeted continual pre-training.

Shokri et al. [11] were the first to introduce MIA in the context of machine learning. Their approach is a black-box, classifier-based method that relies on training a set of shadow models to replicate the behavior of the target model. These models are then used to train an attack model on the output logits, which predicts membership status. Their findings revealed that MIA performance improves significantly when the target model is overfitted, while regularization tends to reduce the model's susceptibility to attacks.

Nasr et al. [8] proposed a white-box classifier-based method that incorporates internal signals such as gradients, logits, and loss values. Their attack model uses convolutional neural networks to process gradients and fully connected layers to handle logits and loss features. This approach improved attack accuracy by up to 20% compared to previous methods. While focused on supervised learning, their method demonstrated that even well-generalized models can leak training data when white-box access is available.

Shi et al. [10] introduced Min-K% Prob, a black-box metric-based method for membership inference in LLMs. The core idea is that unseen text inputs tend to contain rare tokens with extremely low predicted probabilities, whereas seen inputs are more likely to have uniformly high-probability tokens. To reduce sensitivity to such rare token noise, the method averages the log-likelihoods of the top-K% most probable tokens. Notably, this approach requires no access to the pretraining data or a reference model. On the WikiMIA benchmark, Min-K% Prob outperformed prior baselines by 7.4% and proved effective in tasks such as copyright auditing and unlearning validation.

While Min-K% Prob is efficient and easy to apply, it solely relies on output probabilities and ignores internal model behavior. As a result, it may fail to distinguish between genuine memorization and coincidental high-likelihood outputs arising from semantically similar but unseen examples. This motivates the exploration of white-box methods that can leverage parameter-level dynamics for deeper insights.

In this context, Kaneko et al. [4] proposed SaMIA, a black-box method based on Sampling-based Pseudo-Likelihood (SPL). SaMIA estimates token likelihoods without access to internal model scores by sampling multiple continuations from the LLM and comparing them to a ground-truth reference via n-gram overlap. This method enables MIA on proprietary models like ChatGPT, where log-likelihoods are not exposed.

They further introduced an enhanced variant, SaMIA*zlib, which applies compression to generated continuations to quantify redundancy. While both methods perform well—especially on longer text samples—their effectiveness drops for shorter or ambiguous inputs, where n-gram overlap and compression-based metrics become less reliable.

Table 1. Comparison of Proposed Method with Existing MIA Methods

Aspect	Min-K% Prob [10]	SaMIA*zlib [4]	Proposed
Model Access	Black-box	Black-box	White-box
Internal Parameters	✗	✗	✓
Likelihood Dependency	✓	✗	✗
Input Sensitivity	✗	✗	✓
Interpretability	Low	Medium	High
Short Texts	✓	✗	✓

Table 1 compares our proposed method with two representative black-box MIA methods: Min-K% Prob [10] and SaMIA*zlib [4]. The comparison is based on several criteria, including model access level, usage of internal parameters, dependency on output likelihoods, robustness to input variation, interpretability, and performance on short texts. Min-K% Prob is a black-box method that relies on token-level log-likelihoods, which are often unavailable in proprietary models. It performs well on short texts but struggles to differentiate between memorized and semantically similar inputs due to its reliance on output probabilities. SaMIA*zlib addresses the lack of access to likelihoods by relying solely on sampling and compression of generated continuations. While this makes it more applicable in practice, its effectiveness diminishes on short or ambiguous inputs where n-gram overlap becomes unreliable. In contrast, our method assumes a white-box setting and leverages internal parameter drift caused by continual pre-training. This allows it to exploit deeper internal signals that go beyond surface-level output behavior. As a result, it achieves greater robustness against paraphrased inputs and varying input lengths. Moreover, since the metric is derived directly from parameter changes, it provides better interpretability and insight into model behavior.

Table 1 summarizes the qualitative differences among Min-K% Prob, SaMIA*zlib, and our proposed method. While both existing methods operate in a black-box setting—making them applicable even to proprietary models—they are limited in their ability to leverage internal representations of the model. Our method, by contrast, assumes white-box access and directly exploits parameter-level changes to infer data membership. Unlike Min-K% Prob, which relies on token-level log-likelihoods, and SaMIA*zlib, which approximates likelihood using sampling and compression, our approach is likelihood-free and instead uses parameter drift as a membership signal. This design makes our method robust to paraphrased or semantically similar inputs and particularly effective on short or ambiguous texts, where output-based methods tend to degrade. Moreover, parameter drift offers improved interpretability, providing insight into how much the model updates its internal state in response to specific inputs. This facilitates deeper analysis of memorization behavior and enhances the transparency of the inference process.

In summary, although black-box methods remain valuable in restricted-access scenarios, our white-box approach yields higher accuracy, interpretability, and robustness when internal model access is available.

3 Proposed Method

Figure 1 provides an overview of our proposed membership inference method. It consists of two main components: a training step and a feature extraction step.

1. **Training Step** (Sect. 3.1): For each input text sample, we perform continual pre-training on the target LLM. This allows the model to overfit on the given input and reflect its impact on the internal parameters.
2. **Feature Extraction Step** (Sect. 3.2): We extract the model's parameters before and after training, compute the parameter drift, and normalize it to obtain a scalar feature representing how much the model updated in response to the input.

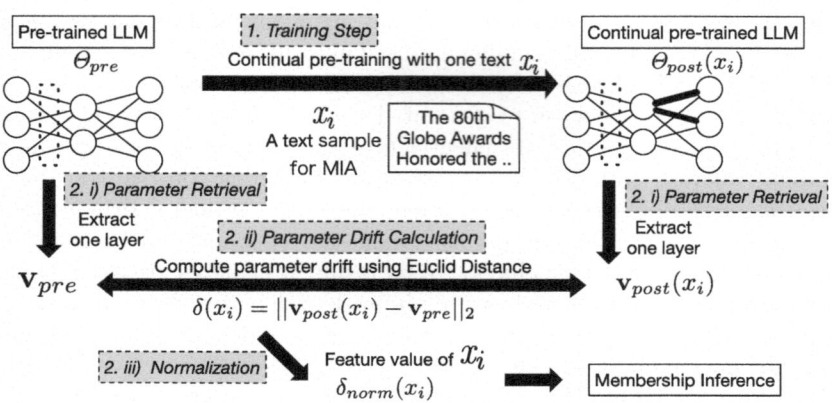

Fig. 1. Overview of Our Proposed Method

3.1 Training Step

Let $\mathcal{D} = \{x_i \mid i = 1, 2, \ldots, N\}$ be a dataset consisting of N text samples. To measure the parameter drift induced by each input, we perform continual pre-training on the target LLM separately for each sample $x_i \in \mathcal{D}$.

The continual pre-training is conducted using gradient descent on the next-token prediction task, with the model trained to predict the next token given a sequence of preceding tokens. To ensure that the observed parameter changes are attributable solely to x_i, we set the batch size to 1 during training. This allows us to capture the input-specific drift without interference from other samples.

We apply early stopping to avoid excessive overfitting or divergence. Training stops if the validation loss does not improve for 50 consecutive epochs, with improvements defined as changes greater than 0.0001. This setup ensures that the model is sufficiently overfitted to x_i to induce measurable drift, while maintaining training stability.

3.2 Feature Extraction Step

To extract features for membership inference, we compute the parameter drift caused by continual pre-training. This process consists of the following three sub-steps:

i) **Parameter retrieval**: We collect the model parameters from each layer of the LLM both before and after continual pre-training. Specifically, we retrieve the weight matrices and bias vectors at each layer for every input sample.

ii) **Parameter drift calculation**: We compute the *parameter drift*, defined as the Euclidean distance between the flattened pre-training and post-training parameter vectors. This drift quantifies how much the model's internal representation has changed in response to a specific input.

iii) **Normalization**: To ensure comparability across inputs and layers, we apply Min-Max normalization to the computed drift values, scaling them to the range $[0, 1]$. The normalized values are used as features for the membership inference task.

These steps are applied independently to each layer of the LLM, allowing the method to capture fine-grained parameter-level changes specific to each input.

i) Parameter retrieval This step retrieves the parameters of the LLM before and after continual pre-training in order to compute the parameter drift for each input x_i.

We denote the parameters of a given layer before pre-training as $\theta_{\mathrm{pre}} = \{\mathbf{W}_{\mathrm{pre}}, \mathbf{b}_{\mathrm{pre}}\}$, where $\mathbf{W}_{\mathrm{pre}}$ and $\mathbf{b}_{\mathrm{pre}}$ are the weight matrix and bias vector, respectively. The weight matrix $\mathbf{W}_{\mathrm{pre}}$ is of size $m \times n$, and the bias vector $\mathbf{b}_{\mathrm{pre}}$ is of size $m \times 1$, where m and n denote the number of output and input units in the layer.

After continual pre-training with input x_i, the updated parameters are denoted as:

$$\theta_{\text{post}}(x_i) = \{\mathbf{W}_{\text{post}}(x_i), \mathbf{b}_{\text{post}}(x_i)\}, \tag{1}$$

where $\mathbf{W}_{\text{post}}(x_i)$ and $\mathbf{b}_{\text{post}}(x_i)$ represent the weight and bias after training on x_i.

Note that θ_{pre} is fixed across all x_i, while $\theta_{\text{post}}(x_i)$ depends on the input x_i. Both parameter sets have the same shape: $m \times (n+1)$. This consistent shape is required to compute the parameter drift via element-wise subtraction in the next step.

This retrieval process is performed on each layer of the target LLM.

ii) Parameter drift calculation Using θ_{pre} and $\theta_{\text{post}}(x_i)$ obtained in step i), this step calculates the parameter drift between the pre-training and post-training states for each input x_i.

First, the parameters before and after pre-training are flattened into one-dimensional column vectors to simplify the distance calculation. Let $\mathbf{v}_{\text{pre}} = \text{vec}(\theta_{\text{pre}})$ and $\mathbf{v}_{\text{post}}(x_i) = \text{vec}(\theta_{\text{post}}(x_i))$, where $\text{vec}(\cdot)$ is an operator that vectorizes the input by concatenating its weight matrix and bias vector column-wise.

If the original parameter θ_{pre} consists of a weight matrix of size $m \times n$ and a bias vector of size $m \times 1$, then both $\mathbf{v}_{\text{pre}}$ and $\mathbf{v}_{\text{post}}(x_i)$ are column vectors of length $m \cdot (n+1)$.

Next, we compute the parameter drift between the pre- and post-training vectors using the Euclidean distance:

$$\delta(x_i) = \|\mathbf{v}_{\text{post}}(x_i) - \mathbf{v}_{\text{pre}}\|_2 \tag{2}$$

Here, $\|\cdot\|_2$ denotes the Euclidean distance.

iii) Normalization To ensure comparability across different samples, we apply Min-Max normalization to the parameter drift values $\delta(x_i)$ computed in step (ii), scaling them to the range $[0,1]$ for all $i \in \{1, \ldots, N\}$. This normalization facilitates stable interpretation across samples and improves the robustness of subsequent analysis.

The normalized parameter drift for input x_i is defined as:

$$\delta_{\text{norm}}(x_i) = \frac{\delta(x_i) - \min_{1 \leq k \leq N} \delta(x_k)}{\max_{1 \leq k \leq N} \delta(x_k) - \min_{1 \leq k \leq N} \delta(x_k)} \tag{3}$$

The resulting normalized values $\delta_{\text{norm}}(x_i)$ serve as the final feature representation used for membership inference in our proposed method.

As described in Sect. 3, our approach computes parameter drift individually for each layer of the target LLM. To assess whether a given input sample was included in the model's pre-training data, we extract features based on the computed drift values. These features can be derived from all layers or from a selected subset, depending on the desired trade-off between inference accuracy and computational efficiency. We empirically investigate the effect of layer selection on classification performance in Sect. 4.2.

4 Experiments

We conduct two experiments to evaluate the effectiveness and generalizability of the proposed membership inference method. **Experiment 1** serves as a proof-of-concept, testing whether parameter drift reliably distinguishes between member and non-member samples using a custom-designed dataset with known ground truth. **Experiment 2** assesses the robustness of our method on a benchmark dataset across multiple publicly available LLMs, and compares its performance against state-of-the-art MIA baselines under varying training and evaluation conditions.

4.1 Experiment 1: Validation on a Self-constructed Dataset

To validate the core hypothesis of our method under controlled conditions, we begin with a small-scale preliminary experiment using a custom-constructed dataset. This dataset is designed to provide explicit ground-truth membership labels, enabling us to directly test whether parameter drift serves as a reliable indicator of whether a given text sample was used during pre-training.

Specifically, we investigate whether samples included in the pre-training corpus exhibit smaller parameter drift than those that were not seen. This experiment provides a proof-of-concept for our method in an idealized setting, laying the groundwork for subsequent large-scale evaluations.

Dataset. To conduct Experiment 1, we constructed two datasets: a *continual pre-training dataset* and a corresponding *evaluation dataset for MIA*, referred to as *myMIA*.

The continual pre-training dataset was created by collecting 30,000 Japanese Wikipedia articles that were newly published after August 2023. Since these articles were unavailable prior to that date, they are guaranteed not to appear in the pre-training data of LLMs released earlier. To ensure textual quality and appropriate input length, we filtered out redirect pages, disambiguation pages, and articles with fewer than 100 or more than 300 words. The resulting corpus was randomly divided into 10,000 training, 10,000 validation, and 10,000 test samples.

We then used the training split to continually pre-train a base LLM, resulting in a continual pre-trained model referred to as *myLLM*. This setup enables full control over the membership status of each sample, making it ideal for white-box MIA evaluation.

To evaluate our method, we constructed the *myMIA* dataset, comprising 10,000 member (i.e., training) and 10,000 non-member (i.e., test) samples from the continual pre-training dataset. This clean separation allows precise and reliable assessment of membership inference performance.

Experimental Settings. In Experiment 1, we used ELYZA-japanese-Llama-2-7b[1] as the base LLM. To construct a model with known training and non-training samples, continual pre-training was performed on this base model using the dataset described in Sect. 4.1. We refer to the resulting model as *myLLM*.

Continual pre-training was conducted using DeepSpeed ZeRO Stage-2[2] with parallel distributed training. The learning rate was set to $3.0 \cdot 10^{-5}$, and the global batch size was 256, achieved via gradient accumulation. The model was trained for one epoch. Each epoch required approximately one minute per input sample, which should be considered when evaluating the practical applicability of the method. Since the exact set of samples used in continual pre-training was known, myLLM was suitable for evaluating membership inference attacks using the myMIA dataset.

In the continual pre-training phase of our proposed method, we set the batch size to 1 to precisely capture the parameter drift induced by each individual input. The AdamW optimizer was used with hyperparameters: $\beta_1 = 0.9$, $\beta_2 = 0.95$, $\epsilon = 1.0 \cdot 10^{-5}$, and weight decay $\lambda = 0.1$. The learning rate remained at $3.0 \cdot 10^{-5}$ and a cosine learning rate scheduler was applied.

Early stopping halted training after 50 epochs without validation loss improvement. An update was considered to have occurred only if the validation loss changed by at least 0.0001 compared to the previous epoch. In preliminary experiments with a learning rate of $3.0 \cdot 10^{-4}$ (as recommended in the original LLaMA 2 paper), training divergence occurred, resulting in NaN losses. This issue was resolved by lowering the learning rate to $3.0 \cdot 10^{-5}$.

Experimental Procedure. We randomly selected 40 samples from the myMIA dataset, consisting of both member (i.e., seen during continual pre-training) and non-member (i.e., unseen) instances. For each sample, we computed the parameter drift across all layers of *myLLM*, as described in Sect. 3.2.

The experiment was carried out in the following three steps:

1. **Continual Pre-Training:** For each input x_i from the myMIA dataset, we performed continual pre-training on *myLLM* using a batch size of 1, following the procedure in Sect. 3.1 and the hyperparameters in Sect. 4.1.
2. **Parameter Drift Measurement:** We retrieved model parameters before and after continual pre-training, and computed the parameter drift as the Euclidean distance between them. The resulting drift values were normalized to the $[0, 1]$ range and used as input features.
3. **Evaluation:** We computed the AUROC to evaluate how well the parameter drift discriminates between member and non-member samples. Additionally, we analyzed the average drift magnitude for each class to assess separability.

Note that this experiment is intended as a small-scale, proof-of-concept study to test the feasibility of our proposed approach under controlled conditions. A

---

[1] https://huggingface.co/elyza/ELYZA-japanese-Llama-2-7b.
[2] https://github.com/microsoft/DeepSpeed.

more comprehensive and large-scale evaluation is presented in Experiment 2 (Sect. 4.2), where we validate our method on multiple pre-trained LLMs and benchmark datasets.

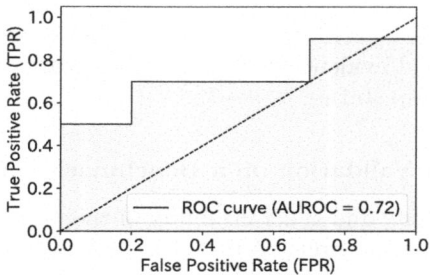

Fig. 2. AUROC curve of Experiment 1.

Results and Discussion. Figure 2 shows the ROC curve of the proposed method on the self-constructed dataset. The curve exhibits a non-linear pattern that reveals distinct behaviors across different false positive rate (FPR) regions.

In the low-FPR range $(0.0 \leq \text{FPR} < 0.2)$, the true positive rate (TPR) remains close to 0.5, indicating that the method behaves almost like random guessing in this region. However, in the moderate-FPR range $(0.2 \leq \text{FPR} < 0.7)$, the TPR rises to approximately 0.7, suggesting that the method can successfully distinguish between member and non-member samples in this region. Finally, in the high-FPR range $(\text{FPR} \geq 0.7)$, the TPR increases sharply and reaches around 0.9, indicating that the most distinguishable cases are captured in this region.

Overall, this ROC profile implies that the proposed method is effective at ranking samples in terms of their membership likelihood, although its discriminative power is limited in the conservative (low-FPR) region. The relatively steep rise in the mid-to-high FPR range contributes significantly to the AUROC score of 0.72 observed in this experiment.

To further analyze this phenomenon, we calculated the average parameter drift separately for member and non-member samples. The mean drift for member samples was 0.1667, whereas non-member samples exhibited a higher average of 0.1904. This aligns with our hypothesis: samples seen during pre-training induce smaller parameter updates, while unseen samples require more substantial adaptation.

We conducted a two-sample t-test to assess the statistical significance of this difference. The test yielded a p-value of 0.0538 under significance level $\alpha = 0.10$, confirming that the difference is statistically significant. These findings support the validity of using parameter drift as a discriminative feature for membership inference.

While the results are promising, it is important to note that this evaluation was performed on a synthetic dataset without paraphrasing or semantic perturbation. Consequently, exact overlaps between training and test samples may have made the task easier compared to real-world scenarios, potentially inflating performance.

Nonetheless, the results confirm the feasibility of the proposed approach and provide strong empirical evidence that parameter drift can serve as a reliable indicator for membership status.

4.2 Experiment 2: Validation on a Benchmark Dataset

To evaluate the scalability and generalizability of the proposed method, we conducted a second experiment using **WikiMIA**[3], a benchmark dataset designed for membership inference attacks (MIA) on LLMs [10].

This experiment assesses the effectiveness of our method across multiple pretrained models in a realistic setting, and compares its performance to state-of-the-art MIA baselines, including Min-K% Prob and SaMIA*zlib. We also investigate how variations in continual pre-training and feature selection influence attack accuracy.

Dataset. We used the 32-token subset of WikiMIA. A key feature of WikiMIA is that membership labels are assigned based on the temporal availability of the content:

- **Member samples**: Sentences describing events that occurred *before 2017*, which are likely to have been included in the pretraining corpora of models released between 2017 and 2023 (e.g., LLaMA-2, OPT).
- **Non-member samples**: Sentences about events that occurred *after 2023*, which could not have been present in any model's pretraining data.

This labeling approach enables a reliable membership ground truth without access to the original pretraining data. Importantly, all WikiMIA samples are paraphrased using ChatGPT, making them lexically distinct from the original corpus and thereby increasing the difficulty of the task. Compared to the myMIA dataset in Experiment 1, which includes exact matches from training data, WikiMIA offers a more realistic and challenging evaluation setting that better reflects real-world auditing scenarios.

Experimental Settings. We evaluated our method on four publicly available pre-trained LLMs of similar scale: GPT-J-6B[4] [13], OPT-6.7B[5] [14], Pythia-6.9B[6] [1], and LLaMA-2-7B[7] [12].

---

[3] https://huggingface.co/datasets/swj0419/wikiMIA.
[4] https://huggingface.co/EleutherAI/gpt-j-6b.
[5] https://huggingface.co/facebook/opt-6.7b.
[6] https://huggingface.co/EleutherAI/pythia-6.9b.
[7] https://huggingface.co/meta-llama/Llama-2-7b.

For each model, we applied continual pre-training using individual samples from the WikiMIA dataset. To ensure that each sample's parameter drift could be independently measured, we used a batch size of 1. Training was performed using DeepSpeed with parallel distributed execution. All hyperparameters (e.g., optimizer settings, learning rate) were aligned with those used in Experiment 1 (see Sect. 4.1).

To mitigate overfitting and ensure stable convergence, we employed early stopping. Training was halted if the validation loss failed to improve for 20 consecutive epochs—a stricter criterion than in Experiment 1, where the threshold was set to 50 epochs.

We applied the proposed method to the four LLMs. For each input sample from the WikiMIA dataset, we performed continual pre-training and computed the corresponding parameter drift as a feature. Membership inference performance was evaluated using the AUROC metric.

First, we report the results when the full set of model parameters is used to compute drift. Next, we investigate the impact of restricting the drift calculation to specific parameter subsets. Finally, we compare our method against state-of-the-art MIA techniques, such as Min-$K\%$ Prob and SaMIA*zlib.

Table 2. AUROC scores of the proposed method using all model parameters under different continual pre-training criteria.

Model	5 epochs	10 epochs	20 epochs	30 epochs	Early Stopping
GPT-J	0.59	0.61	**0.66**	0.64	0.57
OPT	0.51	0.56	0.51	0.51	0.54
Pythia	0.52	0.51	0.64	0.59	**0.68**
LLaMA-2	0.55	0.50	0.53	**0.62**	0.54
Average	0.542	0.545	**0.585**	**0.592**	0.584

Results and Discussion Using All Parameters. Table 2 presents AUROC scores obtained when using the full set of model parameters to compute drift. Each column corresponds to a different pre-training stopping criterion, ranging from fixed numbers of epochs to early stopping (ES).

Across all models and configurations, the AUROC scores consistently exceed 0.5, indicating that the proposed method reliably distinguishes member from non-member samples. Notably, performance improves with increased pre-training epochs. The average AUROC increases from 0.542 at 5 epochs to 0.592 at 30 epochs. A Pearson correlation analysis confirms this trend, with a coefficient of $r = 0.954$ ($p = 0.046$), suggesting that prolonged pre-training enhances signal separation.

These findings align with previous research: Shokri et al. [11] showed that overfitting increases MIA vulnerability, and DeAlcala et al. [2] observed that samples trained for longer exhibit higher membership leakage.

Interestingly, early stopping does not always yield better results. While it prevents overfitting in general-purpose training, in the context of MIA it may halt pre-training before sufficient drift has accumulated—especially for non-member samples. For example, early stopping underperforms compared to 30 epochs in both GPT-J (0.57 vs. 0.64) and LLaMA-2 (0.54 vs. 0.62).

We also observe notable differences across models. Pythia and GPT-J achieve higher AUROC values than OPT, which remains near random guessing across all conditions. This may reflect differences in architecture, training scale, or regularization strength that affect the magnitude of parameter drift.

Overall, these results confirm that parameter drift serves as a strong MIA signal when full model parameters are used. Longer pre-training yields better separation, though practical considerations such as computational cost and inference latency must also be taken into account. In the next section, we evaluate whether focusing on a subset of highly drifting parameters can retain this performance while reducing overhead.

Table 3. AUROC Comparison of the Proposed Method with Partial Parameters and Existing Methods

Target	Attention			MLP			LayerNorm			Top-$K\%$			MinK	SaMIA
	20ep	30ep	ES	20ep	30ep	ES	20ep	30ep	ES	20ep	30ep	ES		
GPT-J	0.63	0.62	0.56	0.68	0.66	0.59	0.62	0.61	0.58	0.65	0.69	0.63	0.71	<u>0.75</u>
OPT	0.53	0.52	0.55	0.51	0.50	0.52	0.54	0.50	0.51	0.70	**0.71**	0.57	0.67	<u>0.81</u>
Pythia	0.64	0.57	0.67	0.64	0.57	0.69	0.54	0.52	0.63	0.65	0.61	**0.76**	0.71	0.75
Llama-2	0.57	0.61	0.55	0.51	0.62	0.53	0.56	0.62	0.53	0.61	**0.64**	0.56	0.58	<u>0.66</u>
Average	0.59	0.58	0.58	0.58	0.59	0.58	0.56	0.56	0.56	0.65	0.66	0.63	0.67	0.74

Results and Discussion Using Parameter Subsets. We analyzed the effect of selecting different subsets of parameters for feature extraction. Table 3 reports the AUROC scores categorized by parameter subsets, specifically focusing on Attention layers, MLP layers, LayerNorm layers, and parameters with Top-$K\%$ highest drift magnitudes. For comparison, the table also includes two existing methods: Min-$K\%$ Prob (abbreviated as "MinK") and SaMIA*zlib (abbreviated as "SaMIA").

When using only parameters from the Attention or MLP layers, the average AUROC was 0.584 in both cases, while LayerNorm parameters yielded a slightly lower average of 0.563. These results suggest that Attention and MLP layers are more informative for MIA than LayerNorm, though the difference is marginal when compared to the 0.592 average AUROC obtained using all parameters (Table 2). This indicates that functional separation alone does not significantly enhance inference performance.

The Attention layer is responsible for information routing via self-attention, while LayerNorm mainly stabilizes activations—neither being highly indicative of data memorization. MLP layers, especially the first feedforward projection, have been shown to encode input-specific representations. However, combining both MLP projections into a single drift measure may dilute the membership signal.

Next, we explored selecting the top-$K\%$ of parameters with the largest drift magnitudes as features. This approach is motivated by our hypothesis (see Sect. 4.1) that untrained inputs induce greater drift. We experimented with various K values and report only the best-performing ones in Table 3. The top-$K\%$ drift strategy achieved AUROC values of 0.652 (20ep), 0.660 (30ep), and 0.634 (ES), surpassing the all-parameter baseline. This confirms that high-drift parameters are particularly discriminative.

Finally, we benchmarked our method against state-of-the-art MIA techniques, Min-$K\%$ Prob and SaMIA*zlib. The bolded values in Table 3 indicate cases where our method outperformed both baselines. We achieved a peak AUROC of **0.76** on Pythia, surpassing both Min-$K\%$ Prob (0.71) and SaMIA*zlib (0.75). Our method also exceeded Min-$K\%$ Prob on OPT and LLaMA-2. However, on GPT-J, it lagged behind both baselines.

Overall, while SaMIA*zlib yields the highest average AUROC (0.74), our method demonstrates competitive or superior performance on select models. It offers a compelling tradeoff between accuracy and interpretability, especially when high-drift parameters are carefully selected. Future work could explore adaptive selection of drift-sensitive parameters and extend evaluation to multilingual or instruction-tuned LLMs.

5   Conclusion

This paper introduces a novel white-box membership inference attack (MIA) for large language models (LLMs) based on internal parameter drift. Unlike prior approaches that rely solely on model outputs, our method quantifies the change in model parameters before and after continual pre-training on a single input. The core hypothesis is that pre-trained inputs require minimal parameter updates, whereas unseen inputs trigger larger changes. This drift signal enables effective membership inference under white-box access.

We validated our approach through two experiments. In Experiment1, we constructed a custom dataset with explicit ground-truth labels, confirming that parameter drift significantly differs between member and non-member samples. In Experiment2, we evaluated the method on WikiMIA using four open-source LLMs and compared it with two strong black-box baselines: Min-$K\%$ Prob and SaMIA*zlib. Our method achieved competitive or superior AUROC scores across several models. Moreover, we showed that selectively using high-drift parameters further improves accuracy, confirming that drift magnitude is a meaningful signal.

While our method did not achieve the best average AUROC across all settings, it offers several distinct advantages. These include interpretability, robustness to paraphrased or ambiguous inputs, and the ability to exploit internal representations unavailable to black-box techniques. When white-box access is permitted, parameter drift offers a powerful and underexplored signal for auditing model training behavior.

For future work, we plan to explore alternative similarity metrics such as cosine distance and Mahalanobis distance to better capture structural differences in parameter updates. Additionally, we aim to apply dimensionality reduction techniques (e.g., PCA or low-rank projection) to mitigate issues associated with high-dimensional parameter spaces and improve scalability. These directions will help extend the applicability of parameter driftâĂŞbased auditing to broader settings.

Overall, our findings establish parameter drift as a promising and interpretable signal for white-box membership inference in LLMs, contributing a new dimension to the landscape of model auditing and transparency.

Acknowledgments. This research was supported by JSPS KAKENHI Grant Number 24K03044 and 23K28383.

References

1. Biderman, S., et al.: Pythia: a suite for analyzing large language models across training and scaling. In: Krause, A., Brunskill, E., Cho, K., Engelhardt, B., Sabato, S., Scarlett, J. (eds.) Proceedings of the 40th International Conference on Machine Learning. Proceedings of Machine Learning Research, vol. 202, pp. 2397–2430. PMLR (2023). https://proceedings.mlr.press/v202/biderman23a.html
2. Dealcala, D., Mancera, G., Morales, A., Fierrez, J., Tolosana, R., Ortega-Garcia, J.: A comprehensive analysis of factors impacting membership inference. In: Proceedings of the IEEE/CVF Conference on Computer Vision and Pattern Recognition (CVPR) Workshops, pp. 3585–3593 (2024)
3. Hu, H., Salcic, Z., Sun, L., Dobbie, G., Yu, P.S., Zhang, X.: Membership inference attacks on machine learning: a survey. ACM Comput. Surv. (CSUR) **54**(11s), 1–37 (2022)
4. Kaneko, M., Ma, Y., Wata, Y., Okazaki, N.: Sampling-based pseudo-likelihood for membership inference attacks. arXiv preprint arXiv:2404.11262 (2024)
5. Karamolegkou, A., Li, J., Zhou, L., Søgaard, A.: Copyright violations and large language models. In: Bouamor, H., Pino, J., Bali, K. (eds.) Proceedings of the 2023 Conference on Empirical Methods in Natural Language Processing, pp. 7403–7412. Association for Computational Linguistics, Singapore (2023). https://doi.org/10.18653/v1/2023.emnlp-main.458. https://aclanthology.org/2023.emnlp-main.458
6. Li, C., Flanigan, J.: Task contamination: language models may not be few-shot anymore. In: Proceedings of the AAAI Conference on Artificial Intelligence, vol. 38, no. 16, pp. 18471–18480 (2024)
7. Meeus, M., Shilov, I., Faysse, M., de Montjoye, Y.A.: Copyright traps for large language models. In: Forty-First International Conference on Machine Learning (2024). https://openreview.net/forum?id=LDq1JPdc55

8. Nasr, M., Shokri, R., Houmansadr, A.: Comprehensive privacy analysis of deep learning: passive and active white-box inference attacks against centralized and federated learning. In: 2019 IEEE Symposium on Security and Privacy (SP), pp. 739–753. IEEE (2019)

9. Sainz, O., Campos, J., García-Ferrero, I., Etxaniz, J., de Lacalle, O.L., Agirre, E.: NLP evaluation in trouble: On the need to measure LLM data contamination for each benchmark. In: Bouamor, H., Pino, J., Bali, K. (eds.) Findings of the Association for Computational Linguistics: EMNLP 2023, Singapore, pp. 10776–10787. Association for Computational Linguistics (2023). https://doi.org/10.18653/v1/2023.findings-emnlp.722. https://aclanthology.org/2023.findings-emnlp.722

10. Shi, W., et al.: Detecting pretraining data from large language models. In: NeurIPS 2023 Workshop on Regulatable ML (2023). https://openreview.net/forum?id=ZLJ6XRbdaC

11. Shokri, R., Stronati, M., Song, C., Shmatikov, V.: Membership inference attacks against machine learning models. In: 2017 IEEE Symposium on Security and Privacy (SP), pp. 3–18. IEEE (2017)

12. Touvron, H., et al.: Llama 2: open foundation and fine-tuned chat models. arXiv preprint arXiv:2307.09288 (2023)

13. Wang, B., Komatsuzaki, A.: GPT-J-6B: A 6 Billion Parameter Autoregressive Language Model (2021). https://github.com/kingoflolz/mesh-transformer-jax

14. Zhang, S., et al.: OPT: open pre-trained transformer language models (2022). https://arxiv.org/abs/2205.01068

MicroSuggest: Kernel-Aware Microservice Decomposition

Harsh Borse[✉], Utkalika Satpathy, Mainack Mondal, and Bivas Mitra

IIT Kharagpur, Kharagpur, India
harshzf2@gmail.com

Abstract. Microservice decomposition typically emphasizes logical or domain-driven boundaries, often overlooking performance bottlenecks from low-level system interactions. We present a system call-aware decomposition method that identifies and separates functions likely to interfere at the kernel level. By defining a collision score based on system call frequency and type, and using a fine-tuned Large Language Model to statically predict syscall behavior, we construct a function interaction graph for clustering. Evaluation on Python-based monoliths shows up to 30% latency reduction and improved scalability compared to traditional approaches, demonstrating the value of kernel-informed microservice design.

1 Introduction

The transition from monolithic architectures to microservices has significantly enhanced the scalability, maintainability, and fault tolerance of modern software systems. However, decomposing a monolithic application into microservices remains a complex engineering challenge. Traditional approaches often rely on domain-driven design (DDD) [2] or runtime profiling of inter-function communication [4], which may overlook deeper system-level interactions, particularly the effects of concurrent system call execution.

In high-throughput systems, concurrent requests frequently trigger similar or overlapping code paths, exerting significant pressure on shared kernel resources. Even without explicit data sharing, processes or threads may experience performance degradation due to contention in the kernel caused by simultaneous system calls such as `fork`, `connect`, `write`, or `open`. These system-level collisions are often subtle and hard to detect through traditional profiling techniques, yet they can severely impact application latency and throughput [1].

We propose a novel decomposition strategy based on analyzing the system call behavior of functions within a monolithic application. For each function, we examine the types and frequencies of system calls it invokes and define a *collision score* that quantifies the potential for performance interference due to concurrent execution. The core hypothesis is that functions which heavily invoke the same types of system calls—especially under concurrent execution—are more likely to contend for kernel resources such as I/O bandwidth, network

© The Author(s), under exclusive license to Springer Nature Switzerland AG 2026
C. K. Leung et al. (Eds.): DaWaK 2025, LNCS 16048, pp. 302–308, 2026.
https://doi.org/10.1007/978-3-032-02215-8_24

sockets, and scheduling queues. By computing pairwise collision scores between all functions in the monolith, we construct a weighted interaction graph that reflects potential syscall-level contention. This graph is then clustered to group functions into microservices in a manner that minimizes intra-service system call interference while preserving logical cohesion (Fig. 1).

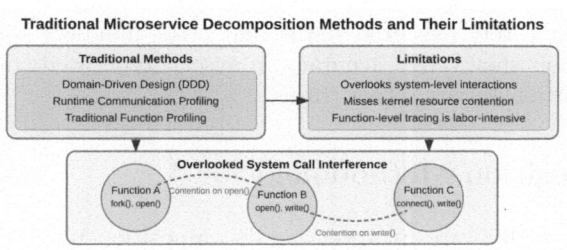

Fig. 1. Traditional Microservice Decomposition

A key challenge in system call-aware decomposition is identifying the syscall behavior of individual functions, especially in large codebases. Traditional tools like `strace`, `perf`, and eBPF-based profilers operate at the process or thread level, making function-level tracing labor-intensive and error-prone [3]. To overcome this, we introduce a Large Language Model (LLM)-based prediction system that infers syscall sequences directly from function-level source code. We curate a dataset of real-world Python functions and corresponding syscall traces from monolithic applications spanning frontend, business logic, and backend layers. A transformer-based model is fine-tuned on this dataset to generalize across unseen code, enabling rapid, automated syscall profile prediction without manual instrumentation [5]. **Key Contributions:** (1) A mathematically rigorous collision score for quantifying kernel-level resource contention between functions; (2) An LLM-based approach for static prediction of syscall behavior from source code; (3) A graph-theoretic optimization framework for performance-aware microservice decomposition; (4) Comprehensive evaluation demonstrating up to 30% latency reduction and improved scalability.

2   System Call Behavior Prediction via LLM

2.1   LLM Architecture and Training

To enable scalable static analysis, we fine-tune a transformer-based, code-oriented LLM on 50,000+ (function, syscall trace) pairs from Python applications in domains like library management, transportation, and e-commerce. Each pair maps [function code] → [syscall profile], encoding syscall types and relative frequencies. The model is trained via supervised learning to minimize syscall prediction error, using cross-entropy loss on next-token generation. It learns associations between code patterns and syscall behaviors—e.g., file I/O maps to `open/read/write`, DB access to `socket`. Self-attention layers capture long-range dependencies such as API and variable usage.

2.2 Inference and Validation

Post-training, the LLM predicts syscall behavior statically for each function in parallel, covering rarely executed paths and error handlers without dynamic tracing. Validation shows 87% accuracy in syscall type prediction and 82% in frequency estimation versus actual traces.

3 Decomposition Methodology

We propose a syscall-aware microservice decomposition pipeline consisting of: (1) fine-tuning a pretrained LLM to predict syscalls from source code, (2) profiling function-level syscall usage and computing collision scores, (3) constructing a weighted function interaction graph, and (4) applying graph clustering to minimize syscall-induced contention. Evaluations on Python-based monoliths (e.g., library and ticketing systems) demonstrate the effectiveness of our approach. The following subsections detail each component.

3.1 System Call Profiling and Collision Score

(A) Function-Level Syscall Instrumentation: We collect fine-grained dynamic syscall profiles for each function in the monolith using static instrumentation combined with dynamic tracing tools such as `strace` and eBPF-based profilers. By augmenting these tools with function context tracking, we attribute each syscall to the highest-level application function on the call stack. This produces a mapping from function names to syscall frequency vectors, indicating how often each syscall type is invoked per execution. For example, a database-accessing function may involve multiple file and socket calls, while a CPU-bound function may primarily issue memory allocation calls.

(B) Defining Syscall Collision: Using the collected traces, we define a syscall collision score to estimate potential resource contention between functions. Functions that frequently execute concurrently and invoke the same syscall types are likely to contend for kernel resources such as CPU, disk, and network, causing performance degradation. Given functions f_i and f_j, and syscall type s, let N_i^s and N_j^s denote the frequency of s by f_i and f_j, respectively. The collision score is defined as:

$$\text{Collision}(f_i, f_j) = \sum_{s \in S} w_s \cdot N_i^s \cdot N_j^s$$

Here, S is the set of syscall categories, and w_s is an optional weight capturing the cost of contention on type s. A high collision score suggests both functions heavily use the same syscall types in overlapping timeframes (e.g., two functions writing to disk or opening sockets), indicating shared kernel resource usage and potential interference. Functions are considered concurrent if they can execute in parallel threads or serve independent user requests.

(C) Collision as a Decomposition Criterion: We use the collision score to guide decomposition by separating high-collision function pairs into distinct microservices. In monolithic architectures, colocated contending functions can lead to kernel-level bottlenecks such as I/O queue contention or CPU scheduling delays. Isolating such functions into separate service processes improves concurrency by leveraging OS-level scheduling and container isolation. This is especially impactful in Python, where the Global Interpreter Lock (GIL) restricts true multi-threaded execution—CPU-bound functions in the same process cannot utilize multiple cores efficiently. By isolating them, we enable parallel execution across cores and reduce resource contention. Thus, the collision score offers a low-level, objective metric that complements high-level coupling/cohesion heuristics for microservice decomposition.

4   Kernel-Aware Decomposition Methodology

4.1   Function Interaction Graph Construction and Clustering

With each function annotated with predicted syscall profiles, we construct a Function Interaction Graph $G = (V, E)$ where vertices V correspond to functions and weighted edges E represent collision scores between function pairs. We compute collision scores for all function pairs, yielding weights $w_{ij} = \text{Collision}(f_i, f_j)$. The graph is sparsely populated with edges only where contention is non-trivial.

We augment the graph with secondary edges based on static call dependencies and module structure. If function f_i invokes f_j, we add edges representing logical coupling with weights lower than collision weights. This ensures strongly cohesive functions are not arbitrarily separated due to syscall behavior alone.

We apply spectral clustering on the weighted graph to derive microservice groupings. Since high collision weights imply dissimilarity (functions should be separated), we invert weights into affinity scores: $a_{ij} = 1/(1 + w_{ij})$. This produces an affinity matrix favoring function pairs with low collision. The spectral clustering procedure optimizes partitioning to minimize intra-cluster collision scores while preserving functional cohesion (Fig. 2 and Table 2).

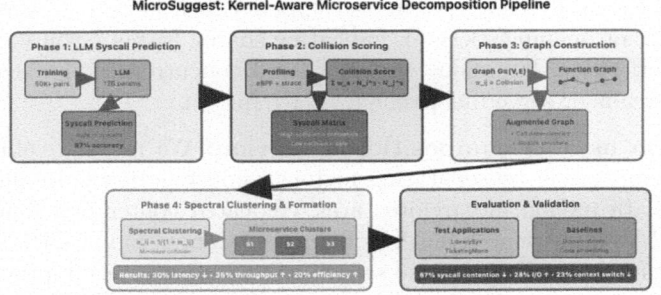

Fig. 2. Function Interaction Graph Construction and Clustering

Table 1. Microservice Decomposition Comparison

System	Approach	#MS Count	Performance Impact
TicketingMono	Traditional	5	Baseline
TicketingMono	Our Approach	8	Optimal collision reduction
LibrarySys	Traditional	8	Baseline
LibrarySys	Our Approach	10	20% improvement in 95th-percentile latency

Table 2. TicketingMonolith Response Time

Approach	MRT (ms)	Improvement
Monolithic	420	Baseline
Traditional	336	20% reduction from monolith
Syscall-aware	235	30% further reduction

5   Implementation and Evaluation

5.1   Experimental Setup and Results

We evaluate on two in-house Python monoliths: *LibrarySys* and *TicketingMonolith*. Each is tested in three configurations: original monolith, expert-driven domain decomposition, and our syscall-aware version. All setups are containerized and deployed with Docker Swarm for consistent benchmarking under simulated user workloads.

Performance Results: Table 1, our syscall-aware decomposition consistently achieves 25–30% reduction in 95th percentile latency compared to traditional approaches (Fig 5). Maximum sustainable throughput increases by 35–42% compared to monolithic baselines. Under 2000 concurrent users, TicketingMonolith showed severe latency degradation in monolithic form; traditional microservices improved this, but still had contention between seat allocation and payment confirmation. Our approach separated these into independent services, resulting in 30% further mean response time reduction (Fig 4).

Function Interaction Graph Construction and Microservice Decomposition

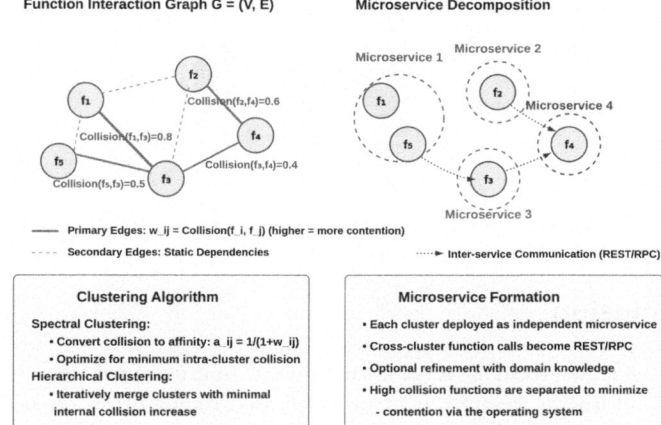

Fig. 3. Function Interaction Graph Construction and Clustering

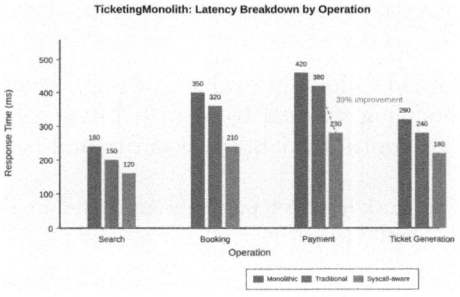

Fig. 4. Component-wise latency analysis.

Resource Utilization: Our decomposition enables more efficient resource utilization with 20% reduction in CPU waste and 15% improvement in memory utilization through better workload isolation. Services handling I/O-intensive operations show 28% improvement in I/O bandwidth utilization when isolated from CPU-intensive services.

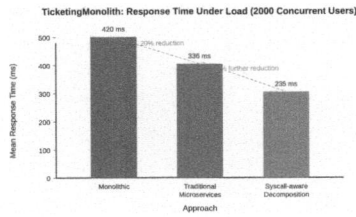

Fig. 5. Analysis of overall mean latency in different decompositions.

6   Conclusion

We introduced MicroSuggest, a kernel-aware microservice decomposition approach that addresses system-level performance issues often missed by domain-driven methods. By combining LLM-based syscall prediction with collision score-based clustering, MicroSuggest improves latency (up to 30%), throughput (35%), and resource efficiency. Our results show that incorporating kernel-level behavior enhances performance without compromising design coherence, offering a new direction for automated, performance-driven microservice decomposition.

Future Work: (1) Broader domain evaluation (e.g., healthcare, finance); (2) Concurrency-aware modeling via real traces; (3) Integration with observability tools; (4) Multi-objective optimization; (5) Security and fault isolation analysis.

Acknowledgment. This work has been partially supported by the DST-SERB funded project with grant CRG/2021/005316.

References

1. Chen, Y., et al.: Characterizing, modeling, and benchmarking RocksDB key-value workloads at Facebook. arXiv preprint arXiv:1709.06666 (2017)
2. Evans, E.: Domain-Driven Design: Tackling Complexity in the Heart of Software. Addison-Wesley Professional (2003)
3. Gregg, B.: Linux Performance: Observability, Analysis, and Tuning. Prentice Hall Professional (2016)
4. Gysel, M., Kölbener, L., Giersche, W., Zimmermann, O.: Service cutter: a systematic approach to service decomposition. In: Aiello, M., Johnsen, E.B., Dustdar, S., Georgievski, I. (eds.) ESOCC 2016. LNCS, vol. 9846, pp. 185–200. Springer, Cham (2016). https://doi.org/10.1007/978-3-319-44482-6_12
5. Vaswani, A., et al.: Attention is all you need. In: Advances in Neural Information Processing Systems, pp. 5998–6008 (2017)

TraceTune: Targeted Fine-Tuning of Attention Heads for Text-to-SQL

Saba Zamankhani[(✉)][iD] and Kai-Uwe Sattler[iD]

TU Ilmenau, Ilmenau, Germany
{saba.zamankhani,kus}@tu-ilmenau.de

Abstract. Large Language Models (LLMs) have demonstrated remarkable performance across various natural language processing tasks. However, adapting them to domain-specific problems like text-to-SQL generation remains computationally expensive, particularly in dynamic environments where schemas frequently evolve. Full fine-tuning for each new database schema is impractical due to high computational and memory costs. To address this challenge, we propose **TraceTune**, a parameter-efficient adaptation method that selectively fine-tunes the most relevant attention heads and bias terms identified through causal tracing. Experiments on multiple Text-to-SQL benchmarks demonstrate that TraceTune achieves comparable accuracy to full fine-tuning while significantly reducing memory and compute overhead. These results highlight TraceTune as an efficient and scalable solution for deploying Text-to-SQL systems in resource-constrained environments.

Keywords: Text-to-SQL · Parameter-Efficient Fine-Tuning · Structured Query Generation · Large Language Models

1   Introduction

Large Language Models (LLMs) have advanced Text-to-SQL generation by enabling natural language queries to be translated into executable SQL [1,2]. However, full fine-tuning remains costly due to complex query structures and schema-specific adaptations. For example, fine-tuning LLaMA-7B [3] on the Spider dataset requires over 12 GPU-hours [4], making frequent retraining impractical.

Parameter-Efficient Fine-Tuning (PEFT) reduces this cost by updating only a small subset of model parameters [5]. Yet, most PEFT methods apply updates uniformly, without prioritizing components most relevant to the task. In contrast, causal tracing techniques [6,7] identify model components—like attention heads—that causally influence outputs, but are typically used for interpretation, not fine-tuning.

We propose **TraceTune**, a hybrid method that combines the efficiency of PEFT with the task-specific focus of causal tracing. TraceTune selectively fine-tunes only the most causally important attention heads—components that control how LLMs align natural language with database schema elements [8]. This

© The Author(s), under exclusive license to Springer Nature Switzerland AG 2026
C. K. Leung et al. (Eds.): DaWaK 2025, LNCS 16048, pp. 309–314, 2026.
https://doi.org/10.1007/978-3-032-02215-8_25

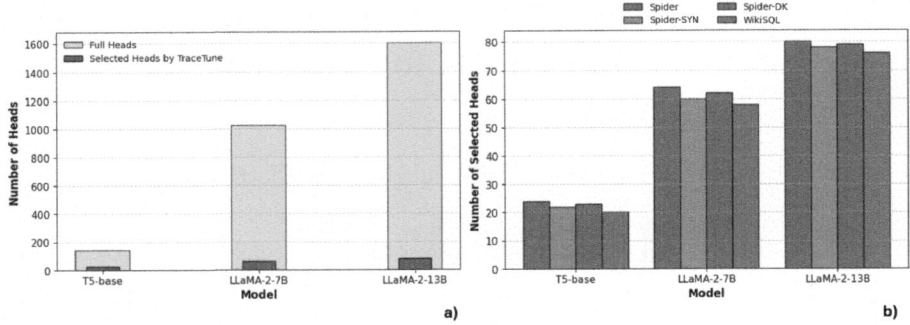

Fig. 1. a: Comparison between the total number of attention heads and the number of heads selected by TraceTune across models averaged over datasets. b: Number of heads selected by TraceTune across different Text-to-SQL benchmarks.

reduces memory and compute overhead while maintaining competitive accuracy. Rather than competing with LoRA and other PEFT methods, TraceTune extends them by focusing adaptation on the most impactful model components. This enables efficient deployment in settings where resources are constrained but accuracy must be preserved. We introduce TraceTune, a hybrid PEFT method guided by causal analysis. We also demonstrate that TraceTune achieves accuracy close to full fine-tuning while significantly reducing parameter, memory, and training costs. finally We show that causally selected heads generalize well across benchmarks, enabling robust adaptation for structured language generation.

2   Related Work

Early Text-to-SQL systems used specialized encoders and grammar-constrained decoders [9,10]. With the emergence of pretrained LLMs like T5, BART, and Codex [11–13], flexible SQL generation became feasible, enabling zero- and few-shot generalization [14]. To better handle structured data, models like SQLova [15], X-SQL [16], TaBERT [17], and StructLM [18] incorporate table-aware representations. However, they typically rely on full fine-tuning, which is resource-intensive for large models. PEFT methods such as adapter tuning [19], prefix tuning [20], and LoRA [5] reduce adaptation costs by updating only a subset of parameters. QLoRA [21] further reduces memory usage through quantization. These methods have been applied to domain-specific Text-to-SQL tasks like FinSQL [22] and ZeroSQL [4]. Additional strategies improve robustness and generalization: PHuber [23] adds regularization for noisy labels, DAIL-SQL [24] benchmarks supervised fine-tuning, and RoE [25] avoids negative transfer via expert routing. Inference-time methods such as candidate sampling [26] and proxy decoding [27] speed up execution without retraining. Parallel work in causal analysis includes tracing [6], activation patching [7], and memory editing [28,29], which intervene on model components based on their causal impact. These methods have mostly focused on interpretability rather than adaptation.

Table 1. Performance comparison of full fine-tuning, LoRA, TraceTune, and random head selection across Text-to-SQL benchmarks. Each entry reports Execution Accuracy/Exact Match (ExAcc/EM).

Model	Method	Spider	Spider-SYN	Spider-DK	WikiSQL
T5-base	Full Fine-Tuning	79.2/74.5	76.5/70.8	72.9/68.3	87.0/84.2
	LoRA	77.0/73.0	73.8/69.5	70.0/65.8	86.0/83.5
	TraceTune	78.1/74.1	74.9/70.2	71.0/66.5	86.5/84.0
	Random Heads	64.8/61.0	61.0/57.0	58.0/53.5	74.5/72.0
LLaMA-2-7B	Full Fine-Tuning	81.5/76.8	78.0/73.5	74.5/69.2	88.0/85.0
	LoRA	77.5/72.8	74.3/68.9	70.2/66.1	85.5/83.0
	TraceTune	78.7/73.9	75.4/70.0	71.5/67.2	86.1/83.5
	Random Heads	64.1/60.5	61.2/57.3	59.0/54.9	73.2/71.0
LLaMA-2-13B	Full Fine-Tuning	83.0/78.0	80.0/75.5	76.0/71.8	89.0/86.0
	LoRA	79.0/76.0	76.0/72.5	72.5/69.0	86.5/84.5
	TraceTune	80.3/77.0	77.2/73.0	74.0/70.5	87.0/85.0
	Random Heads	66.5/63.0	63.5/60.0	60.5/57.0	74.0/72.0

TraceTune bridges these lines of work, combining the efficiency of PEFT with the targeted precision of causal tracing for structured language generation.

3 Methodology

We introduce **TraceTune**, a hybrid fine-tuning method that combines causal analysis with parameter-efficient adaptation to optimize large language models for Text-to-SQL. TraceTune identifies and fine-tunes only the most relevant attention heads in a pretrained decoder-only Transformer. To determine which attention heads are crucial, we apply *causal tracing*. For each head, we blend its activation with a neutral baseline using a mixing factor λ, then measure the drop in execution accuracy ($\Delta_{(l,i)}$) when generating SQL. This drop quantifies the heads causal importance. Heads with the highest $\Delta_{(l,i)}$ are selected for tuning. Selected heads are fine-tuned using *Low-Rank Adaptation (LoRA)* by injecting trainable matrices into their query projections:

$$W_Q^{(l,i)} \leftarrow W_Q^{(l,i)} + A^{(l,i)} B^{(l,i)} \tag{1}$$

Here, $W_Q^{(l,i)}$ is the original query projection matrix for head i in layer l. The matrices $A^{(l,i)} \in \mathbb{R}^{d \times r}$ and $B^{(l,i)} \in \mathbb{R}^{r \times d}$ are small, trainable parameters whose product has the same shape as $W_Q^{(l,i)}$. This low-rank update allows the model to adapt without modifying the original weights directly.

This approach significantly reduces the number of trainable parameters and memory overhead, enabling efficient, targeted adaptation suitable for resource-constrained environments.

Table 2. Comparison of TraceTune with parameter-efficient fine-tuning baselines on the Spider dataset.

Method	Fine-Tuning Type	ExAcc (%)	EM (%)	Trainable Params
Full Fine-Tuning	Full	86	78	100%
LoRA [5]	PEFT (Uniform)	79	76	1–5%
QLoRA [21]	PEFT (Quantized)	70	75	∼1%
PHuber [23]	PEFT + Regularization	77	72	Variable
DAIL-SQL [24]	Supervised Fine-Tuning	79	77	Variable
StructLM [18]	Instruction Fine-Tuning	78	75	Variable
RoE [25]	Expert Retrieval + PEFT	78	76	Varies (per expert)
TraceTune (Ours)	PEFT + Causal Tracing	**83**	**77**	<1% (**top-k heads**)

4 Experimental Setup

We evaluate **TraceTune** on four Text-to-SQL benchmarks: WikiSQL [9], Spider [1], Spider-SYN [30], and Spider-DK [31]. TraceTune is compared against several fine-tuning baselines using the same backbone models: Full Fine-Tuning, LoRA [5], QLoRA [21], PHuber [23], DAIL-SQL [24], StructLM [18], and RoE [25]. Experiments use T5-base, LLaMA-2-7B, and LLaMA-2-13B [3]. Trace-Tune selects top-k causally important attention heads via tracing and fine-tunes their query projections with LoRA (rank $r = 8$). Dropout is set to 0.1. Models are trained for 5 epochs using AdamW (batch size 4, learning rate 3×10^{-5}, linearly decayed), on 80GB A100 GPUs. We use an 80%/10%/10% train/val/test split to evaluate. We evaluate using three metrics: Exact Match (EM) for string-level correctness, Execution Accuracy (ExAcc) for matching query outputs, and Adaptation Efficiency, which measures trainable parameter ratio, memory usage, and training time.

5 Results and Discussion

TraceTune applies causal tracing to identify task-relevant attention heads, adding less than two hours even for LLaMA-2-13B—making it a lightweight pre-adaptation step. As shown in Table 1, TraceTune consistently outperforms LoRA and random head selection across four Text-to-SQL benchmarks, achieving accuracy close to full fine-tuning while updating fewer than 5% of parameters. TraceTune also improves efficiency: it reduces memory usage by over 5× (from 48 GB to 8 GB), training time by nearly 6× (from 12.3 to 2.1 h), and the number of trainable parameters by 20× compared to full fine-tuning, as shown in our efficiency analysis. Figure 1 shows that selected heads are robust (85% overlap across runs), concentrated in deeper layers (72%), and generalize well across datasets (70–80% overlap), validating the causal selection process. Finally, Table 2 shows that TraceTune achieves the highest execution accuracy among PEFT methods on Spider, while requiring fewer updates than LoRA, QLoRA, or expert routing. While promising, TraceTune currently traces only query projections and introduces modest overhead. Future work will explore broader adaptation and faster tracing.

6 Conclusion and Future Work

We presented **TraceTune**, a hybrid fine-tuning method that uses causal tracing to selectively adapt the most relevant attention heads in LLMs for Text-to-SQL. This enables near full fine-tuning accuracy with significantly lower resource requirements. TraceTune outperforms other PEFT methods across four benchmarks in both efficiency and accuracy, demonstrating the value of causality-guided adaptation for structured tasks. Future directions include faster tracing methods and extending TraceTune to tasks like code generation and table-to-text translation.

Acknowledgment. The research is funded by the Carl-Zeiss Foundation via the Project Memwerk and the Deutsche Forschungsgemeinschaft (DFG, German Research Foundation)– Project-ID 434434223 – SFB 1461.

References

1. Yu, T., et al.: Spider: a large-scale human-labeled dataset for complex and cross-domain semantic parsing and Text-to-SQL task. arXiv preprint arXiv:1809.08887 (2018)
2. Wang, B., Shin, R., Liu, X., Polozov, O., Richardson, M.: RAT-SQL: relation-aware schema encoding and linking for Text-to-SQL parsers. In: ACL, pp. 7567–7578 (2020)
3. Touvron, H., et al.: LLaMA 2: open foundation and fine-tuned chat models. arXiv preprint arXiv:2307.09288 (2023)
4. Chang, S., et al.: Zero-shot text-to-SQL learning with auxiliary task. In: AAAI Conference on Artificial Intelligence (2019)
5. Hu, E.J., et al.: LoRA: low-rank adaptation of large language models. ICLR (2022)
6. Meng, K., et al.: Locating and editing factual associations in GPT. In: NeurIPS (2022)
7. Turner, C.L., et al.: Activation patching reveals emergent information in transformers. In: ICLR (2023)
8. Vaswani, A., Shazeer, N., Parmar, N., et al.: Attention is all you need. In: Advances in Neural Information Processing Systems (NeurIPS) (2017)
9. Zhong, V., et al.: Seq2SQL: generating structured queries from natural language using reinforcement learning. In: ACL, pp. 959–968 (2017)
10. Xu, X., Liu, C., Song, D.: SQLNet: generating structured queries from natural language without reinforcement learning. In: EMNLP, pp. 681–691 (2017)
11. Raffel, C., et al.: Exploring the limits of transfer learning with a unified text-to-text transformer. J. Mac. Learn. Res. **21**(140), 1–67 (2020)
12. Lewis, M., et al.: BART: denoising sequence-to-sequence pre-training for natural language generation, translation, and comprehension. In: ACL (2019)

13. Chen, M., et al.: Evaluating large language models trained on code. arXiv preprint arXiv:2107.03374 (2021)
14. Li, Y., Qin, B., Liu, T., et al.: LLMs as a unified database interface. arXiv preprint arXiv:2306.10141 (2023)
15. Hwang, W., et al.: A comprehensive exploration on wikiSQL with table-aware word contextualization. In: EMNLP (2019)
16. Xu, P., Rao, J.R., et al.: Bridging text and tables with multilingual pretraining for structured semantic parsing. arXiv preprint arXiv:2004.08338 (2020)
17. Yin, P., Neubig, G.: TaBERT: pretraining for joint understanding of textual and tabular data. In: ACL (2020)
18. Xiao, C., et al.: StructLM: structure-aware pre-training for language models. In: ACL (2023)
19. Houlsby, N., et al.: Parameter-efficient transfer learning for NLP. In: International Conference on Machine Learning, pp. 2790–2799. PMLR (2019)
20. Li, X.L., Liang, P.: Prefix-Tuning: optimizing continuous prompts for generation. In: Proceedings of the 59th Annual Meeting of ACL and NLP (Volume 1: Long Papers), pp. 4582–4597 (2021)
21. Dettmers, T., et al.: QLoRA: efficient finetuning of quantized LLMs. arXiv preprint arXiv:2305.14314 (2023)
22. Zhang, Y., Chen, W., Wang, W.Y.: FinSQL: enabling financial Text-to-SQL with prompt tuning and data augmentation. In: EACL (2024)
23. Ye, J., et al.: PHuber: robust fine-tuning of language models under label noise. arXiv preprint arXiv:2306.13001 (2023)
24. Zhang, J., et al.: DAIL-SQL: a comprehensive study of supervised fine-tuning for text-to-SQL with LLMs. arXiv preprint arXiv:2310.05090 (2023)
25. Jang, J., et al.: Exploring the benefits of training expert language models over instruction tuning. In: International Conference on Machine Learning, pp. 14702–14729. PMLR (2023)
26. Wang, Y., et al.: Faster decoding for large language models via candidate sampling and rejection. arXiv preprint arXiv:2304.13885 (2023)
27. You, Y., et al.: Accelerating LLM inference with proxy decoding. arXiv preprint arXiv:2308.14508 (2023)
28. Meng, K., et al.: ROME: locating and editing factual knowledge in GPT. In: NeurIPS (2022)
29. Meng, K., Belinkov, Y., Klein, D.: MEMIT: mass editing memory in transformers. In: ACL (2023)
30. Gan, Y., et al.: Huang, towards robustness of text-to-SQL models against synonym substitution (2021)
31. Gan, Y., et al.: Exploring underexplored limitations of cross-domain text-to-SQL generalization (2021)

Neural Networks

ONNYX : Optimized Neural Networks Yielding eXplainable Insights from ECG Signals-Based Data Streams

Sanket Mishra[2](✉) [ID], V. Aravindan[1] [ID], Rajkanwar Singh[1] [ID],
Hasita Chowdary Meka[1] [ID], Sandipan Maiti[1] [ID], and Subhrakanta Panda[3] [ID]

[1] School of Computer Science and Engineering, VIT-AP University, Amaravati, India
sandipan.maiti@vitap.ac.in
[2] Manipal Institute of Technology Bengaluru, Manipal Academy of Higher
Education, Manipal, India
sanket.mishra@manipal.edu
[3] Department of Computer Science and Information Systems, Birla Institute of
Technology and Science, Hyderabad Campus, Hyderabad, India
spanda@hyderabad.bits-pilani.ac.in

Abstract. Deep learning classification models are extensively utilized
for the automated diagnosis of heart disease (HD) by analyzing various
physiological signals, such as electrocardiogram (ECG), magnetocardio-
graphy (MCG), heart sounds (HS) and impedance cardiography (ICG)
signals. In this study, we introduce the ONNYX framework (Optimal
Neural Networks Yielding eXplainable insights from ECG signals-based
data streams), which demonstrates a big data strategy for ECG classification. This framework incorporates several modules, including FastAPI,
MinIO, mlflow, Ray, Kubernetes, and Pulsar. We have developed a high
throughput and low latency system using Kubernetes' distributed architecture and Ray's distributed training to classify ECG signals. The ECG
records of subjects sourced from the MIT-BIH repository are sampled
and input into the classification models to distinguish between normal
and abnormal heart rate patterns in patients. We introduce an innovative
optimal model selection algorithm that assesses classification techniques
according to training efficiency and identifies the most suitable ones for
testing. Our weighted ensemble method attained an overall accuracy
of 99.27% and 99.16% in binary and multiclass classification settings
respectively.

Keywords: ElectroCardiogram signals · Heart Rate Variability ·
Distributed deep learning · Explainable AI

1 Introduction

Cardiovascular diseases (CVDs) are the leading cause of death worldwide, claim-
ing more than 17 million lives each year[1]. These include coronary artery disease,

---

[1] https://www.who.int/health-topics/cardiovascular-diseases.

© The Author(s), under exclusive license to Springer Nature Switzerland AG 2026
C. K. Leung et al. (Eds.): DaWaK 2025, LNCS 16048, pp. 317–331, 2026.
https://doi.org/10.1007/978-3-032-02215-8_26

rheumatic heart disease, and heart failure, with many premature deaths occurring in individuals under 70 years [1]. Conditions such as arrhythmias and valve disorders disrupt cardiac function, often requiring clinical intervention [2]. Manual interpretation of ECGs is time-consuming and error-prone, particularly in low-resource settings. Deep learning offers a promising alternative, with models increasingly used to detect CVDs from ECG signals [3]. These approaches enhance diagnostic speed and accuracy, facilitating personalized care. Advancements in AI have transformed healthcare, excelling in image analysis tasks such as X-rays, CT, MRI, and mammograms, often assisting medical persons in detecting conditions such as tumors and fractures. In deep learning, we have noticed that convolutional networks [4] have proven successful in chest radiographs and ECG data to identify irregular heart patterns (arrhythmias). The capacity of these approaches to identify symptoms of cardiovascular disease may be further enhanced by including transfer learning [5] and federated learning methodologies [6]. On the other hand, the concept of employing an online deep learning system could introduce a vital element for improving the real-time identification of CVDs, which might be essential for quicker diagnostics.

In this work, we propose ONNYX (Optimized Neural Networks Yielding eXplainable insights from ECG signals), a real-time CVD detection framework leveraging a distributed Kubernetes-based architecture. ONNYX concurrently deploys multiple deep learning models, using a multi-armed bandit strategy with Thompson sampling to dynamically select the best-performing model based on incoming data streams. Our key contributions are:

(a) We introduce a microservice-based AI framework for arrhythmia detection from ECG streams using scalable deep learning deployments.
(b) We implement distributed concurrent model execution, with adaptive model selection via probabilistic Thompson sampling; Apache Pulsar is used to simulate real-time data ingestion.
(c) We enhance model interpretability using SHAP and achieve a 99.27% (binary classification) and 99.16% (multiclass classification) accuracy via probabilistic ensembling of top-performing classifiers.

In the subsequent section, we examine various methodologies employed in the domain of ECG classification that have inspired the present study.

2   Related Works

Traditional ECG interpretation systems struggled with intraclass variability and overreliance on supervised datasets, often performing poorly on unseen data [7]. Deep learning methods, particularly CNNs, have since demonstrated cardiologist-level accuracy in multi-label ECG classification, outperforming commercial systems when trained on large datasets [8]. Architectures like 1D ResNet extract features directly from heartbeat sequences, improving diagnostic capability [9]. However, challenges remain in model explainability and bias mitigation, prompting the use of techniques like LIME to highlight important ECG segments [10].

To address class imbalance and enhance generalization, hybrid methods using Bi-LSTM with GANs and SMOTE have been explored, though most studies remain retrospective, lacking scalable, real-time implementations [9]. Big data analytics have enabled ECG-derived predictions of structural abnormalities such as left ventricular mass, yet real-time inference remains underexplored [11]. Variability in ECG formats and annotation standards also impairs model interoperability [12].

Efforts to integrate Complex Event Processing (CEP) with LSTM models show promise for streaming prediction but face scalability issues under fluctuating data loads [13]. While frameworks using Apache Spark and Flink support high-throughput analytics, they either rely on micro-batching or lack native deep learning support, complicating deployment in clinical settings [14]. Clinical decision support systems increasingly incorporate explainable AI, yet face similar integration and scalability challenges when applied to real-time cardiovascular monitoring.

In the next section, we elaborate on the different microservices-based modules that were considered in the ONNYX framework.

3   ONNYX Framework

Figure 1 illustrates the open source and modular containerized framework proposed for the classification of ECG data streams and the early detection of arrhythmias.

To integrate the underlying principles of the algorithms utilized in the proposed ECG classification framework, a concise description and analysis of the performance effects of each module are provided in Table 1.

3.1   Data Ingestion and Preprocessing

Training data is ingested via a FAST API[2] endpoint and stored in MinIO[3], a high-performance, S3-compatible object store used as a data lake. This schema-on-read approach supports scalable, cloud-native storage and efficient handling of large, unstructured deep learning datasets. MinIO enables parallel read/write operations and outperforms traditional databases for such workloads.

The Ray[4] job first ingests this data into Ray Data and applies filtering functions to enhance signal quality:

1. Bandpass Filter: We use a bandpass filter (0.5âÅŞ50 Hz) to remove low and high-frequency noise, like that from muscle contractions or electrical equipment, improving signal clarity.

---

[2] https://fastapi.tiangolo.com/.
[3] https://min.io/.
[4] https://www.ray.io/.

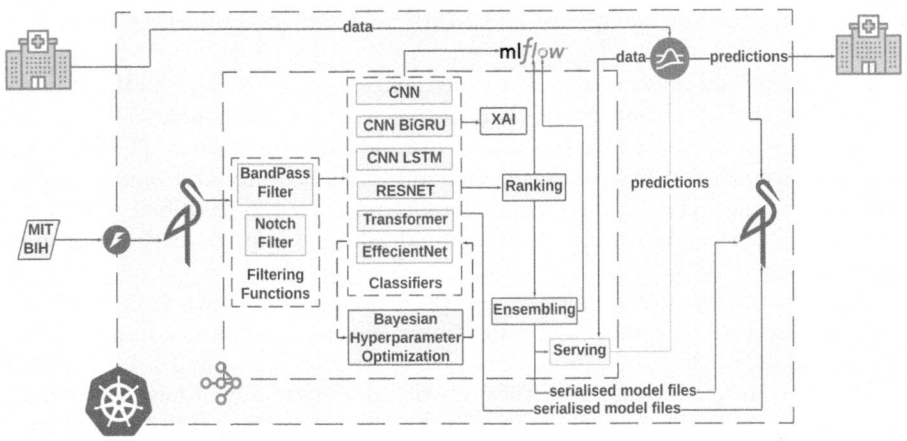

Fig. 1. ONNYX Framework

Table 1. Summary of ONNYX Architecture

Component	Technology	Details
Streaming	Apache Pulsar	**Rationale:** Utilizes a robust event streaming platform for managing real-time data flows **Performance Boost:** Delivers sub-second latency and high throughput for continuous data ingestion
Distributed Computing	Ray	**Rationale:** Leverages a distributed framework to enable concurrent task execution **Performance Boost:** Dynamically scales compute resources to reduce overall processing time
Preprocessing	Ray Data	**Rationale:** Employs distributed data handling to efficiently transform and augment datasets **Performance Boost:** Minimizes I/O and computation bottlenecks to enhance preprocessing throughput
Model Training	Ray Train, PyTorch, TensorFlow	**Rationale:** Integrates scalable deep learning libraries for training on varied hardware **Performance Boost:** Accelerates training cycles by distributing workloads across multiple nodes
HPO (Hyperparameter Optimization)	Ray Tune	**Rationale:** Uses advanced search algorithms to systematically optimize hyperparameters **Performance Boost:** Cuts down tuning iterations, conserving computational resources
Model Serving	Ray Serve	**Rationale:** Provides a modular framework for deploying production-grade ML models **Performance Boost:** Supports autoscaling and asynchronous processing for real-time inference
Explainability (XAI)	SHAP	**Rationale:** Integrates interpretability tools to elucidate model predictions and decisions **Performance Boost:** Enhances transparency and compliance by quantifying feature contributions
Orchestration	Kubernetes	**Rationale:** Automates container orchestration for streamlined service deployment **Performance Boost:** Boosts system resilience and uptime with dynamic scaling and self-healing capabilities
Model Tracking	MLflow	**Rationale:** Enables systematic logging and tracking of experiments and metadata **Performance Boost:** Facilitates reproducibility and comparative analysis of model versions
Storage	MinIO	**Rationale:** Employs a high-performance object storage system compatible with S3 APIs **Performance Boost:** Supports scalable, schema-on-read storage for efficient data access in ML workloads

2. Notch Filter: Notch filter removes narrow frequency interference, specifically powerline noise (60 Hz in the US, 50 Hz elsewhere).

Ray Data facilitates efficient I/O and preprocessing operations that scale across multiple nodes, overcoming data bottlenecks that typically arise when using in-memory single-machine compute via Pandas[5].

3.2   Model Training

Our system uses Ray Train to parallel-train multiple deep learning models (built with PyTorch[6]) across distributed CPUs and a dedicated GPU. This accelerates training and optimizes resource use by dynamically scaling tasks across clusters. We elaborate on the various classifiers, their respective architecture and the rationale behind using them in Table 2.

Table 2. Architectural Summary of Deep Learning Models for ECG Classification

Model	Architecture	Rationale
Simple1DCNN [15]	Stacked 1D Convolutional layers → Max Pooling → Fully Connected layer	Automatically extracts spatial hierarchies from raw signals, excelling in the detection of ECG arrhythmias *Generalizes well across signal qualities, robust to noise*
CNN-BiGRU	Convolutional layers → Max Pooling → Bidirectional GRU layer → Fully Connected layer	Combines spatial and temporal modeling. CNN detects morphology, BiGRU captures bidirectional dependencies *Improves sequence awareness in arrhythmia classification*
CNN-LSTM [16]	Convolutional layers → Max Pooling → Bidirectional LSTM layer → Fully Connected layer	Extracts spatial features via CNN and models temporal dependencies via LSTM *Effective for long ECG windows where rhythm and morphology matter*
EfficientNet1D [17]	Scaled 1D Convolutional blocks → Batch Normalization and Activation → Adaptive Average Pooling → Fully Connected layer	Balances depth, width, and resolution for optimal performance with fewer parameters *Efficient and accurate for real-time ECG applications*
ResNet1D [18]	Initial Convolution → Residual Blocks with skip connections → Deep stacking → Fully Connected layer	Residual connections prevent vanishing gradients *Captures fine-grained ECG variations while training deeper networks*
Transformer1D [19]	Initial Convolution → Transformer Encoder blocks → Multi-head Attention + Feed-forward layers → Fully Connected layer	Uses self-attention to capture global dependencies *Ideal for modeling long ECG sequences without recurrence*

---

[5] https://pandas.pydata.org/.

[6] https://pytorch.org/.

Hyperparameter Optimization. Ray Tune[7] uses Bayesian Hyperparameter Optimization for parallel execution to quickly find optimal classifier hyperparameters. It is more efficient than grid or random search since it uses a probabilistic model to explore promising options and prunes underperforming models early, saving computational resources.

Explainability (XAI). Integrating XAI is crucial for us when working with models trained on the MIT-BIH arrhythmia dataset, given the critical nature of medical AI. SHapley Additive exPlanations (SHAP) significantly enhances our model's interpretability, which is vital for achieving clinical acceptance and building trust. It helps us identify if our models are learning incorrect patterns (for example, from irrelevant ECG features) and also aids in amending any biases. SHAP also provides explanations for each patient, offering clinicians valuable insight into the rationale behind our model's specific ECG classifications.

3.3   Model Ranking

Instead of static weighting, we rank classifiers using a multi-armed bandit approach with Thompson Sampling. This allows us to dynamically balance exploring new ranking possibilities with exploiting previously successful ones, ensuring we always select the optimal classifier for predictions.

Multi-bandits Algorithm for Model Ranking. The algorithm begins with **Initialization** (line 2), setting success and failure counters $s_i = 0$ and $f_i = 0$ for each classifier i.

In the **Main Loop** (lines 3–14), repeated for $t = 1$ to T, it first performs **Sampling** (line 5) by drawing $\theta_i \sim \text{Beta}(s_i + 1, f_i + 1)$, modeling uncertainty using the Beta-binomial framework.

Then, it executes **Selection** (line 7), choosing $j = \arg\max_i \theta_i$, balancing exploration and exploitation.

The chosen classifier yields a binary **Reward** $r \sim \text{Binomial}(1, p_j)$ (line 8), and the algorithm **Updates Counters** (lines 9–12): incrementing s_j if $r = 1$, otherwise f_j.

Finally, after T rounds, the estimated success probability for each classifier is computed as

$$\hat{p}_i = \frac{s_i}{s_i + f_i + \epsilon}$$

with ϵ preventing division by zero. Classifiers are then ranked by $\hat{p}_i$ (line 15).

Each step systematically integrates new information, ensuring an optimal selection process through a balance of exploration and exploitation.

The rationale behind ranking models using Thompson sampling than just using metrics like F1 Score or Accuracy is that they are a point estimate and inform us how well a model did on a fixed test set, ignoring variance and lacking dynamic adjustment to account for possible data distributions drift.

---

[7] https://docs.ray.io/en/latest/tune/index.html

Algorithm 1. Thompson Sampling for Classifier Selection

1: **Input:** Number of classifiers n, number of rounds T, true reward probabilities $\{p_1, p_2, \ldots, p_n\}$, classifier names.
2: **Initialize:** For each classifier $i = 1, \ldots, n$, set

$$s_i \leftarrow 0 \quad \text{(successes)} \quad \text{and} \quad f_i \leftarrow 0 \quad \text{(failures)}.$$

3: **for** $t = 1$ **to** T **do**
4:     **for** $i = 1$ **to** n **do**
5:         Sample $\theta_i \sim \text{Beta}(s_i + 1, f_i + 1)$.
6:     **end for**
7:     Select classifier $j = \arg\max_i \theta_i$.
8:     Obtain reward $r \sim \text{Binomial}(1, p_j)$.
9:     **if** $r = 1$ **then**
10:         $s_j \leftarrow s_j + 1$.
11:     **else**
12:         $f_j \leftarrow f_j + 1$.
13:     **end if**
14: **end for**
15: **Output:** For each classifier i, compute the posterior mean:

$$\hat{p}_i = \frac{s_i}{s_i + f_i + \epsilon},$$

with a small $\epsilon > 0$ to avoid division by zero. Sort classifiers based on $\hat{p}_i$ in descending order.

3.4   Model Ensembling

In our approach, we selected the top $n = 3$ models from the Thompson Sampling Multi-Armed Bandit ranking, as supported by the ablation study in Table 5. To ensure numerical stability during weight computation, we introduced a small constant $\epsilon = 1 \times 10^{-6}$. The weight for each model was calculated using:

$$w = \frac{1}{(1 - \text{Accuracy}) + \epsilon}$$

This formulation assigns greater importance to more accurate models. We obtained class probabilities from each selected model on the test set X_{test}, and combined them using a weighted average. Final binary predictions were derived by applying a 0.5 threshold to the aggregated probabilities.

The PWPAE framework inspired our method [20], which emphasises dynamic model selection and performance-based weighting. By prioritising models with lower error rates, we enhanced the robustness and generalizability of the ensemble. We evaluated its performance using accuracy, precision, recall, and F1 score.

The rationale to use this ensembling technique over top-ranked classifier or a non weighted ensemble is that it gives higher weight to better models while still retaining contributions from others, thereby mitigating overconfidence and reducing the impact of model-specific bias and overfitting tendencies.

3.5   Model Management and Experiment Tracking

MLflow[8] handles model logging, tracking performance metrics, and managing serialized model files. It also allows data scientists to perform comparisons with various visualization tools.

3.6   Model Serving

ONNYX serves models via Ray Serve over an HTTP endpoint to enable low-latency, autoscalable inference as a Prediction-as-a-Service framework. To manage burst traffic and cold-start delays, Apache Pulsar[9] is used as a message queue between the client and the model endpoint. Predictions are published back to the client and forwarded to the data lake via Pulsar.

Unlike Apache Kafka, which uses polling and disk-based storage that can increase latency and overwhelm downstream services, Pulsar offers native queueing, in-memory caching, and multitopic routing—enabling controlled, efficient, and parallel data flow to clients and storage.

3.7   Kubernetes

Kubernetes[10] orchestrates ONNYX's distributed model services, offering scalability, reliability, and cloud-agnostic deployment across local or cloud infrastructures (AWS, GCP, Azure). It ensures dynamic load balancing, autoscaling, and fault tolerance. For local development, we used Minikube[11] with the Docker[12] driver, ensuring consistent testing across environments.

4   Experimental Results and Discussion

In this section, we examine the various experimental results obtained during the ECG classification process.

4.1   MIT-BIH Arrhythmia Dataset

The MIT-BIH Arrhythmia Database [21], a widely recognized resource for ECG analysis, contains 48 half-hour, two-channel ambulatory ECG recordings collected from 47 subjects between 1975 and 1979. The researchers digitized these recordings at 360 samples per second with 11-bit resolution, selecting a mix of random excerpts and clinically significant cases of arrhythmia.

In our study, we adopted a 70:30 intra-patient train-test split strategy, where ECG beats from the same patient may appear in both training and testing sets. This is consistent with several prior works that benchmark ECG arrhythmia classification performance on the MIT-BIH dataset using similar splitting strategies [3,4].

---

[8] https://mlflow.org/.
[9] https://pulsar.apache.org/.
[10] https://kubernetes.io/.
[11] https://minikube.sigs.k8s.io/docs/.
[12] https://www.docker.com/.

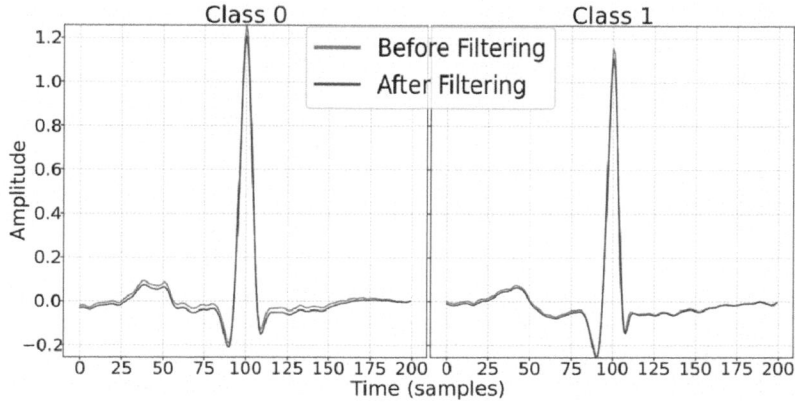

Fig. 2. Signal Before and After Filtering

Figure 2 demonstrates the impact of signal processing using a bandpass and notch filter on each class within the dataset. This filtering procedure enhances the signal by mitigating baseline drift, attenuating high-frequency noise, and diminishing power line interference, while maintaining critical electrocardiogram characteristics.

Table 3. Thompson Ranked Classifier Metrics on Binary Classification

Rank	Name	Estimated Reward	Accuracy	Precision	Recall	F1 Score	Hyperparameters
1	ResNet	0.9927	0.9920	0.9689	0.9558	0.9623	lr = 0.0016
2	CNN-LSTM	0.9821	0.9810	0.9515	0.8673	0.9075	conv = 8.4945, lstm = 25.5679, lr = 0.0095
3	EfficientNet	0.9779	0.9816	0.9623	0.8618	0.9093	lr = 0.0038
4	Transformer	0.9730	0.9766	0.9315	0.8440	0.8856	lr = 0.0007
5	CNN-BiGRU	0.7778	0.9823	0.9427	0.8888	0.9150	conv = 8.4945, gru = 30.8171, lr = 0.0073
6	1D-CNN	0.5000	0.9725	0.9522	0.7832	0.8595	conv = 9.1833, lr = 0.0030

Table 3 illustrates the results of various models along with their respective rankings. Our analysis covers six deep learning methodologies, executed concurrently within the distributed ONNYX framework. We observed that ResNet requires a substantially longer duration for training in comparison to other models, whereas the 1D-CNN model demonstrates the shortest training time. Additionally, we recognize that sophisticated models like Transformers also require considerable training time. The implemented ranking method positions ResNet, CNN-LSTM, and EfficientNet as the top three models. This ranking strategy facilitates the identification of models that may be integrated with probabilities to advance in the ensemble process.

Table 4. Thompson Ranked Classifier Metrics on Multiclass Classification

Rank	Name	Estimated Reward	Accuracy	Precision	Recall	Hyperparameters
1	ResNet	0.9914	0.9911	0.9911	0.9911	lr = 0.001644585
2	Transformer	0.9869	0.9868	0.9864	0.9868	lr = 0.000675028
3	1D-CNN	0.9863	0.9804	0.9795	0.9804	conv = 9.1833, lr = 0.002983168
4	EfficientNet	0.9757	0.9838	0.9831	0.9838	lr = 0.003807947
5	CNN-LSTM	0.9388	0.9326	0.8986	0.9326	conv = 8.4945, lstm = 25.5679, lr = 0.009512072
6	CNN-BiGRU	0.9369	0.9115	0.8768	0.9115	conv = 8.4945, gru = 30.8171, lr = = 0.00734674

To demonstrate the framework's capability, we also perform multiclass classification, and the respective metrics are depicted in Table 4. While we do see a
shift in the rankings of the classifiers, given the dynamic nature of the pipeline,
the best-performing ensemble will ultimately be put into production. We have
presented upcoming results on binary and multiclass classification to depict the
capability of ONNYX in detecting arrhythmia on ECG data streams.

Figure 3 illustrates the accuracy of deep learning models, demonstrating
marked improvements in classification performance in various epochs for both
Binary and Multi Class classifiers (Table 6).

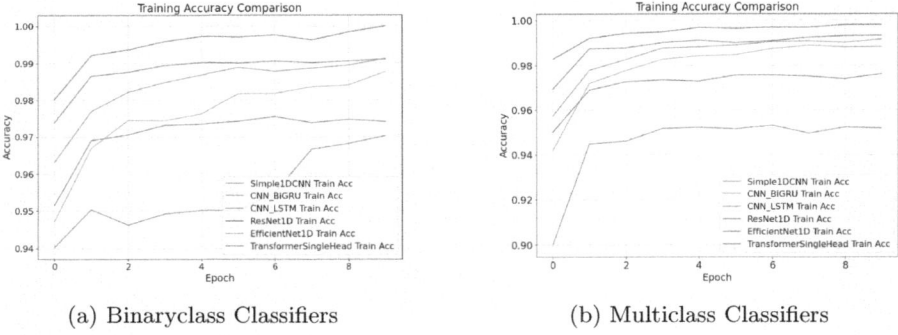

(a) Binaryclass Classifiers   (b) Multiclass Classifiers

Fig. 3. Change in Accuracy Epoch in Binary and Multiclass Classifiers

Table 5. Thompson Ranked Weighted Ensemble Binary Classifier Metrics

Ensemble Size	Models	Model Weights	Accuracy	Precision	Recall	F1-Score
2	ResNet,EfficientNet	0.7096,0.2904	0.9925	0.9761	**0.9533**	0.9646
3	**ResNet,EfficientNet,CNN-LSTM**	0.5713,0.2338,0.1950	**0.9927**	**0.9804**	0.9509	**0.9654**
4	ResNet,EfficientNet,CNN-LSTM,1DCNN	0.4868,0.1992,0.1661,0.1479	0.9923	0.9785	0.9490	0.9635

Table 6. Thompson Ranked Weighted Ensemble Multiclass Classifier Metrics

Ensemble Size	Models	Model Weights (Normalized)	Accuracy	Precision	Recall	F1-Score
2	**ResNet, Transformer**	0.5983, 0.4017	**0.9916**	**0.9915**	**0.9916**	**0.9915**
3	ResNet, Transformer, 1D-CNN	0.4707, 0.3161, 0.2132	0.9909	0.9907	0.9909	0.9907
4	ResNet, Transformer, 1D-CNN, EfficientNet	0.3741, 0.2512, 0.1694, 0.2053	0.9912	0.9910	0.9912	0.9910

Table 5 shows the results of the different ensembles created from the top-ranked models identified in Table 3 for binary classification. We conducted an ablation analysis using two, three and four models to determine the optimal ensemble combination that may improve classification performance. In addition, custom weights were formulated and integrated into each model to facilitate improved convergence. We observed that the ensemble comprising three methodologies, namely ResNet, EfficientNet, and CNN-LSTM, demonstrated superior performance across all metrics, achieving an exceptional accuracy of 99.27%. Performance variations among the various ensembles fluctuated within a margin of 0.01% to 0.03%, indicating enhanced performance through the majority voting ensembles. In addition, Table 6 presents the same analysis for a multiclass scenario. In this table, we observe that the ensemble size of 2 outperforms all other ensembles and thus is put to production.

Table 7. Inference Metrics for Top 3 Binary Ensemble Classifier

Signals	Total Time (s)	Throughput (SPS)	Avg Latency (ms)	Min Latency (ms)	Max Latency (ms)	95% Latency (ms)
50	1.5083	33.1490	30.1626	14.3369	60.9237	58.9972
100	3.9701	25.1880	39.6973	13.7745	65.6122	65.3230
200	7.7110	25.9371	38.5503	13.8197	73.2426	65.5490
500	19.8225	25.2239	39.6403	13.9686	72.7405	65.2281
1000	40.3844	24.7621	40.3795	14.1674	67.4932	65.5553

Table 7 presents the inference metrics for the classifier of the highest performing binary ensemble, while being served on Ray, evaluated using increasing request sizes from 50 to 1000. We observe that the average latency ranges from 30.16 ms (50 requests) to 40.38 ms (1000 requests). At the 95th percentile latency also remains consistent at 65 ms. The maximum latency increases slightly with load, peaking at 73 ms, but does not indicate exponential degradation. Throughput (signals per second, SPS) falls slightly as the number of requests increases from 33.15 SPS, at 50 requests to 24.76 SPS, at 1000 requests. Similarly Table 8 presents the top-performing multiclass ensemble classifier. We observe that the classifier maintains stable performance under increasing load, with average latency remaining around 32–35 ms and 95th percentile latency consistently below 57 ms, supporting real-time inferencing. Compared to the binary classifier (Table 7), the multiclass classifier (Table 8) exhibits slightly lower 95th percentile latency and more stable throughput under load.

Table 8. Inference Metrics for Top 2 Multiclass Ensemble Classifier

Signals	Total Time (s)	Throughput (SPS)	Avg Latency (ms)	Min Latency (ms)	Max Latency (ms)	95% Latency (ms)
50	1.7333	28.8462	34.6614	10.3165	57.7038	57.4033
100	3.2563	30.7096	32.5583	10.2194	58.3754	57.2018
200	6.7294	29.7203	33.6428	10.5564	57.7880	56.3047
500	15.9278	31.3917	31.8512	9.8152	58.0600	56.3062
1000	34.3374	29.1228	34.3327	9.6526	61.5697	56.5906
2000	69.6609	28.7105	34.8254	9.5381	59.2717	56.9662

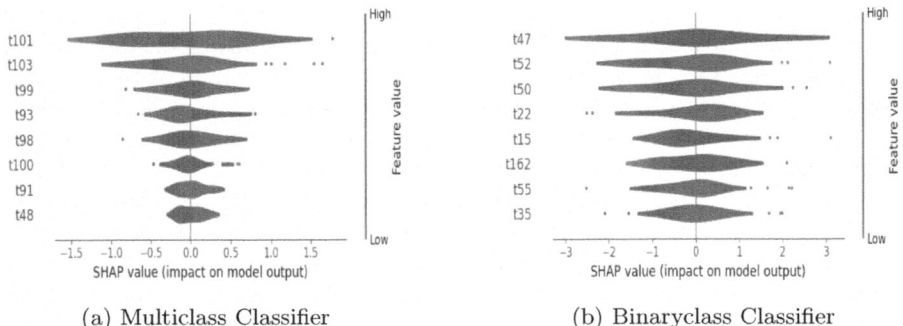

(a) Multiclass Classifier   (b) Binaryclass Classifier

Fig. 4. XAI Plot for RESNET

Figure 4a and Fig. 4b are violin summary plots for RESNET which offers compact representation of the distribution and variability of SHAP values for each feature. The plots above illustrate the top 8 features. Shap values on the X-axis correspond to the importance of the element in the prediction of the model. From Fig. 4a, we learn that on the multi-class model, features t101, t103, and t99 dominate the influence of the model. While from Fig. 4b we observe that on the binary class model, features t47, t52, t50 dominate the influence of the model. The binary model shows a wider shap value spread, indicating stronger feature influence on predictions, while the multi class model showcases a narrow spread, indicating lower variability in feature influence across different predictions. The contrasting SHAP distributions suggest that the binary model relies more heavily on specific features for decision-making, while the multi-class model distributes importance more uniformly across features.

We compare ONNYX to existing benchmarks in the field of ECG classification in Table 9.

Table 9. Performance Comparison of ONNYX with Benchmark Models

Study/Model	Accuracy	Type	Remarks
ONNYX	99.27%	Binary & Multiclass	**Approach:** Ensemble of DL models with adaptive selection (Thompson sampling)
			Key Features: Real-time Kubernetes deployment, SHAP explainability, Apache Pulsar
			Limitations: High computational complexity
EBM [22]	96.84%	Binary	**Approach:** Explainable Boosting Machine
			Key Features: Uses a lightweight, interpretable EBM on resource-constrained edge devices
			Limitations: Lower accuracy, relying on handcrafted feature extraction
CNN-LSTM-SE [23]	98.5%	Multiclass	**Approach:** CNN + LSTM + SE attention
			Key Features: Uses ensemble empirical mode decomposition (EEMD) preprocessing; SE attention over channels
			Limitations: Lower accuracy, slower LSTM, sensitive to imbalance
1D ResNet [24]	98.63%	Multiclass	**Approach:** Residual 1D CNN with SMOTE
			Key Features: 6 conv layers + 3 pooling; high specificity (99.06%)
			Limitations: Lower sensitivity (92.4%), fixed architecture
MB-MHA-TCN [25]	98.75%	Multiclass	**Approach:** Multi-branch Temporal Convolutional Networks + Multi-head Attention
			Key Features: Parallel dilated branches, focal loss, Bayesian tuning
			Limitations: High memory/computation

5   Limitations and Future Work

This study introduces a modular and reproducible framework intended for the binary classification of ECG signals. We recognize the potential for integrating this framework within a real-time system, thereby simulating its applicability in real-world environments. However, several limitations or challenges encountered during deployment are elaborated on in the following points:

– ONNYX has been deployed on Kubernetes using the Docker driver for local development for distributed setup. We noted that the platform extensively uses RAM for storing the intermediate results which claims much of memory resources. This does not affect the predictions as the training and testing times are maintained well.
– Adapting the system to handle heterogeneous data sources will need writing of custom pre-processing jobs.

Our future initiatives include the integration of tools for scheduling and event-based job orchestration, as well as the enhancement of observability tools to improve system administration. Moreover, we intend to incorporate components that define our framework into an Infrastructure as Code (IaC) tool, thereby rendering deployment processes more adaptable to future advancements.

While ONNYX is designed to process diverse ECG datasets, as noted in the limitations, modifications to preprocessing jobs might be necessary depending on the structure of incoming data. By modifying ONNYX to avoid storing intermediate results extensively in RAM, we can accommodate even larger volumes of data in memory-constrained setups, leading to a more cost-effective solution. For end-to-end integration with a hospital in a real-world setup, since the model is offered via an HTTP API endpoint, a hospital would just need to provide a way for the system to access their data and get results. The modularity offered by ONNYX also facilitates the setup of a custom frontend.

6   Conclusion

We presented ONNYX, a big data analytics framework for the classification of ECG data streams with an open source modular architecture. Our framework used deep learning to detect heart arrhythmias and allows components to be interchanged for various ECG use cases. We integrate Thompson Sampling with probabilistic weights to find optimal methods. Using a ranking strategy, we achieved 99.27% accuracy in binary and 99.16% accuracy in multiclass with a majority voting ensemble of top-ranked approaches selected by the multibandit approach. We presented architectural details, hyperparameters, and comparisons with state-of-the-art models, including multiclass performance. ONNYX supports adaptive ensembling, enabling model combinations to evolve dynamically during real-world deployment, with the architecture accommodating these changes. XAI with Shapley plots provides insight into the model's predictions.

References

1. Li, Z., et al.: Comparative analysis of atherosclerotic cardiovascular disease burden between ages 20–54 and over 55 years: insights from the global burden of disease study 2019. BMC Med. **22**(1), 303 (2024)
2. Rabadia, J.P., Thite, V.S., Desai, B.K., Bera, R.G., Patel, S.: Cardiovascular system, its functions and disorders. In: Pullaiah, T., Ojha, S. (eds.) Cardioprotective Plants, pp. 1–34. Springer, Singapore (2024). https://doi.org/10.1007/978-981-97-4627-9_1
3. Shoughi, A., Dowlatshahi, M.B., Amiri, A., Kuchaki Rafsanjani, M., Batth, R.S.: Automatic ECG classification using discrete wavelet transform and one-dimensional convolutional neural network. Computing **106**(4), 1227–1248 (2024)
4. Prem Narayan Singh and Rajendra Prasad Mahapatra: A novel deep learning approach for arrhythmia prediction on ECG classification using recurrent CNN with GWO. Int. J. Inf. Technol. **16**(1), 577–585 (2024)
5. Peimankar, A., Ebrahimi, A., Wiil, U.K.: xECG-beats: an explainable deep transfer learning approach for ECG-based heartbeat classification. Netw. Model. Anal. Health Inform. Bioinform. **13**(1), 45 (2024)
6. Ying, Z., Zhang, G., Pan, Z., Chu, C., Liu, X.: FedECG: a federated semi-supervised learning framework for electrocardiogram abnormalities prediction. J. King Saud Univ. Comput. Inf. Sci. **35**(6), 101568 (2023)
7. Zhang, X., et al.: Automated detection of cardiovascular disease by electrocardiogram signal analysis: a deep learning system. Cardiovas. Diagn. Ther. **10**(2), 22735–22235 (2020)
8. Weston Hughes, J., et al.: Performance of a convolutional neural network and explainability technique for 12-lead electrocardiogram interpretation. JAMA Cardiol. **6**(11), 1285–1295 (2021)
9. Khan, F., Yu, X., Yuan, Z., Rehman, A.: ECG classification using 1-d convolutional deep residual neural network. PLOS ONE **18**(4), e0284791 (2023)
10. Chung, C.T., et al.: Clinical significance, challenges and limitations in using artificial intelligence for electrocardiography-based diagnosis. Int. J. Arrhythmia **23**(1), 24 (2022)

11. Tison, G.H., Zhang, J., Delling, F.N., Deo, R.C.: Automated and interpretable patient ECG profiles for disease detection, tracking, and discovery. Circ. Cardiovas. Qual. Outcomes, **12**(9) (2019)

12. Somani, S., et al.: Deep learning and the electrocardiogram: review of the current state-of-the-art. EP Europace **23**(8), 1179–1191 (2021)

13. Mishra, S., Jain, M., Siva Naga Sasank, B., Hota, C.: An ingestion based analytics framework for complex event processing engine in internet of things. In: Mondal, A., Gupta, H., Srivastava, J., Reddy, P.K., Somayajulu, D.V.L.N. (eds.) BDA 2018. LNCS, vol. 11297, pp. 266–281. Springer, Cham (2018). https://doi.org/10.1007/978-3-030-04780-1_18

14. Mishra, S., Hota, C.; A rest framework on IoT streams using apache spark for smart cities. In: 2019 IEEE 16th India Council International Conference (INDICON), pp. 1–4 (2019)

15. LeCun, Y., Bottou, L., Bengio, Y., Haffner, P.: Gradient-based learning applied to document recognition. Proc. IEEE **86**(11), 2278–2324 (1998)

16. Shi, X., Chen, Z., Wang, H., Yeung, D.Y., Wong, W.K., Woo, W.C.: Convolutional LSTM network: a machine learning approach for precipitation nowcasting. Adv. Neural Inf. Proces. Syst. **28** (2015)

17. Tan, M., Le, Q.: EfficientNet: rethinking model scaling for convolutional neural networks. In: Chaudhuri, K., Salakhutdinov, R., (eds.) Proceedings of the 36th International Conference on Machine Learning. Proceedings of Machine Learning Research, vol. 97 pp. 6105–6114. PMLR (2019)

18. He, K., Zhang, X., Ren, S., Sun, J.: Deep residual learning for image recognition. In: Proceedings of the IEEE Conference on Computer Vision and Pattern Recognition (CVPR) (2016)

19. Vaswani, A., et al.: Attention is all you need. In: Guyon, I., et al (eds.) Advances in Neural Information Processing Systems, vol. 30. Curran Associates, Inc., (2017)

20. Yang, L., Manias, D.M., Shami, A.: PWPAE: an ensemble framework for concept drift adaptation in IoT data streams. In: 2021 IEEE Global Communications Conference (globecom), pp. 01–06. IEEE (2021)

21. Goldberger, A.L., et al.: The MIT-BIH long term database (1992)

22. Xiaolin, L., Qingyuan, W., Panicker, R.C., Cardiff, B., John, D.: Binary ECG classification using explainable boosting machines for IoT edge devices. In: 2022 29th IEEE International Conference on Electronics, Circuits and Systems (ICECS), pp. 1–4 (2022)

23. Sun, A., Hong, W., Li, J., Mao, J.: An arrhythmia classification model based on a CNN-LSTM-se algorithm. Sensors **24**(19), 6306 (2024)

24. Cretu, I., Tindale, A., Abbod, M., Khir, A., Balachandran, W., Meng, H.: Reliable multimodal heartbeat classification using deep neural networks (2023)

25. Bi, S., Rongjian, L., Qiang, X., Zhang, P.: Accurate arrhythmia classification with multi-branch, multi-head attention temporal convolutional networks. Sensors **24**(24), 8124 (2024)

SpaPool: Soft Partition Assignment Pooling for Graph Neural Networks

Rodrigue Govan[1]([✉])(iD), Romane Scherrer[1](iD), Philippe Fournier-Viger[2](iD),
and Nazha Selmaoui-Folcher[1](iD)

[1] Institute of Exact and Applied Sciences (EA7484), University of New Caledonia,
BP R4, 98851 Nouméa, New Caledonia
`rodrigue.govan@gmail.com, nazha.selmaoui@unc.nc`
[2] Big Data Institute, College of Computer Science and Software Engineering,
Shenzhen University, Shenzhen, China
`philfv@szu.edu.cn`

Abstract. This paper introduces SpaPool, a novel pooling method that combines the strengths of both dense and sparse techniques for a graph neural network. SpaPool groups vertices into an adaptive number of clusters, leveraging the benefits of both dense and sparse approaches. It aims to maintain the structural integrity of the graph while reducing its size efficiently. Experimental results on several datasets demonstrate that SpaPool achieves competitive performance compared to existing pooling techniques and excels particularly on small-scale graphs. This makes SpaPool a promising method for applications requiring efficient and effective graph processing.

Keywords: attributed graphs · neural networks · graph pooling

1 Introduction

Graph Neural Networks (GNNs) offer great flexibility for processing unstructured data, making them well-suited to domains such as chemistry, economics, and epidemiology, where instances are often interdependent. These instances typically exhibit attribute correlations with neighboring vertices, forming a topological structure that must be captured in networks of interrelated entities. While graphs are effective for modeling complex data, processing high-dimensional graphs poses significant challenges in terms of computational efficiency and structural preservation. To address this, pooling methods such as DiffPool [13] and TopKPool [2] have been introduced. These approaches reduce graphs by selecting representative vertices while retaining structural information. However, they show limitations when applied to heterogeneous graphs, which vary widely in size and topology. Most existing pooling methods fall into one of two categories: dense pooling, which clusters vertices into a fixed number of supernodes defined by the user, and sparse pooling, which retains a user-defined fraction $k \in (0, 1]$ of

© The Author(s), under exclusive license to Springer Nature Switzerland AG 2026
C. K. Leung et al. (Eds.): DaWaK 2025, LNCS 16048, pp. 332–340, 2026.
https://doi.org/10.1007/978-3-032-02215-8_27

vertices based on importance scores, discarding the rest. To overcome these limitations, we propose SPAPOOL (Soft Partition Assignment Pooling), a novel pooling method that integrates the strengths of both paradigms. SPAPOOL performs vertex clustering with an adaptive number of representative vertices, dynamically adjusted to each graph's specific structure.

The article is organized as follows. The next section reviews a brief state of the art on graph neural networks and pooling methods. Section 3 details the proposed SPAPOOL method, while Sect. 4 describes the experimental validation against existing pooling techniques. Section 5 presents results from multiple datasets. Finally, the last section concludes and discusses future perspectives.

2 Related Work

Let $G = (V, E, X)$ be an attributed graph such that $V = \{v_1, \ldots, v_N\}$ is the set of N vertices, $E \subseteq \{(v_i, v_j) \in V^2 \mid \forall\, i, j \in \{1, \ldots, N\} \wedge i \neq j\}$ is the set of m edges, and $X = \{x_i \in \mathbb{R}^F \mid \forall\, i \in \{1, \ldots, N\}\}$ is the attribute matrix where x_i represents the vector of F attributes associated to vertex v_i.

Graph Convolution. To our knowledge, the first formal graph convolutional network (GCN) was introduced near the end of the last decade [4]. This model applies deep learning directly to graph-structured data, avoiding information loss that can occur when transforming graphs into tabular formats. Inspired by the first-order Laplacian, the GCN layer-wise propagation rule is defined as:

$$H^{(\ell+1)} = \sigma\left(\tilde{D}^{-\frac{1}{2}}\,\tilde{A}\,\tilde{D}^{-\frac{1}{2}}\,H^{(\ell)}\,W^{(\ell)}\right) \tag{1}$$

where $\tilde{A} = A + I_N$ is the adjacency matrix with self-loops added via the identity matrix I_N, $H^{(\ell)}$ is the vertex embedding matrix at layer ℓ, and $W^{(\ell)}$ is a trainable weight matrix applying a linear transformation to $H^{(\ell)}$. The diagonal degree matrix $\tilde{D}$ is defined by $\tilde{D}_{i,i} = \sum_{j=1}^{N} \tilde{A}_{i,j}$. The function $\sigma(\cdot)$ denotes an activation function, such as $\mathrm{ReLU}(\cdot) = \max(0, \cdot)$. Initially, at layer $\ell = 0$, we have $H^{(0)} = X$. In the literature, numerous GNN models have appeared, such as GRAPHSAGE [3] which proposed a convolutional layer that aggregates a subset of neighbor vertices, GAT [11] which aggregates neighbor vertices according to an importance score, and GIN [12] which employs an injective aggregation function in order to capture a richer structural information. GNNs are designed to process graph-structured data and typically address three main tasks: vertex prediction (labeling vertices within a graph), edge prediction (identifying relationships between vertices), and graph prediction (classifying entire graphs). Although GNNs have been applied to tasks such as graph matching [6], this paper focuses on graph classification. As large graphs often pose computational challenges, it is essential to reduce their size while preserving the most relevant information during training.

Graph Pooling. This paper focuses on hierarchical pooling approaches, which preserve graph topology, with particular emphasis on trainable pooling designed

specifically for GNNs. Pooling in GNNs involves three steps: (1) selecting significant vertices via an operator S, which can be a matrix or vector—often a sparse
assignment matrix defining supernodes (centroids of vertex groups); (2) reducing
the vertex embedding matrix H using S; and (3) adjusting the adjacency matrix
A to remain consistent with the reduced H.

Among trainable hierarchical methods, pooling approaches can be divided
into dense and sparse techniques. Dense methods such as DIFFPOOL [13], MIN-
CUT [1] and DMoN [10], aim to group subsets of vertices into a fixed number
of supernodes whose cardinality is $O(N)$. These supernodes represent an aggregation of both local and global information. Although our proposed method is
dense, this paper focuses on sparse methods from the literature, as our method
reduces the input graph adaptively as sparse methods.

Unlike dense pooling methods, sparse pooling directly selects a subset of vertices based on criteria such as importance scores, producing supernodes with
cardinality $O(1)$. This requires setting a ratio $k \in (0, 1]$ to retain approximately
$\lceil kN \rceil$ vertices, reducing the graph while preserving its local structure. Several sparse methods have been proposed, such as TOPKPOOL [2] which defines
the importance score according to a trainable normalized weight vector, SAG-
POOL [5] which employs a self-attention mechanism to calculate an importance
score vector, and ASAPOOL [9] which uses attention mechanisms to dynamically
identify and group coherent local substructures. After computing importance
scores, most sparse pooling methods select the top $\lceil kN \rceil$ vertices with the highest scores and remove the corresponding rows and columns from the adjacency
matrix [2, 5, 9].

3   SPAPOOL: A Dense but Adaptive Pooling Approach

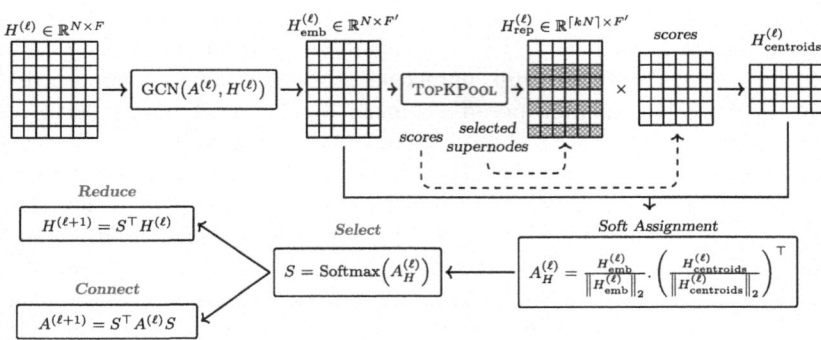

Fig. 1. Illustration of the proposed graph pooling method (SPAPOOL).

In this article, we propose SPAPOOL (Soft Partition Assignment Pooling). It
combines adaptive vertex selection and grouping to preserve graph information

(Fig. 1) by selecting significant vertices as centroids, as sparse methods do, and grouping vertices to maintain graph structure, as dense methods do.

Select. To define our operator S, we proceed as follows. At layer ℓ, the attributed graph is represented by the adjacency matrix $A^{(\ell)} \in \mathbb{R}^{N \times N}$ and the vertex embedding matrix $H^{(\ell)} \in \mathbb{R}^{N \times F}$. Each vector in row $h_i^{(\ell)}$ in the vertex embedding matrix denotes embeddings of vertex v_i in the graph. The layer-wise process of SpaPool at layer ℓ proceeds as follows:

$$H_{\text{emb}}^{(\ell)} = \text{GCN}\left(A^{(\ell)}, H^{(\ell)}\right) \; ; \; H_{\text{rep}}^{(\ell)}, H_{\text{scores}}^{(\ell)} = \text{TopKPool}\left(H_{\text{emb}}^{(\ell)}, k\right)$$

$$H_{\text{centroids}}^{(\ell)} = H_{\text{rep}}^{(\ell)} H_{\text{scores}}$$

$$S = \text{Softmax}\left[\frac{H_{\text{emb}}^{(\ell)}}{\left\|H_{\text{emb}}^{(\ell)}\right\|_2} \cdot \left(\frac{H_{\text{centroids}}^{(\ell)}}{\left\|H_{\text{centroids}}^{(\ell)}\right\|_2}\right)^{\top}\right] \tag{2}$$

where S applies a Softmax defined by $\text{Softmax}(z)_i = \exp(z_i) \cdot \left(\sum_{j=1}^{K} \exp(z_j)\right)^{-1}$.

The operator S in SpaPool (Eq. 2) is a Softmax applied to the cosine similarity between representatives and vertex embeddings from the GCN layer.

Reduce and Connect. In reduction and connection steps, SpaPool readjusts the attribute and adjacency matrices as existing pooling methods [1, 8, 13], respectively defined by:

$$H^{(\ell+1)} = S^{\top} H^{(\ell)} \tag{3} \qquad\qquad A^{(\ell+1)} = S^{\top} A^{(\ell)} S \tag{4}$$

Auxiliary Loss. As SpaPool is based on a dense approach (i.e., grouping vertices into supernodes), we considered an auxiliary loss function that we added into the classification loss function, i.e., the cross-entropy function (Eq. 5). Because we group vertices into supernodes, it can become difficult for our pooling method to avoid local minima issues. Therefore, in the classification loss function (cross-entropy), we added the same auxiliary losses used in DiffPool [13]:

$$\mathcal{L}_{\text{CE}} = -\sum_{i=1}^{N} y_i \log(\hat{y}_i) \tag{5}$$

$$\mathcal{L}_{\text{DiffPool}} = \mathcal{L}_{\text{LP}} + \mathcal{L}_{\text{E}} = \left[\left\|A^{(\ell)}, SS^{\top}\right\|_F\right] + \left[\frac{1}{N}\sum_{i=1}^{N} E(S_i)\right] \tag{6}$$

where y_i and $\hat{y}_i$ are respectively the observed and predicted classes of graph i, $\|\cdot\|_F$ denotes the Frobenius norm, and $E(\cdot)$ is the entropy function.

4 Experimental Validation

This section outlines the methodology used to evaluate SPAPOOL against existing pooling methods. A GNN model was designed with only the pooling layer varied across experiments. We then present the datasets used for evaluation.

Graph Neural Network. To isolate the impact of the pooling layer, the GNN model (Fig. 2) is composed of two MLP blocks and two GCN blocks. Only one layer concerns the pooling layer allowing us to compare pooling methods.

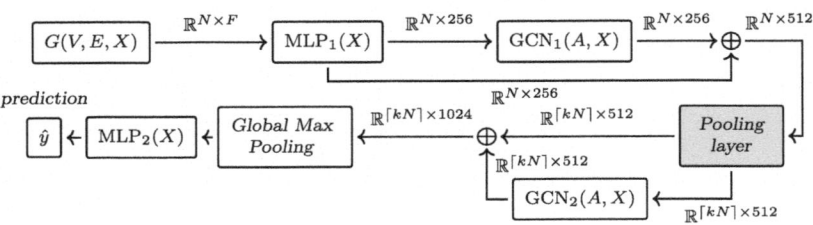

Fig. 2. Graph Neural Network model used in the experiments.

Datasets. Each dataset (Table 1) was randomly split into 80% training, 10% validation, and 10% test sets. These datasets were chosen for their variety in terms of number of graphs, vertices, attributes and classes. Models were trained with a batch size of 64, learning rate of 5×10^{-4}, pooling ratio $k = 0.5$, using SGD for up to 500 epochs with early stopping after 100, and conducted using Python 3.8 with `PyTorch` 1.13.1 and `PyTorch-Geometric` 2.6.1 libraries.

Table 1. Summary of datasets [7] used in our graph classification experiments.

Id	Dataset	Graphs	Vertices (avg)	Edges (avg)	Attributes	Classes
[#01]	PROTEINS	1,113	$39.06_{\pm 45.76}$	$72.82_{\pm 84.6}$	4	2
[#02]	ENZYMES	600	$32.63_{\pm 15.28}$	$62.14_{\pm 25.5}$	21	6
[#03]	DD	1,178	$284.32_{\pm 272}$	$715.66_{\pm 693.91}$	89	2
[#04]	Mutagenicity	4,337	$30.32_{\pm 20.12}$	$30.77_{\pm 16.82}$	14	2
[#05]	github_stargazers	12,725	$113.79_{\pm 164}$	$234.64_{\pm 427.23}$	1	2
[#06]	reddit_threads	203,088	$23.93_{\pm 16.55}$	$24.85_{\pm 19.14}$	1	2
[#07]	OHSU	79	$82.01_{\pm 43.44}$	$199.66_{\pm 165.08}$	190	2
[#08]	twitch_egos	127,094	$29.67_{\pm 11.1}$	$86.59_{\pm 70.37}$	1	2
[#09]	COLLAB	5,000	$74.49_{\pm 62.3}$	$2457.22_{\pm 6438.92}$	1	3
[#10]	IMDB-BINARY	1,000	$19.77_{\pm 10.06}$	$96.53_{\pm 105.6}$	1	2

5   Results

This section presents results on the tested datasets (Table 1) and an ablation study of SPAPOOL's components, a common approach in GNN pooling [2,9].

Performance Comparison. A GNN model (Fig. 2) was trained ten times using ten different random splits to assess accuracy variability across datasets displayed in Table 1. All experiments were conducted using one NVIDIA Tesla V100 GPU with 32 GB of dedicated memory. In terms of time performance, on the `twitch_egos` dataset, SPAPOOL required approximately 21 h to complete ten training runs, which is comparable to ASAPOOL, which took 23 h. In contrast, TOPKPOOL and SAGPOOL completed the same number of runs in 10 and 13 h, respectively. This difference stems from the pooling architectures: SPAPOOL/ASAPOOL involve heavier matrix operations than TOP-KPOOL/SAGPOOL. Specifically, SPAPOOL performs matrix computations to determine vertex centroids, while ASAPOOL applies a graph convolution that aggregates each vertex's neighbors to compute importance scores. In contrast, TOPKPOOL and SAGPOOL rely solely on a trainable weight vector to compute vertex importance scores [2,5].

Table 2. Accuracies on the graph classification experiments. The numbers in the header refer to the corresponding dataset in Table 1.

	[#01]	[#02]	[#03]	[#04]	[#05]	[#06]	[#07]	[#08]	[#09]	[#10]
TOPKPOOL	$68_{\pm 11}$	$63_{\pm 5}$	$75_{\pm 5}$	$75_{\pm 3}$	$65_{\pm 2}$	$76_{\pm 2}$	$61_{\pm 17}$	$67_{\pm 3}$	$66_{\pm 4}$	$64_{\pm 5}$
ASAPOOL	$\mathbf{72_{\pm 9}}$	$66_{\pm 4}$	$76_{\pm 4}$	$76_{\pm 2}$	$\mathbf{68_{\pm 1}}$	$\mathbf{77_{\pm 1}}$	$55_{\pm 19}$	$70_{\pm 1}$	$67_{\pm 4}$	$69_{\pm 3}$
SAGPOOL	$71_{\pm 10}$	$55_{\pm 6}$	$\mathbf{77_{\pm 4}}$	$75_{\pm 4}$	$67_{\pm 2}$	$77_{\pm 1}$	$\mathbf{63_{\pm 15}}$	$\mathbf{70_{\pm 0}}$	$\mathbf{68_{\pm 3}}$	$\mathbf{69_{\pm 5}}$
SPAPOOL	$71_{\pm 8}$	$\mathbf{68_{\pm 5}}$	$72_{\pm 0}$	$\mathbf{77_{\pm 2}}$	$65_{\pm 4}$	$76_{\pm 1}$	$54_{\pm 20}$	$69_{\pm 1}$	$55_{\pm 2}$	$56_{\pm 4}$

Although SPAPOOL outperformed existing methods only on 2 of the 10 datasets, it consistently achieved comparable results with low variability across ten training runs. Notably, since SPAPOOL is TOPKPOOL-based for vertex scoring, it outperformed this baseline on 5 of the 10 datasets. Its advantage was more pronounced on datasets with smaller graphs (averaging 30 vertices), where TOP-KPOOL removes less significant vertices, while SPAPOOL preserves structure by grouping vertices into supernodes. In such cases, where every vertices contribute crucial information, dense pooling methods like SPAPOOL prove more effective than sparse methods.

Effect of Vertex Selection. We assessed the impact of the vertex selection method in SPAPOOL by comparing TOPKPOOL and SAGPOOL. SAGPOOL applies a GCN layer (Eq. 1) to the vertex embeddings before selecting important vertices, while TOPKPOOL uses a normalized weight vector for this purpose. Table 3 shows that using TOPKPOOL appeared to optimize performance compared to SAGPOOL, as it decreased accuracies in the tested datasets.

Effect of Vertex Aggregation. We further assessed the importance of the vertex aggregation method by comparing the cosine similarity with the scalar product and the attention mechanism, respectively defined as:

$$A_H^{(\ell)} = \text{Softmax}\left(H^{(\ell)} \, H_{\text{rep}}^{(\ell)^\top}\right)$$

$$A_H^{(\ell)} = \text{Softmax}\left(\frac{H_q^{(\ell)} \, H_k^{(\ell)^\top}}{\sqrt{\alpha}}\right) \; ; \; H_q^{(\ell)} = \text{MLP}\left(H^{(\ell)}\right) \text{ and } H_k^{(\ell)} = \text{MLP}\left(H_{\text{rep}}^{(\ell)}\right)$$

where we set $\alpha = 16$, $H_q^{(\ell)} \in \mathbb{R}^{N \times \alpha}$, and $H_k^{(\ell)} \in \mathbb{R}^{N \times \alpha}$.

In most datasets (Table 3), cosine similarity outperformed both scalar product and attention mechanism, with lower variability.

Effect of the Auxiliary Loss Function. While SPAPOOL employs the auxiliary loss from DIFFPOOL (Eq. 6), we also tested auxiliary losses from DMON and MINCUT, respectively defined by:

$$\mathcal{L}_{\text{DMoN}} = \mathcal{L}_M + \mathcal{L}_{\text{CR}} = \left[-\frac{1}{2m} \text{Tr}(S^\top B S)\right] + \left[\frac{\sqrt{C}}{n}\left\|\sum_{i=1}^{n} S_i^\top\right\|_F - 1\right] \quad (7)$$

$$\mathcal{L}_{\text{MinCUT}} = \mathcal{L}_c + \mathcal{L}_o = \left[-\frac{\text{Tr}(S^\top \tilde{A} S)}{\text{Tr}(S^\top \tilde{D} S)}\right] + \left[\left\|\frac{S^\top S}{\|S^\top S\|_F} - \frac{I_C}{\sqrt{C}}\right\|_F\right] \quad (8)$$

where $\|\cdot\|_F$ is the Frobenius norm. In $\mathcal{L}_{\text{DMoN}}$, $B = A - \frac{dd^T}{2m}$ where d is the vertex degree vector, m is the number of edges, and C is the user-set number of supernodes following the pooling layer. Since SPAPOOL is an adaptive method, we modified $\mathcal{L}_o$ by replacing C by k_i which represents the number of supernodes in the graph i. As each graph was pooled according to their own k_i and S_i, $\mathcal{L}_o$ in $\mathcal{L}_{\text{MinCUT}}$ was computed for each pooled graph and the final $\mathcal{L}_o$ was aggregated according to the mean value. The same modification was applied on $\mathcal{L}_{\text{DMoN}}$. In our SPAPOOL ablation study, the auxiliary loss had the least consistent impact, with no tested variant clearly outperforming the others

Table 3. Graph classification accuracies according to the ablation study.

		PROTEINS	ENZYMES	Mutagenicity	OHSU	IMDB-BINARY
Selection method	TOPKPOOL	$71.24_{\pm 7.53}$	$68.29_{\pm 5.31}$	$76.58_{\pm 1.91}$	$\mathbf{53.75_{\pm 20.19}}$	$\mathbf{55.9_{\pm 3.56}}$
	SAGPOOL	$68.21_{\pm 12.24}$	$66.25_{\pm 4.96}$	$\mathbf{77.72_{\pm 1.91}}$	$51.25_{\pm 18.92}$	$55.7_{\pm 3.61}$
Aggregation method	COSINE	$\mathbf{71.24_{\pm 7.53}}$	$68.29_{\pm 5.31}$	$76.58_{\pm 1.91}$	$\mathbf{53.75_{\pm 20.19}}$	$\mathbf{55.9_{\pm 3.56}}$
	SCALAR	$71.07_{\pm 9.78}$	$\mathbf{69.64_{\pm 6.99}}$	$77.35_{\pm 1.89}$	$47.5_{\pm 20.77}$	$55.7_{\pm 4.15}$
	ATTENTION	$66.16_{\pm 12.26}$	$69.09_{\pm 5.98}$	$\mathbf{77.6_{\pm 2.2}}$	$43.75_{\pm 21.1}$	$54.1_{\pm 3.33}$
Loss function	$\mathcal{L}_{\text{DiffPool}}$	$\mathbf{71.24_{\pm 7.53}}$	$68.29_{\pm 5.31}$	$76.58_{\pm 1.91}$	$53.75_{\pm 20.19}$	$\mathbf{55.9_{\pm 3.56}}$
	$\mathcal{L}_{\text{DMoN}}$	$66.07_{\pm 12.32}$	$68.81_{\pm 5.93}$	$\mathbf{78.34_{\pm 2.1}}$	$\mathbf{53.75_{\pm 17.72}}$	$54.5_{\pm 3.38}$
	$\mathcal{L}_{\text{MinCUT}}$	$68.75_{\pm 10.68}$	$\mathbf{70.82_{\pm 5.39}}$	$76.8_{\pm 1.79}$	$52.5_{\pm 18.37}$	$55.9_{\pm 3.21}$

across datasets (Table 3). We evaluated SPAPOOL on diverse datasets varying in graph count and attribute dimensionality. Defining a universal auxiliary loss for arbitrary graph data remains a key challenge in GNNs.

6 Conclusion

While most GNN pooling methods are either dense and fixed or sparse and adaptive, this paper proposed SPAPOOL, a dense yet adaptive approach that combines the advantages of both. Although SPAPOOL did not outperform existing methods on all datasets, it achieved comparable results, demonstrating strong potential. Notably, it performed well on datasets with small graphs (around 30 vertices per graph) by grouping vertices into supernodes to preserve both local and global structure. Next steps include adjusting the auxiliary loss function to optimize results on large graphs. Future work will also includes explainability components to identify which attribute features the trained GNN considers important.

Acknowledgments. This work was funded by the French National Research Agency as part of the SPIraL program (grant number ANR-19-CE35-0006-02).

Disclosure of Interests. The authors declare that no competing interests exist.

References

1. Bianchi, F.M., Grattarola, D., Alippi, C.: Spectral clustering with graph neural networks for graph pooling. In: Proceedings of the 37th International Conference on Machine Learning, pp. 874–883 (2020)
2. Gao, H., Ji, S.: Graph u-nets. In: Proceedings of the 36th International Conference on Machine Learning, vol. 97, pp. 2083–2092 (2019)
3. Hamilton, W.L., Ying, R., Leskovec, J.: Inductive representation learning on large graphs. In: Proceedings of the 31st International Conference on Neural Information Processing Systems, pp. 1025–1035 (2017)
4. Kipf, T.N., Welling, M.: Semi-supervised classification with graph convolutional networks. In: International Conference on Learning Representations (2017)
5. Lee, J., Lee, I., Kang, J.: Self-attention graph pooling. In: Proceedings of the 36th International Conference on Machine Learning, pp. 3734–3743 (2019)
6. Li, Y., Gu, C., Dullien, T., Vinyals, O., Kohli, P.: Graph matching networks for learning the similarity of graph structured objects. In: Proceedings of the 36th International Conference on Machine Learning, pp. 3835–3845 (2019)
7. Morris, C., Kriege, N.M., Bause, F., Kersting, K., Mutzel, P., Neumann, M.: Tudataset: a collection of benchmark datasets for learning with graphs. In: ICML 2020 Workshop on Graph Representation Learning and Beyond (2020)
8. Noutahi, E., Beaini, D., Horwood, J., Giguère, S., Tossou, P.: Towards interpretable sparse graph representation learning with Laplacian pooling. arXiv preprint arXiv:1905.11577 (2019)
9. Ranjan, E., Sanyal, S., Talukdar, P.: Asap: adaptive structure aware pooling for learning hierarchical graph representations. In: Proceedings of the 34th AAAI Conference on Artificial Intelligence, pp. 5470–5477 (2020)

10. Tsitsulin, A., Palowitch, J., Perozzi, B., Müller, E.: Graph clustering with graph neural networks. J. Mach. Learn. Res. **24**(127), 1–21 (2023)
11. Veličković, P., Cucurull, G., Casanova, A., Romero, A., Liò, P., Bengio, Y.: Graph attention networks. In: International Conference on Learning Representations (2018)
12. Xu, K., Hu, W., Leskovec, J., Jegelka, S.: How powerful are graph neural networks? In: International Conference on Learning Representations (2019)
13. Ying, Z., You, J., Morris, C., Ren, X., Hamilton, W., Leskovec, J.: Hierarchical graph representation learning with differentiable pooling. In: Proceedings of the 32nd International Conference on Neural Information Processing Systems (2018)

Prediction of Iterative Solvers' Convergence Using Pretraining by Natural Images

Yuki Oba[1]([✉]), Taro Tezuka[2][iD], and Hidehiko Hasegawa[3,4][iD]

[1] Graduate School of Science and Technology, University of Tsukuba,
Tsukuba, Japan
s2230178@u.tsukuba.ac.jp
[2] Institute of Systems and Information Engineering, University of Tsukuba,
Tsukuba, Japan
tezuka@iit.tsukuba.ac.jp
[3] Institute of Library, Information and Media Science, University of Tsukuba,
Tsukuba, Japan
hasegawa@slis.tsukuba.ac.jp
[4] Kogakuin University, Shinjuku, Japan

Abstract. Many problems in scientific computing can be formalized as
solving a large system of sparse linear equations. Iterative methods are
usually employed for solving the problem, but many coefficient matrices
exist where a given iterative method does not converge within a realistic
time frame. When an iterative method fails, one must try another, and
the time used for the first method is wasted. Suppose a method exists
to predict whether a given iterative method will likely converge for a
given matrix. In that case, one can save much time without running the
iterative method. We trained a deep learning model (EfficientNetV2-S)
to enable such classification after transforming matrices into grayscale
images by scaling their components into a limited range. The proposed
method achieved high predictive performance in discriminating whether
the iterative method converges on a matrix. Furthermore, we found a considerable performance improvement when we introduced the pretrained
neural network using a large-scale dataset of natural images (ImageNet).

Keywords: a real non-symmetric system of linear equations ·
convolutional neural network · iterative solver · sparse matrix ·
scientific data · pretraining

1 Introduction

Artificial Intelligence (AI) is increasingly used to support scientific investigations. Our current work introduces deep learning for predicting the convergence
of an iterative solver. In scientific computing, the problem is usually expressed
by a partial differential equation. To numerically simulate the process, the differential equation is discretized into a large-scale linear equation $Ax = b$ using

© The Author(s), under exclusive license to Springer Nature Switzerland AG 2026
C. K. Leung et al. (Eds.): DaWaK 2025, LNCS 16048, pp. 341–348, 2026.
https://doi.org/10.1007/978-3-032-02215-8_28

the finite element method, the boundary element method, or the finite difference method. The linear equation is solved either by a direct or an iterative method. When A is a large sparse matrix, an iterative method, most notably the Krylov subspace method, is commonly used to solve the equation. When A is a real non-symmetric matrix, the BiConjugate Gradient (BiCG) method and the BiConjugate Gradient STABilized (BiCGSTAB) method are among the first choices. However, there are many matrices where these methods fail to converge within a practical time. In such a case, one must try a different method, and the time used for running the first method is wasted.

We used deep learning to train a classifier that distinguishes whether the iterative methods, BiCG and BiCGSTAB, converge within a given time frame. Such a classifier can significantly benefit numerical simulation users since if they can check in advance if an iterative method does not converge for a given matrix, they can try an alternative approach. It can reduce the time needed to solve a large-scale sparse linear equation. Imaging and classifying sparse matrices can yield results faster than actually performing iterative methods.

We applied modern pretrained image classification models, ResNet-50 and EfficientNetV2-S, to this classification task. Pretraining in image classification uses large datasets to learn image features. Training a target task on a pretrained model can achieve efficient learning based on features already learned. Using a pretrained model on a large image dataset can efficiently perform better than starting training only on the current task dataset. Our code used for the experiments is available at https://github.com/itumizu/pred_conv_ pretrained_dawak2025.

2   Method

2.1   Target Data

We target sparse non-symmetric matrices for the following reasons. If the coefficient matrix of a system of linear equations is dense and can be stored in memory, it can be solved using the direct method, and there is no need to use an indirect method such as the BiCG method. For symmetric matrices, the Conjugate Gradient (CG) method usually converges, and there are no other popular alternatives. For non-symmetric matrices, the algorithm is not perfect unless all intermediate vectors are stored. However, doing so requires as much storage space as the direct method. Therefore, most algorithms restrict the number of vectors to be stored, resulting in various arrangements. Since there are many options, judging whether an algorithm converges without running it saves time.

2.2   Matrix Transformation

We used 875 sparse real non-symmetric matrix data from the SuiteSparse Matrix Collection [1]. Table 1 shows the tasks in scientific computing and optimization from which the coefficient matrices originate. We converted the coefficient matrices into gray-scale images following Ota et al. [5]. The absolute value of each coefficient is normalized, and then the whole matrix is down-sampled. We set d to 224, so the sizes of the images were all normalized to 224×224 pixels. The reason for converting the coefficient matrices into images is to make them manageable by down-sampling. Since the computation time depends on the size of the matrix, down-sampling significantly reduces the necessary time. The dataset consists of different matrix sizes. Since a convolutional neural network can only accept matrices of one size, down-sampling is necessary in most cases.

Table 1. Most frequent domains from which coefficient matrices originate.

Domain	BiCG	BiCGSTAB
subsequent circuit simulation problem	122 (non-conv.) / 60 (conv.)	123 (non-conv.) / 59 (conv.)
computational fluid dynamics problem	44/52	54/42
chemical process simulation problem	64/3	64/3
economic problem	56/6	56/6
directed graph	45/5	48/2
circuit simulation problem	27/19	30/16
2D/3D problem	18/21	22/17
directed weighted graph	26/13	24/15
eigenvalue/model reduction problem	24/11	19/16
subsequent computational fluid dynamics problem	22/8	22/8
power network problem	24/0	24/0
electromagnetics problem	9/13	11/11
semiconductor device problem	10/6	6/10
structural problem	7/8	7/8
materials problem	8/6	10/4
subsequent semiconductor device problem	11/2	11/2
subsequent optimization problem	13/0	13/0
directed multigraph	12/0	12/0
combinatorial problem	9/2	7/4
subsequent 2D/3D problem	5/5	4/6
thermal problem	3/7	3/7

2.3   Model

We trained two deep-learning classification models whose structures are based on a convolutional neural network, namely ResNet-50 [2] and EfficientNetV2-S [8]. We used CNN because it is among the first choices for 2-dimensional data such as images. As the baseline models, we trained the simple CNN model based on Ota et al. [5].

EfficientNetV2-S is a model based on a convolutional neural network that achieves faster training speeds by optimizing its network structure. Compared to others with equivalent performance, this model has fewer parameters while still delivering high performance.

For pretraining, we used ImageNet-1k and ImageNet-21k, subsets of ImageNet [6]. Specifically, we used implementation and pretrained weights obtained from the timm library [10]. For ImageNet-21k, we could only get a pretrained model for EfficientNetV2-S, not ResNet-50.

2.4 Experiment Settings

Calculating Linear Algebra and Labeling. The task is to classify whether a specific iterative method converges for a given coefficient matrix. We used BiCG and BiCGSTAB as iterative methods. The model is trained using ground-truth labels obtained by running an iterative method up to a set number of maximum iterations. We label the matrix convergent if the relative residual 2-norm goes below the threshold before the maximum iteration count. Otherwise, it is considered to be non-convergent. The convergence of the iterative methods was determined using SciPy [9].

The iterative methods aim to solve a large system of linear equations, $Ax = b$. We used the following procedure used by SciPy to set the initial values of the iterative methods of both BiCG and BiCGSTAB. In what follows, A is a sparse matrix. For set b to Ax where $x := \begin{bmatrix} 1 & -1 & 1 & -1 & \cdots & (-1)^{N-1} \end{bmatrix}^{\mathsf{T}}$. We used $x_0 := \begin{bmatrix} 0 & 0 & \cdots & 0 \end{bmatrix}^{\mathsf{T}}$ as the initial solution for the iterative method. The convergence tolerance was set to 10^{-6}, and the maximum iterations were set to N, which is the size of the coefficient matrix.

Training Models. We conducted 5-fold cross-validation. Three folds were used for the training phase, one fold for the validation phase, and the remaining fold served as the test set. All models used binary cross-entropy as the loss function. We trained each model for a maximum of 100 epochs, employing an early stopping strategy that halts training if the validation loss does not improve for 20 consecutive epochs. We selected the best weights where the model shows the lowest validation loss during the training epochs. We used the AdamW optimizer and a cosine annealing learning-rate scheduler to set the hyperparameters with a warm-up phase [3]. We used the CosineLRScheduler implemented by the timm library and set the following parameters. We set the initial learning rate to 10^{-5}, which linearly warms up to 10^{-3} over three epochs. Afterward, the learning rate was decreased to 0 using cosine decay over a period equal to the maximum number of epochs. The process had no restart, and the minimum learning rate was set to 0. The batch size was 16, and the weight decay was set to 10^{-8}. To compare with an existing method, we trained the CNN-based model [5]. In this experiment, we applied a learning rate of 0.001 and no learning rate scheduler and used the Adam optimizer.

3   Results

We evaluated the classification task using three metrics: accuracy, AUROC, and Matthews correlation coefficient (MCC) [4]. High AUROC and MCC ensure that convergent and non-convergent matrices are correctly classified. Tables 2 - 3 indicate the classification performance of each model. The table presents the average evaluation results of test data using 5-fold cross-validation.

Figure 1 visualizes the responding areas of a trained network using Grad-CAM [7]. Grad-CAM is a visualization method that focuses on the final convolution layer. Its output is a localization map showing regions on which the network is more dependent when making predictions.

Table 2. Predictive performance of classification task for the BiCG method.

Model	Pretraing type	Accuracy	Recall	Precision	AUROC	MCC
CNN (baseline)	–	0.7909	0.6634	0.6698	0.8610	0.5163
		±0.0254	±0.0952	±0.0804	±0.0268	±0.0581
ResNet-50		0.8000	0.5919	**0.7449**	0.8471	0.5217
		±0.0225	±0.1680	**±0.1470**	±0.0326	±0.0649
	✓ (ImageNet-1k)	0.7851	0.5813	0.7136	0.8267	0.4932
		±0.0257	±0.1456	±0.1489	±0.0303	±0.0650
EfficientNetV2-S		0.7737	0.5627	0.6347	0.7993	0.4504
		±0.0805	±0.3362	±0.2089	±0.1092	±0.2630
	✓ (ImageNet-1k)	0.8057	**0.7469**	0.6708	0.8578	0.5629
		±0.0408	**±0.0409**	±0.0609	±0.0202	±0.0662
	✓ (ImageNet-21k)	**0.8240**	0.6890	0.7213	**0.8691**	**0.5790**
		±0.0306	±0.0995	± 0.0736	**±0.0376**	**±0.0930**

4   Discussion

Tables 2–3 show that pretrained EfficientNetV2-S models beat baseline CNN models on predictive performance for both tasks of BiCG and BiCGSTAB. The tables indicate that pretrained models perform better than non-pretrained models. For example, the predictive performance of the pretrained EfficientNetV2-S models improved by about 4.5% on the BiCG task and 5.0% on the BiCGSTAB task compared to the non-pretrained models.

In general, pretraining improved performance in the more difficult domains of the non-pretrained models. In contrast, performance improvement was small for tasks where the non-pretrained models performed well already. The grayscale images exhibit many light-colored lines running parallel to their diagonal elements. Pretraining may increase the classifier's ability to detect such lines since

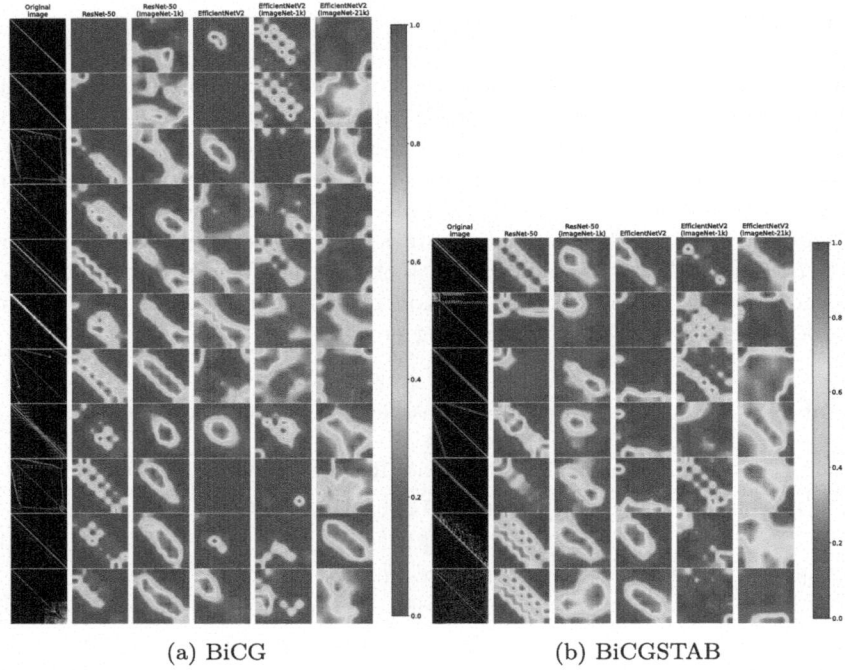

(a) BiCG   (b) BiCGSTAB

Fig. 1. Visualization of the responding areas using Grad-CAM. The left-most column is a grayscale image converted from a coefficient matrix. The rest are the outputs of Grad-CAM in the following order: ResNet-50 (non-pretrained, pretrained on ImageNet-1k), EfficientNetV2-S (non-pretrained, pretrained on ImageNet-1k, pretrained on ImageNet-21k). The images are of coefficient matrices that pretrained EfficientNetV2-S predicted correctly, but other models didn't. They show what spatial features were helpful for pretrained EfficientNetV2-S when making predictions for difficult samples for other models.

long and straight lines are frequently found in natural images. The Grad-CAM output shown in Fig. 1 also shows that EfficientNetV2-S is indeed focusing on the diagonal elements.

It is often said that CNN's structure is designed to capture the hierarchical structure of locality in natural images of real-world objects. In contrast, the coefficient matrices originating from partial differential equations and optimization are usually very sparse. It has not been clear until now if CNN is effective for this task because we cannot suppose such a hierarchy in coefficient matrices. Despite the difference, the experiments showed that the method is effective. In addition, pretraining using natural images improved the performance, indicating that common features are shared between coefficient matrices and natural images.

The main feature of CNN is its convolutional layers that act as filters for extracting image features. Pretraining the filters may help them capture local

Table 3. Predictive performance of classification task for the BiCGSTAB method.

Model	Pretraining type	Accuracy	Recall	Precision	AUROC	MCC
CNN (baseline)	–	0.7817	0.6399	0.6746	0.8321	0.4979
		±0.0247	±0.1317	±0.0648	±0.0390	±0.0803
ResNet-50		0.7737	0.6461	0.6471	0.8329	0.4796
		±0.0472	±0.1416	±0.0814	±0.0576	±0.1287
	✓ (ImageNet-1k)	0.7829	0.6483	0.6670	0.8265	0.4998
		±0.0733	±0.1621	±0.1421	±0.1043	±0.1888
EfficientNetV2-S		0.7851	0.6550	0.6834	0.8398	0.5100
		±0.0227	±0.0646	±0.0597	±0.0383	±0.0206
	✓ (ImageNet-1k)	0.8103	0.7051	0.7218	0.8690	0.5717
		±0.0491	±0.0360	±0.0996	±0.0340	±0.0912
	✓ (ImageNet-21k)	**0.8309**	**0.7405**	**0.7409**	**0.8867**	**0.6157**
		±0.0354	**±0.0657**	**±0.0792**	**±0.0368**	**±0.0836**

features in grayscale images, such as diagonal lines. Notably, EfficientNetV2-S employs progressive learning [8] in its pretraining, dynamically adjusting image scales, while ResNet does not. The matrices were scaled to the same image size (224×224), irrespective of their original sparse matrix dimensions. EfficientNetV2-S's architecture will likely be more adept at extracting features from images scaled from their original sparse matrix. One explanation for our results is that without pretraining, CNN may be consuming much of its training time to capture any form of local spatial structure in the original coefficient matrices. Since most coefficient matrices originate from discretizing differential equations, they have strong locality properties. Combining CNN and pretraining may efficiently capture and utilize these properties.

5   Conclusion

We developed a deep learning model that predicts the convergence of iterative methods for solving a system of linear equations. It was shown that converting the coefficient matrix into a gray-scale image and training a CNN can predict the convergence with high accuracy. The result is surprising because it was previously unknown that the hierarchical and locality-based structure of the CNN is effective for classifying images representing highly sparse matrices. Our results showed that the inductive bias represented by the local kernels of the CNN also applies to coefficient matrices in scientific computing and optimization, probably due to their inherent spatial nature.

Another interesting outcome of our work is the effectiveness of pretraining using natural images. It was already known that pretraining using natural images improves the classification accuracy of medical images. Our results further generalize this phenomenon to matrices originating from scientific computing and optimization. At first sight, these matrices appear to be vastly different from natural images consisting of large-scale objects, but it seems that pretraining makes models learn common locality structures across a wide range of datasets.

Acknowledgment. This work was supported by JSPS KAKENHI Grant Number 22K12056.

References

1. Davis, T.A., Hu, Y.: The university of Florida sparse matrix collection. ACM Trans. Math. Softw. **38**(1) (2011)
2. He, K., Zhang, X., Ren, S., Sun, J.: Deep residual learning for image recognition. In: 2016 IEEE Conference on Computer Vision and Pattern Recognition (CVPR). pp. 770–778 (2016)
3. Loshchilov, I., Hutter, F.: SGDR: stochastic gradient descent with warm restarts. arXiv preprint arXiv:1608.03983 (2016)
4. Matthews, B.W.: Comparison of the predicted and observed secondary structure of t4 phage lysozyme. Biochimica et Biophysica Acta (BBA)-Protein Struct. **405**(2), 442–451 (1975)
5. Ota, R., Hasegawa, H.: Predicting the convergence of BiCG method from grayscale matrix images. JSIAM Lett. **12**, 45–48 (2020)
6. Russakovsky, O., et al.: Imagenet large scale visual recognition challenge. Int. J. Comput. Vis. **115**, 211–252 (2015)
7. Selvaraju, R.R., Cogswell, M., et al.: Grad-cam: visual explanations from deep networks via gradient-based localization. In: Proceedings of the IEEE International Conference on Computer Vision, pp. 618–626 (2017)
8. Tan, M., Le, Q.: EfficientNetV2: smaller Models and Faster Training. In: Proceedings of the 38th International Conference on Machine Learning, vol. 139, pp. 10096–10106. PMLR (2021)
9. Virtanen, P., Gommers, R., Oliphant, T.E., et al.: SciPy 1.0: fundamental algorithms for scientific computing in python. Nat. Methods **17**, 261–272 (2020)
10. Wightman, R.: Pytorch image models (2019). https://github.com/rwightman/pytorch-image-models

Local-Aware Convolutional Modulation for Short-Term Sequential Recommendation

Tianxing Wang[1]([✉]), Can Wang[1]([✉]), Hui Tian[1], and Hong Shen[2]

[1] Griffith University, Gold Coast 4215, Australia
tianxing.wang@griffithuni.edu.au, {can.wang,hui.tian}@griffith.edu.au
[2] Central Queensland University, Brisbane 4000, Australia
h.shen@cqu.edu.au

Abstract. Sequential recommendation models have predominantly relied on self-attention mechanisms in recent years. However, beyond self-attention, other deep neural architectures such as convolutional neural networks (CNNs) offer promising alternatives for capturing sequential patterns. In this paper, we explore the CNN-based architecture and propose **L**ocal-aware **C**onvolutional **M**odulation for Short-Term Sequential **Rec**ommendation (LCMRec). Like other convolutional neural network-based models, LCMRec benefits from strong local modelling capabilities through its convolutional architecture. By introducing the multi-head convolutional modulation (MHCM) unit, which applies convolutions with varying kernel sizes across multiple heads locally, LCMRec dynamically captures short-term dependencies at multiple scales and keeps a linear computational complexity. In experiments, LCMRec outperforms baseline models, demonstrating the efficacy of the convolutional architecture and validating the effectiveness of our approach in balancing multi-scale dependency modelling with computational efficiency.

Keywords: Sequential Recommendation · Convolutional Neural Network · Local Modelling

1 Introduction

The self-attention (SA)-based recommender has significantly improved recommendation performance by capturing intricate long-term dependencies within user historical sequences [3,7,15]. Despite its effectiveness in capturing users potential interests via the historical records, the SA-based models still face the notable challenges in sequential recommendation tasks. For example, SA calculates pairwise item interactions across the sequence, assigning attention weights uniformly to all positions. While this enables effective long-term dependency modeling, it weakens the influence of recent interactions, which are often more indicative of immediate user preferences [1]. Consequently, while SA is capable of modelling short-term dependencies, it lacks an explicit mechanism for localized information exchange [4]. Unlike SA, which models short-term dependencies

© The Author(s), under exclusive license to Springer Nature Switzerland AG 2026
C. K. Leung et al. (Eds.): DaWaK 2025, LNCS 16048, pp. 349–355, 2026.
https://doi.org/10.1007/978-3-032-02215-8_29

through learned attention weights over the entire sequence, convolutional neural networks (CNNs) inherently capture localized patterns through hierarchical feature modelling. CNN-based design is particularly advantageous in sequential recommendation, especially in the short-term scenario, where recent interactions often carry more weight in determining future preferences. Inspired by this property, we propose **L**ocal-aware **C**onvolutional **M**odulation for Short-Term Sequential **Rec**ommendation (LCMRec), a novel CNN-based model enhancing short-term pattern modeling. Compared with previous CNN approaches, such as Caser and NextItNet [14,18], which use fixed-size convolutional kernels, our design introduces **M**ixed **H**ead **C**onvolutional **M**odulation (MHCM) Unit, a mechanism that applies variable kernel sizes to capture short-term dependencies across multiple scales [6,12].

The contributions of our paper can be summarized as follows:

- We propose a novel **L**ocal-aware **C**onvolutional **M**odulation for Short—Term Sequential **Rec**ommendation (LCMRec), which leverages the local modelling abilities of CNN to effectively capture the short-term dependencies.
- We introduce the Multi-head Convolutional Modulation (MHCM) unit, which applies variable kernels across multiple heads to capture short-term dependencies at multiple scales.
- In experiments, LCMRec consistently outperforms baseline models, demonstrating the potential of the convolutional architecture and validating the effectiveness of our approach in balancing multi-scale dependency modelling with computational efficiency.

2   Related Work on Sequential Recommendation

Early sequential recommendation models are based on Markov chains, such as FPMC [5]. With the advent of deep learning, researchers began exploring recurrent neural networks (RNNs) [9] and convolutional neural networks (CNNs) [14]. These models are good at capturing temporal dependencies, enabling better predictions of user behavior. The introduction of self-attention (SA) mechanisms has further enhanced sequential recommendations by effectively modelling item-item relevance within sequences [15]. SASRec was the first model that purely utilizes the SA architecture [7]. Following this, several models have further explored the potential of the SA mechanism [3,11]. Recently, there has been a trend toward lightweight architectures that achieve performance comparable to SA. Models like MLP4Rec leverages MLP structures to capture sequential patterns at both the feature and sequence levels [10]. EDM incorporates an external attention mechanism to model local and global preferences [16]. LRURec introduces the linear recurrent unit to accelerate training and inference processes [19]. Despite these advancements, CNN-based models still hold significant potential for efficiently modelling short-term dependencies. Our approach adopts the idea of the convolutional modulation and multi-kernel technology [6,12] to capture multi-level dependencies across sequences.

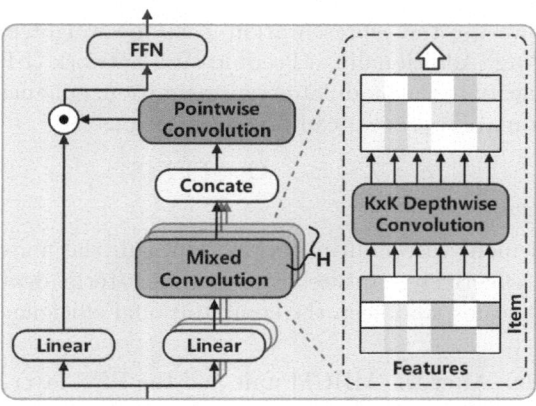

Fig. 1. The structure of the MHCM unit.

3   Our Proposed Model

The goal of sequential recommendation is to predict a user's next interaction based on their historical interaction sequence. Figure 1 illustrates the structure of the MHCM unit.

Multi-Head Convolutional Modulation Unit. We propose the Multi-Head Convolutional Modulation (MHCM) unit, which is designed to capture short-term dependencies across multiple scales. The MHCM adopts a convolutional modulation structure, a SA-style network, where the learned convolutional features are used as weight maps to modulate the input embeddings [6]. Firstly, the input is separated to multiple heads. Each head is then fed into the mixed convolution layer, which allocates a distinct kernel size to each head [12]. Specifically, each head applies a depthwise separable convolution with a specific K by K kernel. We set the K start from 3, and gradually increase by 2 per head. Such design creates the variable receptive fields to capture short-term dependencies at different ranges. After that, the outputs from each head are concatenated and passed through a pointwise convolution, integrating information across various temporal scales. The formulas of the discussed two layers are:

$$\mathbf{X}_{\text{conv}}^{(h)} = \text{DepthwiseConv}_{K_h}(\mathbf{X}^{(h)}), \tag{1}$$

$$\mathbf{Y} = \text{Concat}(\mathbf{X}_{\text{conv}}^{(1)}, \mathbf{X}_{\text{conv}}^{(2)}, \dots, \mathbf{X}_{\text{conv}}^{(H)}), \tag{2}$$

$$\mathbf{Z} = \text{PointwiseConv}(\mathbf{Y}), \tag{3}$$

where $\mathbf{X}^{(h)} \in R^{L \times \frac{d}{H}}$ is h-th head of input embedding $\mathbf{E}_l$, H is the total number of head, and $K \in \{3, 5, \cdots, (3+2(H-1))\}$ represents the kernel size which increases by 2 for each successive head. After the above two convolutional layers, the representation Z encapsulates the multi-scale information of the input sequence.

Subsequently, we use the representation Z as the weight map to modulate the input embedding. Additionally, a Feedforward Network (FFN) is applied to introduce non-linearity to the modulated representation, enhancing its expressive capacity. The modulation process can be formulated as:

$$\mathbf{S} = \mathbf{Z} \odot \mathbf{E}_l, \quad \mathbf{O} = \text{FFN}(\mathbf{S}) \tag{4}$$

where $\odot$ is the element-wise multiplication. Through the above processes, the MHCM unit can effectively capture dynamic short-term dependencies across different temporal scales, and keep the computational efficiency.

Prediction Layer. After the MHCM unit and the FFN layer, we get the final output embedding $\mathbf{O}$. In the prediction layer, we use Euclidean distance to estimate the similarity between the modeled representation $\mathbf{O}$ and each candidate item embedding c_i. The lower distances indicate the higher relevance that the item might be interacted with the user. The prediction score for candidate item i on timestamp $t + 1$ can be formulated as:

$$r_{i,t+1} = ||Avg(\mathbf{O}) - c_i||_2, \tag{4}$$

where $Avg()$ is the average operation that aggregate the information of $\mathbf{O}$, and $||\cdot||_2$ denotes the Euclidean norm. By ranking items based on their prediction scores, the model identifies the most relevant recommendations.

Model Training. In the training session, we adopt the sliding window strategy with size $|L+T|$ to the user's sequence: for each user, we take each $|L|$ successive item interactions as the input, and the next $|T|$ records as the targets to predict. Meanwhile, our model is trained with Bayesian Personalized Ranking (BPR) loss [13], which optimizes the ranking of positive items over negative ones for each user. The BPR loss is defined as:

$$\mathcal{L}_{\text{BPR}} = - \sum_{(i,j)\in\mathcal{D}} \log \sigma(r_{i,t} - r_{j,t}), \tag{5}$$

where $\sigma(\cdot)$ is the sigmoid function, and (i, j) is the positive-negative item pair.

4 Experiment

4.1 Experiment Settings

Datasets and Baselines. We evaluate our model on two widely-used datasets: Foursquare [17] and Gowalla [2], which is composed of the implicit feedback data. Foursquare contains 1083 users, 9989 items, and 227,428 interactions, while Gowalla includes 3785 users, 32338 items, and 474391 interactions.

The baselines used to validate our model can be grouped into two categories: (1) Self-attention-based sequential recommendation models: **SASRec** [8], **TiSASRec** [8], and **STOSA** [3]; (2) Lightweight sequential recommendation models: **AutoMLP** [10], **EDM** [16], and **LRURec** [19].

Model Configuration. For baseline models, we follow the original parameter settings in their papers. We perform grid search to determine the optimal parameters for our proposed model: the number of heads in the mixed convolution layer is set to 4, the embedding dimension to 256, and the dilation rate to 1. We also set the number of stacking blocks M = 2M=2, each consisting of one MHCM unit followed by one dilated MHCM unit. During training, we set a short-term learning procedure: the batch size to 512, learning rate to 0.001, embedding dimension to 256, and input sequence length L to 5 with target length T to 3.

Table 1. Our experimental results on Precision (Pre@5) and Normalize Discounted Cumulative Gain (nD@5).

Metrics	Foursquare		Gowalla	
	Pre@5	nD@5	Pre@5	nD@5
SASRec	0.066	0.069	0.056	0.064
TiSASRec	0.070	0.066	0.058	0.061
STOSA	0.062	0.064	0.060	0.063
AutoMLP	0.071	0.077	0.061	0.073
EDM	0.075	0.080	0.068	0.082
LRURec	0.077	0.076	0.065	0.078
LCMRec	**0.081**	**0.091**	**0.076**	**0.095**

4.2   Performance Analysis

Table 1 presents the performance of our proposed LCMRec alongside all baseline models. LCMRec demonstrates robust results across both datasets. On Foursquare, it achieves a precision@5 of 0.081 and an nDCG@5 of 0.091, while on Gowalla, LCMRec leads with a precision@5 of 0.076 and an nDCG@5 of 0.095. Notably, LCMRec's nDCG shows over a 15% improvement, highlighting its effectiveness in capturing short- and long-term dependencies through the MHCM design, and its stacking structure. For the baseline models, we observe that lightweight models generally outperform SA-based models, suggesting that simpler architectures also possess strong capabilities for sequential modelling.

5   Conclusion

We proposed LCMRec, the local-aware convolutional modulation for short-term sequential recommendation that captures short-term dependencies through a multi-head convolutional modulation (MHCM) structure. Our experiments demonstrate that LCMRec outperforms several state-of-the-art baselines, achieving notable improvements in a short-term input scenario. These results highlight the effectiveness of combining multi-scale convolutions with efficient modulation for sequential modelling. In the future, we will extend this concept from short-term recommendation to long-sequence and multi-modal recommendations.

References

1. Beltagy, I., Peters, M.E., Cohan, A.: Longformer: the long-document transformer. arXiv preprint arXiv:2004.05150 (2020)
2. Cho, E., Myers, S.A., Leskovec, J.: Friendship and mobility: user movement in location-based social networks. In: Proceedings of the 17th ACM SIGKDD International Conference on Knowledge Discovery and Data Mining, pp. 1082–1090 (2011)
3. Fan, Z., et al.: Sequential recommendation via stochastic self-attention. In: Proceedings of the ACM Web Conference 2022, pp. 2036–2047 (2022)
4. Guo, M.H., Lu, C.Z., Liu, Z.N., Cheng, M.M., Hu, S.M.: Visual attention network. Comput. Visual Media $9(4)$, 733–752 (2023)
5. He, R., McAuley, J.: Fusing similarity models with Markov chains for sparse sequential recommendation. In: 2016 IEEE 16th International Conference on Data Mining (ICDM), pp. 191–200. IEEE (2016)
6. Hou, Q., Lu, C.Z., Cheng, M.M., Feng, J.: Conv2former: a simple transformer-style convnet for visual recognition. IEEE Trans. Pattern Anal. Mach. Intell. (2024)
7. Kang, W.C., McAuley, J.: Self-attentive sequential recommendation. In: 2018 IEEE International Conference on Data Mining (ICDM), pp. 197–206. IEEE (2018)
8. Li, J., Wang, Y., McAuley, J.: Time interval aware self-attention for sequential recommendation. In: Proceedings of the 13th International Conference on Web Search and Data Mining, pp. 322–330 (2020)
9. Li, J., Ren, P., Chen, Z., Ren, Z., Lian, T., Ma, J.: Neural attentive session-based recommendation. In: Proceedings of the 2017 ACM on Conference on Information and Knowledge Management, pp. 1419–1428 (2017)
10. Li, M., et al.: Automlp: automated MLP for sequential recommendations. In: Proceedings of the ACM Web Conference 2023, pp. 1190–1198 (2023)
11. Li, Y., Chen, T., Zhang, P.F., Yin, H.: Lightweight self-attentive sequential recommendation. In: Proceedings of the 30th ACM International Conference on Information & Knowledge Management, pp. 967–977 (2021)
12. Lin, W., Wu, Z., Chen, J., Huang, J., Jin, L.: Scale-aware modulation meet transformer. In: Proceedings of the IEEE/CVF International Conference on Computer Vision, pp. 6015–6026 (2023)
13. Rendle, S., Freudenthaler, C., Gantner, Z., Schmidt-Thieme, L.: BPR: Bayesian personalized ranking from implicit feedback. arXiv preprint arXiv:1205.2618 (2012)
14. Tang, J., Wang, K.: Personalized top-n sequential recommendation via convolutional sequence embedding. In: Proceedings of the Eleventh ACM International Conference on Web Search and Data Mining, pp. 565–573 (2018)
15. Vaswani, A.: Attention is all you need. In: Advances in Neural Information Processing Systems (2017)
16. Wang, T., Wang, C.: Global-aware external attention deep model for sequential recommendation. In: Pacific-Asia Conference on Knowledge Discovery and Data Mining, pp. 335–347. Springer (2023)
17. Yang, D., Zhang, D., Zheng, V.W., Yu, Z.: Modeling user activity preference by leveraging user spatial temporal characteristics in LBSNs. IEEE Trans. Syst. Man Cybern. Syst. $45(1)$, 129–142 (2014)

18. Yuan, F., Karatzoglou, A., Arapakis, I., Jose, J.M., He, X.: A simple convolutional generative network for next item recommendation. In: Proceedings of the Twelfth ACM International Conference on Web Search and Data Mining, pp. 582–590 (2019)
19. Yue, Z., Wang, Y., He, Z., Zeng, H., McAuley, J., Wang, D.: Linear recurrent units for sequential recommendation. In: Proceedings of the 17th ACM International Conference on Web Search and Data Mining, pp. 930–938 (2024)

Author Index

© The Editor(s) (if applicable) and The Author(s), under exclusive license
to Springer Nature Switzerland AG 2026
C. K. Leung et al. (Eds.): DaWaK 2025, LNCS 16048, pp. 357–358, 2026.
https://doi.org/10.1007/978-3-032-02215-8

The manufacturer's authorised representative in the EU is Springer
Nature Customer Service Centre GmbH, Europaplatz 3, 69115 Heidelberg,
Germany. If you have any concerns regarding our products, please
contact ProductSafety@springernature.com

Printed and bound by CPI Group (UK) Ltd, Croydon, CR0 4YY
01/05/2026
02101080-0004